Developments and Approaches in Science Diplomacy:

Latin America and the Caribbean

Luisa Fernanda Echeverría-King
Universidad Simón Bolivar, Colombia

Branislav Pantović
Universidad Nacional de Río Negro, Argentina

Rafael Enrique Pineros-Ayala
Universidad Externado de Colombia, Colombia

Pedro Figueroa
DiploCientifica, Chile

Alejandra Fabiola Flores-Zamora
Universidad de Guadalajara, Mexico

A volume in the Advances in Electronic Government, Digital Divide, and Regional Development (AEGDDRD) Book Series

IGI Global
PUBLISHER of TIMELY KNOWLEDGE

Published in the United States of America by
IGI Global
Information Science Reference (an imprint of IGI Global)
701 E. Chocolate Avenue
Hershey PA, USA 17033
Tel: 717-533-8845
Fax: 717-533-8661
E-mail: cust@igi-global.com
Web site: http://www.igi-global.com

Library of Congress Cataloging-in-Publication Data

CIP DATA PROCESSING

2024 Information Science Reference
hcISBN: 9798369327463
scISBN: 9798369349236
eISBN: 9798369327470

British Cataloguing in Publication Data
A Cataloguing in Publication record for this book is available from the British Library.

Table of Contents

Detailed Table of Contents

Emerging technologies in LATAM are challenging for decision-makers, characterized by rapid evolution, limited expertise in the region, and a need for individual skills. To make impactful and resilient science diplomacy, sustainable thinking is required in current decision-making processes, often due to insufficient anticipation of future scientific developments. This chapter examines the existing status of emerging technologies and the current decision-making process. A significant focus is placed on the impact of individual cognitive processes of science diplomats presenting an anticipatory and sustainable decision-making framework designed to enhance critical thinking skills, targeting the root causes of prevalent issues of today. More informed, future-oriented policy development in science diplomacy are required. The chapter aims to offer actionable insights and a strategic direction for navigating the complex domain of emerging technologies, emphasizing the need for foresight and inference awareness in individual decision-makers.

The prospect for the solution of the most urgent challenges in Latin America, and also the projection of the countries and the region, is associated with the strength of the relation between science, technology, innovation, foreign policy, and decision making. This nexus requires a special, tailor-made science diplomacy that takes into account the history, culture's characteristics, and particularity of governance. In other words, it is beneficial to start addressing issues from local perspectives, knowing its contexts and needs, and on the other hand, knowing how to calculate and predict diplomatic-political maneuvering and productivity. Therefore, an appropriate response to a challenge via science diplomacy should include the design of science diplomacy taxonomy, a tree of interconnected concepts and practices.

This chapter explores the role of science communication in advancing science diplomacy in the Latin America and the Caribbean (LAC) region. It examines communication initiatives and collaborative approaches among LAC institutions through science diplomacy amidst challenges posed by the COVID-19 pandemic and global environmental change (GEC). Case studies featuring organizations such as the National Secretariat of Science, Technology, and Innovation from Panama (SENACYT), the Science Diplomacy Network in Latin America and the Caribbean (DiploCientifica), the Inter-American Institute for Global Change Research (IAI), Puerto Rico Science Policy Action Network (PRSPAN), and Ciencia Puerto Rico (CienciaPR) highlight their efforts in disseminating science, technology, and innovation (STI). These organizations serve as intermediaries, translating scientific insights for decision-makers and facilitating connections between scientists, diplomats, policymakers, and society. By analyzing the organizations' missions, visions, and contributions, the chapter aims to identify key factors for effective scientific communication in science diplomacy and essential skills.

German and Latin American science diplomacy fosters institutional partnerships and academic exchanges, bridging cultural and geographical divides to enhance productivity. This analysis highlights Germany's advanced research infrastructure and Berlin's dynamic innovation ecosystem, marked by strategic multi-actor frameworks, effective science communication, and robust public-private partnerships. It addresses Germany's challenges and opportunities in strengthening its economic position as a key player in scientific and technological progress. The network between Germany and Latin America is examined, showcasing initiatives that enhance international networking, drive innovation, stimulate economic activity, and generate mutual benefits. Collaboration between Germany and Latin American countries advances scientific knowledge, fostering bilateral progress and joint growth. These efforts address contemporary challenges, promote sustainability, and underscore the importance of science diplomacy.

Science can shape beliefs, behaviors, and societal structures globally. Recently, the concept of science diplomacy (SD) has gained prominence. However, SD has faced criticism for overlooking power dynamics inherent in diplomatic agendas. Countries from the Global South have begun integrating SD strategies. Yet, there is a gap in understanding the diplomacy for science dimension captured by the formal agreements these nations are pursuing. This chapter addresses this gap by examining SD initiatives from Argentina, Chile, and Colombia with major economic powers like the United States, the European Union, and China. Using a comparative case study research design this chapter identifies and categorizes formal agreements from leading scientific agencies between 2018 and 2023. It also builds a taxonomy of science diplomacy instruments. It concludes that despite a historical subordination to the science diplomacy agendas of the economic powers, Argentina, Chile, and Colombia are emerging as proactive partners, transitioning from recipients to viable collaborators in training and research.

More than 60% of the world's flow of fresh water is shared, of the 33 countries that make up the Latin American and Caribbean region, 22 share watercourses, lakes and aquifers. The basin is considered the ideal unit to carry out the sustainable management of water resources, that contributes to territorial management and social balance at the local, national and cross-border level. The Puyango-Tumbes river basin (Peru-Ecuador) it constitutes one of the most important binational basins in South America. As the quality of water degrades or the available quantity of water has to meet increasing demands, competition among water users intensifies; nowhere is this situation more destabilizing than in river basins that cross political borders. The most critical situation began in 1980s due to mining activity in Portovelo and Zaruma area, with the presence of informal companies that were later relocated to the Calera and Amarillo rivers, the main tributaries of the Puyango-Tumbes river, dumping the mining effluents there, focusing the contamination problem and compromising the populations settled downstream of the basin. This water source presents a serious environmental problem. On the one hand, it is the common recipient of mining effluents and on the other hand of wastewater, urban solid waste and contamination from the excessive use of agrochemicals in the agricultural sector. In this context, it is required to consolidate water security through diplomacy and geopolitics based on strengthening hydrodiplomacy, considering the institutional frameworks for governance in transboundary watercourses and the fundamental principles of international water law.

The COVID-19 pandemic caused an important global change in political, financial, and health sectors to face the generated crisis. Therefore, new strategies have been developed, assumed, and implemented, with the support of governments and international organizations, on innovative capabilities that make it sustainable to invest in science to advance human development improvement of our nations. The objective of this chapter is to present in three different sections, before, during, and after the pandemic, a comprehensive overview of this topic in LATAM, which has generated close links among vaccine development and foreign policy strategies considering health and vaccine diplomacy as paramount branches included on science diplomacy actions in this region. Addressing the challenges and benefits of those strategies constitutes an element to be considered in foreign policy by governments interested in allocating efforts and resources into science, health, and sustainable development.

Preface

It is with great pride and anticipation that we present *Developments and Approaches in Science Diplomacy: Latin America and the Caribbean*. This reference book emerges at a critical juncture in the evolution of science diplomacy, particularly within the context of Latin America and the Caribbean, a region characterized by dynamic scientific advancements and complex socio-economic challenges.

Science diplomacy as a field has gained considerable relevance in Latin America, serving as a nexus between science, international relations, and public policy. In emerging economies, science diplomacy aims to address development objectives, strengthen ties with the scientific diaspora, and bridge gaps in science, technology, and innovation through international collaborations. While our primary focus is on Latin America, the insights and approaches discussed will resonate across the Global South, offering valuable perspectives on how emerging economies can leverage science diplomacy to foster sustainable development.

This book addresses a significant gap in the literature by providing a comprehensive examination of science diplomacy cases within Latin America and the Caribbean. It is designed to serve as a vital resource for researchers, policymakers, and diplomats who are engaged in or interested in the development of science, technology, and innovation systems in the region. By bringing together diverse voices and expertise, this collective work aims to deepen our understanding of the varied meanings, instruments, and scopes of science diplomacy across different regions and countries, with a particular emphasis on Latin America and the Global South.

Our objectives with this book are fourfold:

1. **Identify Science Diplomacy Strategies in Latin America:** We aim to uncover the specific strategies employed across the region to harness science diplomacy for developmental goals.
2. **Examine Institutional Instruments and Thematic Axes:** We explore the institutional frameworks, thematic focuses, and characteristics of science diplomacy utilized by states and other actors in Latin America.

3. **Compare Strategies Through Case Studies:** By presenting detailed case studies, we compare the effectiveness of various strategies implemented by different actors, assessing their utility in the Latin American context.
4. **Generate Proposals for Strengthening Governance:** We propose ways to enhance multi-level and multi-actor governance to address shared and common challenges more effectively.

Latin America and the Caribbean are at a pivotal moment of scientific and technological disruption, presenting both challenges and opportunities for their societies. This book seeks contributions from diplomats, decision-makers, researchers, scholars, private sector representatives, and organized civil society members. We believe this volume will be invaluable to those interested in the role of science diplomacy in addressing the region's challenges and opportunities.

The chapters cover a broad array of topics, including public diplomacy, technology diplomacy, innovation diplomacy, scientific heritage, global challenges, environmental diplomacy, science diplomacy education, knowledge diplomacy, health and vaccine diplomacy, scientific diasporas, multilateral diplomacy, evidence-based public policies, science communication, innovation ecosystems, and paradiplomacy. Each chapter is meticulously crafted to offer in-depth analysis and practical insights, contributing to the broader narrative of how science diplomacy can drive progress in Latin America and the Caribbean.

We are deeply grateful to our esteemed contributors for their dedication and expertise, which have been instrumental in bringing this project to fruition. It is our hope that *Developments and Approaches in Science Diplomacy: Latin America and the Caribbean* will not only inform but also inspire continued exploration and innovation in this burgeoning field. We look forward to seeing the advancements and collaborations that this work will undoubtedly inspire, as we collectively strive towards a more interconnected and scientifically advanced future for the region.

Chapter 1: A Critical Thinking Approach for Sustainable and Anticipatory Science Diplomacy

Emerging technologies in Latin America present significant challenges for decision-makers due to their rapid evolution, the region's limited expertise, and the necessity for individual skill development. This chapter delves into the current state of emerging technologies and decision-making processes in the region, highlighting the importance of sustainable thinking and future anticipation. Claudia Alarcon Lopez emphasizes the impact of individual cognitive processes on science diplomats and proposes a framework for anticipatory and sustainable decision-making. This framework aims to enhance critical thinking skills, targeting the root causes of current issues and fostering more informed, future-oriented policy development. The chapter offers actionable insights and strategic directions for navigating the complex landscape of emerging technologies, underscoring the need for foresight and inference awareness among individual decision-makers.

Chapter 2: A Taxonomy of Science Diplomacy From a Latin American and Caribbean Perspective

The chapter explores the essential relationship between science, technology, innovation, foreign policy, and decision-making in Latin America and the Caribbean. The authors argue that a specialized, context-sensitive science diplomacy is necessary, taking into account the region's history, cultural characteristics, and governance particularities. By addressing issues from local perspectives and predicting diplomatic and political maneuvers, the authors propose a taxonomy of science diplomacy—a tree of interconnected concepts and practices. This taxonomy serves as a framework for designing effective science diplomacy strategies tailored to the unique challenges and opportunities in Latin America and the Caribbean.

Chapter 3: The Necessity of Innovation Diplomacy in the LAC Region

With the rapid globalization of innovation ecosystems, the chapter highlights the emerging importance of Innovation Diplomacy in the Global South. The authors discuss how Innovation Diplomacy, focused on financial and commercial growth through collaborative innovation, can enhance international relations and improve other forms of diplomacy, such as economic and public diplomacy. They analyze the "Sao Paulo School of Advanced Sciences on Science and Innovation Diplomacy" (InnSciD-SP), emphasizing its relevance to the Latin American and Caribbean region. The chapter argues for a stronger focus on Innovation Diplomacy to drive mutual benefits and international cooperation.

Chapter 4: Synergies between Science Communication and Science Diplomacy to Address Global Challenges: Lessons From the LAC Region

This chapter examines the critical role of science communication in advancing science diplomacy in Latin America and the Caribbean. Through case studies on COVID-19 and climate change efforts, the authors highlight organizations like SENACYT, DiploCientifica, IAI, PRSPAN, and CienciaPR that act as intermediaries, translating scientific insights for policymakers. The chapter identifies key factors for effective science communication in science diplomacy and essential skills for building connections between scientists, diplomats, and policymakers. It underscores the importance of clear communication strategies to address global challenges collaboratively.

Chapter 5: Boosting Productivity through Science Diplomacy: German and Latin American Networks

The chapter showcases how science diplomacy fosters innovation and productivity by strengthening international networks through strategic science and technology initiatives. It highlights Germany's robust research infrastructure and Berlin's vibrant innovation ecosystem, emphasizing strategic academic frameworks, effective science communication, and public-private partnerships. The chapter explores how German science diplomacy in Latin America promotes institutional partnerships and academic exchanges, bridging cultural and geographical divides. These efforts enhance the scientific community and contribute to joint growth, addressing global challenges and promoting sustainable development through international cooperation.

Chapter 6: Geopolitics and Science and Technology Cooperation in Argentina, Chile, and Colombia: An Analysis of Diplomacy for Science Dynamics

This chapter addresses the diplomacy for science dynamics in Argentina, Chile, and Colombia, focusing on formal agreements with major economic powers such as the United States, the European Union, and China. Using a comparative case study approach, the authors identify and categorize science diplomacy instruments from 2018 to 2023. They argue that despite historical subordination to the agendas of economic powers, these countries are emerging as proactive partners, transitioning from recipients to viable collaborators in training and research. The chapter provides a taxonomy of science diplomacy instruments and highlights the evolving role of these countries in the global science diplomacy landscape.

Chapter 7: Environmental Science Diplomacy and Traditional Knowledge: Diplomacy Within National Borders in Latin America and the Caribbean

This chapter explores environmental science diplomacy within national borders in Latin America and the Caribbean, emphasizing the integration of traditional ecological knowledge. The authors discuss the challenges of overcoming cultural barriers and promoting equitable collaboration between diplomats, scientists, and local communities. They highlight the importance of inclusiveness, intercultural dialogue, and mutual respect in addressing global environmental challenges. The chapter underscores the need for diplomacy that acknowledges diverse epistemologies and approaches, fostering cooperation between Western science and traditional knowledge to develop effective, evidence-based environmental policies.

Chapter 8: Strategic Actions and Prospective Hydrodiplomacy in the Context of Transboundary Waters

The chapter focuses on hydrodiplomacy in the management of transboundary waters, with a case study of the Binational Puyango-Tumbes basin (Peru-Ecuador). The authors discuss the critical situation caused by mining activities and the need for sustainable water management. They argue for the consolidation of water security through hydrodiplomacy, considering fundamental principles of International Water Law. The chapter emphasizes the importance of strategic actions and prospective planning to address competition for water resources, ensuring sustainable development and cooperation between countries sharing watercourses.

Chapter 9: Traditional Knowledge Diplomacy: A Latin American Case Study on FILAC

This chapter presents a case study on the diplomatic activities of the Fund for the Development of the Indigenous Peoples of Latin America and the Caribbean (FILAC) from COP 15 onwards. The authors introduce the concept of Traditional Knowledge Diplomacy (TKD) and analyze how FILAC integrates Indigenous Peoples' perspectives into its diplomatic practices. Through semi-structured interviews and documentary reviews, they explore the roles of traditional knowledge in diplomacy, for diplomacy, and diplomacy for traditional knowledge. The chapter provides a methodological framework for investigating TKD and highlights its significance in complementing Science Diplomacy (SD).

Chapter 10: Vaccine and Health Diplomacy in Latin America and the Caribbean in the Pre- and Post-COVID-19 Context: Past and Present Learned Lessons as a Contribution to Sustainable Development in Our Region

The chapter examines the evolution of vaccine and health diplomacy in Latin America and the Caribbean before, during, and after the COVID-19 pandemic. The author discusses how the pandemic prompted new strategies supported by governments and international organizations to enhance scientific capabilities and human development. The chapter provides a comprehensive overview of vaccine development and foreign policy strategies, highlighting the importance of Health and Vaccine Diplomacy as key branches of Science Diplomacy. It addresses the challenges and benefits of these strategies, offering insights into their role in sustainable development and foreign policy in the region.

As editors of *Developments and Approaches in Science Diplomacy: Latin America and the Caribbean*, we are honored to conclude this pivotal work, which underscores the profound impact and potential of science diplomacy within the region. The chapters compiled in this volume paint a comprehensive picture of the multifaceted nature of science diplomacy, highlighting both its challenges and its transformative capabilities.

In Chapter 1, Claudia Alarcon Lopez's exploration of sustainable and anticipatory science diplomacy sets the stage by emphasizing the importance of foresight and critical thinking in navigating the fast-evolving landscape of emerging technologies. Her proposed framework for decision-makers underscores the need for cognitive agility and strategic foresight in fostering resilient policies.

Chapter 2's taxonomy of science diplomacy from a Latin American and Caribbean perspective, articulated by Luisa F. Echeverría King and her colleagues, provides a foundational structure for understanding the unique interplay between science, technology, innovation, and foreign policy in the region. This taxonomy is essential for tailoring science diplomacy strategies to local contexts and cultural specificities.

The focus on Innovation Diplomacy in Chapter 3, discussed by Almas Awan and co-authors, brings to light the critical role of collaborative innovation in enhancing international relations and economic growth. The insights from the Sao Paulo School of Advanced Sciences on Science and Innovation Diplomacy (InnSciD-SP) offer a valuable blueprint for leveraging innovation ecosystems in the Global South.

Branislav Pantovic and his team's examination of the synergies between science communication and science diplomacy in Chapter 4 highlights the essential role of effective communication in addressing global challenges. Their case studies provide practical examples of how science diplomacy can facilitate meaningful dialogue and cooperation among diverse stakeholders.

Chapter 5, authored by Hannia Bridg, presents a compelling case for the productivity-boosting potential of science diplomacy through German and Latin American networks. The analysis of Germany's robust research infrastructure and its application in Latin America underscores the importance of strategic international partnerships in fostering scientific and technological advancement.

In Chapter 6, Julian Prieto and his colleagues offer a nuanced analysis of the geopolitics and science diplomacy dynamics in Argentina, Chile, and Colombia. Their comparative case study approach reveals how these nations are transitioning from subordinate roles to proactive collaborators in the global science diplomacy arena.

Juan Reina Rozo and Angie Vargas Junco's exploration of environmental science diplomacy and traditional knowledge in Chapter 7 underscores the necessity of inclusive and intercultural dialogue in addressing environmental challenges. Their emphasis on the integration of traditional ecological knowledge highlights the importance of diverse epistemologies in formulating effective policies.

The strategic actions and hydrodiplomacy discussed in Chapter 8 by Bertha Garcia Cienfuegos and Felix Enrique Alcocer Torres provide crucial insights into the management of transboundary waters. Their case study of the Binational Puyango-Tumbes basin illustrates the need for cooperative and sustainable water management practices.

Chapter 9's case study on FILAC by Martha Isabel Gómez Lee and Santiago Gómez R introduces the novel concept of Traditional Knowledge Diplomacy (TKD). Their methodological framework for investigating TKD offers a valuable complement to scientific diplomacy, emphasizing the integration of Indigenous perspectives in diplomatic practices.

Finally, Danev Ricardo Perez Valerino's examination of vaccine and health diplomacy in Chapter 10 provides a timely analysis of the lessons learned from the COVID-19 pandemic. His discussion of the strategic importance of Health and Vaccine Diplomacy underscores its critical role in sustainable development and international cooperation.

As we conclude this volume, we are filled with optimism for the future of science diplomacy in Latin America and the Caribbean. The diverse perspectives and expertise presented herein offer a rich tapestry of knowledge and insights that we hope will inspire continued innovation, collaboration, and progress in this ever-expanding field. We extend our deepest gratitude to all our contributors for their dedication and scholarship, and we look forward to the advancements and collaborations that this work will undoubtedly inspire. Together, we can strive towards a more interconnected, scientifically advanced, and sustainable future for the region and beyond.

Luisa Echeverría-King
Universidad Simón Bolivar, Colombia

Branislav Pantović
Universidad Nacional de Río Negro, Argentina

Rafael Enrique Pineros Ayala
Universidad Externado de Colombia, Colombia

Pedro Figueroa
DiploCientifica, Chile

Alejandra Fabiola Flores-Zamora
Universidad de Guadalajara, Mexico

Chapter 1
A Critical Thinking Approach for Sustainable and Anticipatory Science Diplomacy

Claudia Alarcon Lopez
https://orcid.org/0000-0001-7026-1247
Swiss Federal Institute of Technology, Lausanne, Switzerland

ABSTRACT

Emerging technologies in LATAM are challenging for decision-makers, characterized by rapid evolution, limited expertise in the region, and a need for individual skills. To make impactful and resilient science diplomacy, sustainable thinking is required in current decision-making processes, often due to insufficient anticipation of future scientific developments. This chapter examines the existing status of emerging technologies and the current decision-making process. A significant focus is placed on the impact of individual cognitive processes of science diplomats presenting an anticipatory and sustainable decision-making framework designed to enhance critical thinking skills, targeting the root causes of prevalent issues of today. More informed, future-oriented policy development in science diplomacy are required. The chapter aims to offer actionable insights and a strategic direction for navigating the complex domain of emerging technologies, emphasizing the need for foresight and inference awareness in individual decision-makers.

DOI: 10.4018/979-8-3693-2746-3.ch001

INTRODUCTION

The rapidly evolving landscape of new and emerging technologies presents both unprecedented opportunities and significant challenges to global sustainability. As regions like Latin America and the Caribbean undergo significant technological transformation, it is imperative to approach these changes with robust critical thinking and forward-thinking strategies. By investing in the critical thinking abilities of both current professionals and future generations, we can ensure a sustainable and prosperous future for these regions.

Latin America and the Caribbean are at the forefront of leveraging technology for sustainable development. According to the Inter-American Development Bank, the adoption of digital technologies in Latin America has the potential to increase GDP by an estimated 3.7% by 2025, provided there is substantial investment in digital infrastructure and skills development (Inter-American Development Bank, 2021). The United Nations Economic Commission for Latin America and the Caribbean (ECLAC) also highlights that emerging technologies can play a crucial role in addressing key sustainability challenges (United Nations Economic Commission for Latin America and the Caribbean, 2021).

To navigate this complex terrain, this chapter provides an extensive examination of science diplomacy for emerging technologies. It introduces what emerging technologies are, how they are adopted, and the role of critical thinking in the science diplomats' decision-making process. The chapter uses Facione's critical thinking framework, established in 1990, to provide a structured approach to understanding critical thinking through its six dimensions: interpretation, analysis, inference, explanation, evaluation, and self-regulation. While Facione's framework focuses on critical thinking, this chapter extends its application to the realms of sustainability and anticipation.

The six dimensions of Facione's critical thinking framework are essential for comprehensively addressing the challenges posed by emerging technologies. Interpretation involves understanding and clarifying meaning, which is crucial for grasping complex technological concepts. Analysis refers to examining ideas and identifying arguments, helping policymakers dissect technological impacts and benefits. Inference is the process of drawing conclusions based on evidence, a vital skill for anticipating future technological scenarios. Explanation entails articulating findings and justifying decisions, necessary for communicating policy choices effectively. Evaluation assesses the credibility and relevance of information, ensuring that decisions are based on sound evidence. Finally, self-regulation involves reflecting on and adjusting one's own thinking processes, essential for continuous learning and improvement in decision-making.

While all these dimensions are crucial, this chapter places particular emphasis on inference. Inference is vital for drawing logical conclusions from evidence and making reasoned judgments, which are essential for effective decision-making in the context of science diplomacy. Inference allows science diplomats to navigate uncertainties and anticipate future scenarios, making it a cornerstone of strategic planning and policy formulation.

This work delves into the cognitive processes related to inference, with a particular focus on "inference to the best explanation." This approach is vital for generating hypotheses and anticipating future scenarios based on limited or incomplete data. By fostering these skills through targeted educational interventions and practical applications, we can equip current and future decision-makers with the tools they need to create more resilient, inclusive, and forward-thinking policies.

In addition to inference, the concept of anticipation plays a significant role in this chapter. Anticipation complements critical thinking skills by enabling science diplomats to proactively prepare for and shape future developments rather than simply reacting to them as they occur. By integrating foresight into the decision-making process, anticipation allows diplomats to predict and prepare for the long-term impacts of emerging technologies. This proactive approach ensures that decisions made today are aligned with sustainable development goals and can adapt to future challenges and opportunities.

Anticipation involves not only understanding current trends but also foreseeing future technological and environmental changes. When paired with the other dimensions of critical thinking, anticipation equips decision-makers to better navigate the uncertainties of technological progress. This enhanced skill set emphasizes the necessity of predicting and preparing for the long-term impacts of emerging technologies, ensuring that these advancements contribute positively to sustainable development goals.

Science anticipation and critical thinking are deeply intertwined, with anticipation providing a strategic outlook and critical thinking offering the tools to evaluate and implement these strategies effectively. This chapter will explore how these skills can be cultivated among science diplomats, emphasizing the importance of educational initiatives and continuous professional development. By investing in the development of these skills, we can ensure that science diplomats are well-equipped to address the complex and dynamic challenges posed by emerging technologies.

The hype cycle is another critical concept explored in this chapter. Developed by Gartner, the hype cycle describes the pattern of expectations and adoption of new technologies. It comprises several phases: the Innovation Trigger, Peak of Inflated Expectations, Trough of Disillusionment, Slope of Enlightenment, and Plateau of Productivity. Understanding where a technology lies within this cycle helps science diplomats manage expectations and strategize accordingly. It aids in recognizing the initial excitement and subsequent challenges that new technologies face before they reach stable productivity.

The hype cycle not only informs science diplomats about the stages of technology adoption but also emphasizes the importance of critical thinking at each stage. For instance, during the Innovation Trigger and Peak of Inflated Expectations, diplomats must critically evaluate the hype surrounding new technologies, separating realistic potential from exaggerated claims. As technologies enter the Trough of Disillusionment, critical thinking helps in assessing which technologies have genuine long-term potential and deserve continued support. During the Slope of Enlightenment and Plateau of Productivity, critical thinking ensures that strategies are adjusted to leverage the technologies' mature benefits fully.

Furthermore, the chapter will discuss the role of inference awareness training in enhancing the critical thinking capacity of science diplomats. By becoming aware of their own cognitive biases and the limitations of their inferential processes, diplomats can make more informed and objective decisions. This awareness is particularly important in the context of emerging technologies, where the pace of change and the novelty of innovations can easily lead to biased judgments and decisions.

In conclusion, this chapter will provide a comprehensive examination of the role of critical thinking in science diplomacy's emerging technologies decision-making. By integrating Facione's critical thinking framework with the principles of science anticipation and the hype cycle, we can equip science diplomats with the tools they need to navigate the complexities of technological advancements. Through targeted educational interventions and practical applications, we can foster a culture of continuous learning and improvement, ensuring a sustainable and prosperous future for Latin America and the Caribbean.

REGIONAL CONTEXT: LATIN AMERICA AND THE CARIBBEAN

Technological transformation in Latin America and the Caribbean is driven by various initiatives aimed at sustainable development. The region's leap into the future is marked by the integration of key emerging technologies such as quantum computing, biotechnology, clean energy, nanotechnology, and artificial intelligence. These technologies hold the key to addressing many pressing sustainability challenges in the region, which has a unique set of challenges and opportunities that influence the adoption and impact of these technologies. For instance, while there is significant potential for renewable energy, the infrastructure required to harness and distribute this energy is still developing. Similarly, the rich biodiversity in Latin America presents both opportunities for biotechnological advancements and challenges in ensuring sustainable and ethical practices. The digital divide is another critical issue. Despite advancements, there are significant disparities in access to technology and the internet, particularly between urban and rural areas. Addressing this divide is essential for ensuring that all communities can benefit from technological advancements. Initiatives such as increasing broadband access, promoting digital literacy, and supporting local technology startups are crucial steps toward bridging this gap. The region's socio-political landscape also plays a significant role in the adoption and implementation of emerging technologies. Political stability, regulatory frameworks, and government support are critical factors that influence the success of technological initiatives. Policies that promote innovation, protect intellectual property, and encourage foreign investment are essential for fostering a conducive environment for technological growth.

In order to promote decision-making processes in science diplomats leading the required multilateral approaches to promote the development, integration, and regulation of emerging technologies in the Latin American region, ensuring international cooperation for sustainable, ethical, and inclusive practices, critical thinking is crucial. Sustainable decision support systems exist to guide decision making based on science (Hersh, 1999). Even though the capacity building of science diplomats train them in decision-making practices, an individual approach to the cognitive aspects of thinking critically will be discussed in this chapter.

Critical Thinking

What is critical thinking? Even though we can all define it in our own way, a consortium of experts gathered in 1990 to define what dimensions could define the critical thinking construct in the most integral way. It is known as Facione's critical thinking framework, developed at the American Philosophical Association, and it provides a structured approach to understanding critical thinking through its six dimensions: interpretation, analysis, inference, explanation, evaluation, and self-regulation. These dimensions are important for navigating the complexities of science diplomacy and making informed decisions regarding emerging technologies. Each section is described here:

Interpretation

Interpretation involves understanding the subtle nuances of information and identifying underlying assumptions. Decision-makers must sift through a deluge of data and scientific findings, discerning relevant patterns and trends. This dimension extends beyond mere understanding to uncover biases and assumptions that influence judgment.

Analysis

Analysis involves breaking down complex information into its constituent parts to understand relationships and patterns. In the context of science diplomacy, this skill is vital for assessing the impacts of new technologies on various stakeholders, including governments, industries, and communities. Systematic exploration of the interactions between technological advancements, socio-economic factors, and geopolitical environments provides decision-makers with insights into potential synergies and conflicts, enabling more strategic and holistic policy responses.

Inference

Inference pertains to drawing logical conclusions from evidence and making reasoned judgments. Each innovation has the potential to shift power dynamics and alter the geopolitical landscape. Decision-makers must navigate uncertainty and ambiguity, often relying on probabilistic reasoning to make informed decisions. Evidence-based policymaking allows them to apply their inferential skills to mitigate risks and enhance resilience and adaptive capacity in the face of emerging issues.

Explanation

Explanation involves articulating coherent and persuasive rationales for decisions and proposals. Effective science diplomacy relies on the ability to communicate complex scientific concepts and policy advice to diverse audiences, including policymakers, scientists, and the public. Building consensus through compelling, evidence-based narratives fosters trust, which is essential for the acceptance and implementation of sustainable and inclusive policies.

Evaluation

Evaluation encompasses assessing the validity and reliability of information and arguments. In an era of rampant misinformation and disinformation, decision-makers must be vigilant and skeptical, sorting fact from fiction to ground their policies in solid empirical evidence and expert consensus. This vigilance protects the integrity of the decision-making process.

Self-Regulation

Self-regulation involves self-monitoring and reflecting on one's cognitive processes to recognize and address biases and shortcomings. In the fast-paced world of science diplomacy, practitioners must continuously foster intellectual humility and openness to alternative perspectives. Adopting a culture of continuous learning enhances the resilience and flexibility of decision-making frameworks, promoting agile and responsive governance.

Critical thinking is indispensable for science diplomacy today and in the future. The various dimensions of CT offer a comprehensive view of the personal thinking skills necessary in the field, enabling individuals to manage the complexities of emerging technologies with foresight and precision. Developing these skills through targeted capacity building and training in science diplomacy is crucial, especially in regions where such initiatives are relatively new (Mauduit et al., 2020). Additionally, anticipating the future impacts of emerging technologies is essential, and it must be considered in the science diplomacy processes to proactively prepare for and shape future developments rather than merely reacting to them, specially when talking about emerging technologies, which is why anticipation has been added in this chapter's framework (see figure 1).

Figure 1. Critical Thinking Skillset of a Science Diplomat Tackling Emerging Technologies

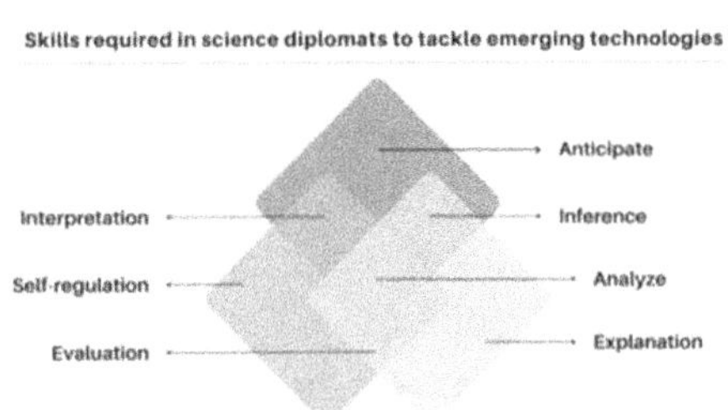

The skill of anticipation in science diplomacy may complement critical thinking by enabling science diplomats to proactively prepare for and shape future developments rather than merely reacting to them. This involves integrating a forward-thinking approach to predict and prepare for the long-term impacts of emerging technologies, ensuring alignment with sustainable development goals. Scenario planning and risk management are essential components of this approach. Anticipation in the context of sustainability involves imagining probable future scenarios and preparing for various possible outcomes. This forward-thinking approach means that LATAM can plan for and shape its future rather than just reacting to changes as they happen. For example, understanding the potential for growth in clean energy allows for the creation of policies that support sustainable energy projects, while in biotechnology, it helps anticipate new treatments or agricultural methods that could benefit the region's people and its environment. In order to foster forward-thinking, scientific development, and science diplomacy in an accessible manner, especially in the Global South, open science is required (Camacho Toro et al, 2024). An integral approach is required to tackle the challenges and opportunities scientific and technological advances bring to the region. In the next section, we discuss more about the emerging technologies that have been highlighted as relevant for both the academic and industrial communities.

Emerging Technologies' Hype Cycle and Science Diplomacy

Understanding the hype cycle is crucial for navigating the adoption and implementation of emerging technologies. The hype cycle, a concept introduced by Gartner, describes the pattern of expectations and adoption of new technologies. It comprises five key phases: Innovation Trigger, Peak of Inflated Expectations, Trough of Disillusionment, Slope of Enlightenment, and Plateau of Productivity. Additionally, it integrates stages of technology adoption, including early adopters, early majority, late majority, and laggards. Together, these elements provide a comprehensive framework for understanding how technologies evolve and are adopted over time (See figure 2).

1. **Innovation Trigger**: This phase marks the beginning of a technology's journey. A breakthrough or a potential technological innovation creates significant interest and publicity. Initial proof-of-concept stories and media interest trigger excitement, but commercial viability is not yet proven, and there are no usable products. Early adopters, who are technology enthusiasts, are typically the first to engage with the technology during this phase.
2. **Peak of Inflated Expectation**s: During this phase, early publicity produces a number of success stories—often accompanied by scores of failures. Some companies take action; many do not. The high expectations often lead to a bubble of unrealistic forecasts and predictions about the technology's potential. Early adopters drive the initial adoption, but their experiences may vary significantly.
3. **Trough of Disillusionment**: As time progresses, interest wanes as experiments and implementations fail to deliver. Producers of the technology shake out or fail. Investments continue only if the surviving providers improve their products to the satisfaction of early adopters. The initial excitement gives way to

skepticism and disillusionment. It is during this phase that the early majority begins to show interest, cautiously observing the technology's evolution.

4. **Slope of Enlightenment**: Although the technology has largely fallen out of public favor, some businesses continue through the Trough of Disillusionment and begin to understand the benefits and practical applications of the technology. More instances of how the technology can benefit the enterprise start to crystallize and become more widely understood. The early majority becomes more engaged as the technology's practical benefits become clearer.
5. **Plateau of Productivity**: Finally, the technology reaches the Plateau of Productivity, where mainstream adoption starts to take off. The benefits of the technology become widely demonstrated and accepted. Tools and methodologies become stable, and second- and third-generation products emerge. The technology's broad market applicability and relevance are clearly paying off. The late majority and laggards start to adopt the technology as it becomes more stable and proven.

Figure 2. Hype Cycle for Emerging Technologies and Adoption Rate in Science Diplomacy

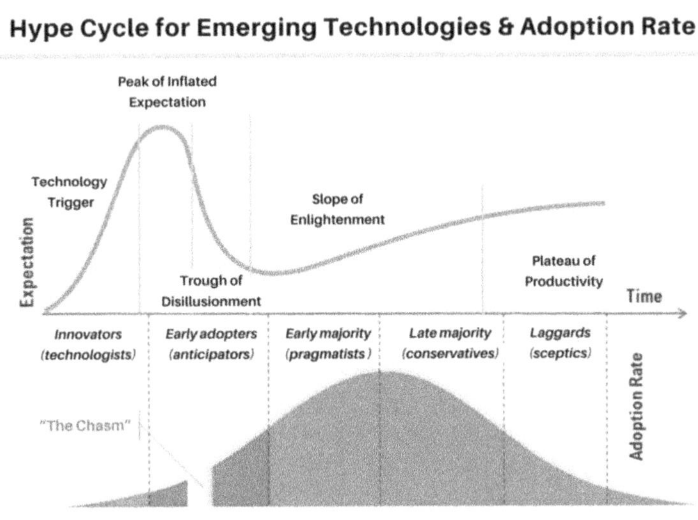

ADOPTION PROFILE AND THE DIFFUSION OF INNOVATION

Understanding the diffusion of innovation is critical for science diplomats, as it provides a roadmap for how new technologies spread through societies. This diffusion process is characterized by the adoption profile, which segments the population into different categories based on their willingness and speed to adopt new technologies. These segments include innovators (2.5%), early adopters (13.5%), early majority (34%), late majority (34%), and laggards (16%). Each segment has unique psychographic characteristics that influence their behavior and decision-making processes. For science diplomats, understanding these characteristics is essential to effectively promote and manage the adoption of emerging technologies.

Innovators (2.5%)

Psychographic Characteristics: Technologists

Innovators are the trailblazers in the adoption curve, representing 2.5% of the population. They are technology enthusiasts who are always on the lookout for the latest advancements. Innovators are characterized by their willingness to take risks and their passion for new ideas. They have the technical expertise to understand and experiment with new technologies, often being the first to test and provide feedback on these innovations. Innovators are driven by a desire to be at the forefront of technology, and they thrive on the excitement and challenges that come with exploring uncharted territories.

Role in Science Diplomacy: Innovators provide invaluable insights into the technical feasibility and potential applications of new technologies. They help identify early challenges and opportunities, which can form the basis for more comprehensive adoption strategies. For science diplomats, collaborating with innovators is crucial as they can offer firsthand knowledge about the capabilities and limitations of new technologies. Innovators' feedback can guide the development of initial regulatory frameworks and policies, ensuring that emerging technologies are aligned with broader societal goals.

Early Adopters (13.5%)

Psychographic Characteristics: Visionaries

Early adopters represent 13.5% of the population and are visionaries who see the potential of new technologies beyond their immediate applications. They are influential within their communities and often act as opinion leaders. Early adopters are motivated by the strategic advantages that new technologies can bring, and they are willing to invest time and resources to explore these possibilities. They are not just interested in the technology itself but also in its ability to drive innovation and change.

Role in Science Diplomacy: Early adopters play a pivotal role in validating new technologies and demonstrating their practical applications. Their endorsement can significantly accelerate the adoption process, as they provide social proof that the technology is valuable and viable. Science diplomats should engage with early adopters to gather detailed case studies and success stories that can be used to persuade the early majority. By leveraging the insights and experiences of early adopters, science diplomats can craft compelling narratives that highlight the benefits and address potential concerns associated with new technologies. Early adopters are also essential for science anticipation, as their visionary approach aligns closely with the principles of preparing for future technological trends.

Early Majority (34%)

Psychographic Characteristics: Pragmatists

The early majority consists of 34% of the population and includes pragmatists who adopt new technology once it has been proven and has become more widely accepted. Pragmatists are practical and deliberate in their decision-making, often seeking reassurance and validation from the experiences of early adopters. They are motivated by the practical benefits that new technologies can offer and prefer to see tangible results before making a commitment.

Role in Science Diplomacy: The early majority is critical for transitioning new technologies from niche applications to mainstream use. Science diplomats must focus on addressing the specific concerns and needs of this group, providing concrete evidence of the technology's benefits and reliability. Engaging the early majority requires a comprehensive approach that includes robust demonstrations, pilot projects, and detailed evaluations. By providing clear and practical information, science diplomats can facilitate the early majority's adoption of new technologies, ensuring that these innovations are integrated into broader societal and economic frameworks. The early majority's acceptance signals a move towards broader societal acceptance and integration of the technology, making it essential for diplomats to address their needs comprehensively.

Late Majority (34%)

Psychographic Characteristics: Conservatives

The late majority also represents 34% of the population and comprises conservatives who are skeptical about new technologies. They adopt new innovations only after they have been established and widely accepted by the majority. This group requires substantial evidence and social proof before changing their practices. They are often driven by a need for stability and prefer to adhere to traditional methods until change becomes inevitable.

Role in Science Diplomacy: Engaging the late majority involves building strong, evidence-based narratives and leveraging the social proof provided by the early majority. Science diplomats need to address the specific concerns of this group, emphasizing the long-term benefits and reliability of new technologies. This may involve extensive outreach and education efforts, as well as collaboration with community leaders and stakeholders who can influence this group's adoption decisions. By providing robust support and clear communication, science diplomats can help the late majority overcome their resistance to change and adopt new technologies in a way that ensures societal benefits. The late majority's adoption is crucial for achieving widespread societal integration of new technologies.

Laggards (16%)

Psychographic Characteristics: Skeptics

Laggards are the most resistant to change and represent 16% of the population. They are typically the last to adopt new technologies, preferring traditional methods and only adopting new innovations when it becomes absolutely necessary. Laggards are often characterized by a high degree of skepticism and a preference for the status quo.

Role in Science Diplomacy: While engaging laggards can be challenging, their eventual adoption is necessary for the complete diffusion of technology. Science diplomats should focus on providing compelling evidence and addressing specific barriers that prevent laggards from adopting new technologies. This involves understanding their motivations and concerns and tailoring communication strategies to address these issues. By demonstrating the inevitability and necessity of adopting new technologies, science diplomats can help laggards transition and ensure that no segment of the population is left behind in technological advancement. Engaging laggards requires patience and persistence, as well as targeted strategies that address their specific needs and reservations.

The critical transition point in the adoption process has been explained by Maloney's 16% Rule. It mentions that once a technology reaches 16% adoption, the strategy must shift from emphasizing scarcity to leveraging social proof to accelerate through the chasm to the tipping point. The chasm represents the gap between early adopters and the early majority. Crossing the chasm is crucial for achieving widespread adoption. That moment is where understatement of science anticipation is crucial. Dialogue with technologists and expert scientists of the area, monitoring on the advances, and forward-thinking are required to anticipate as accurately as possible, understanding the peak of inflated expectations and the trough of disillusionment stages will happen.

LINKING THE HYPE CYCLE TO SCIENCE DIPLOMACY DECISION-MAKING

For science diplomats, understanding where an emerging technology falls within the hype cycle is essential for making informed decisions. The hype cycle provides a roadmap for navigating the often tumultuous journey of technological adoption, highlighting the importance of tempering expectations during the early stages and recognizing the potential for significant advancements once the technology matures.

During the Innovation Trigger and Peak of Inflated Expectations phases, science diplomats must exercise caution. It is crucial to critically evaluate the claims and potential impacts of new technologies, ensuring that decisions are grounded in sound evidence rather than hype. This is where critical thinking skills, particularly interpretation and analysis, are invaluable. Diplomats must sift through the initial excitement to identify realistic opportunities and challenges.

In the Trough of Disillusionment, the role of science diplomats becomes even more critical. As interest wanes and skepticism rises, it is essential to maintain a balanced perspective. This phase requires robust evaluation and self-regulation skills to avoid abandoning potentially transformative technologies prematurely. By staying the course and continuing to support promising technologies through this phase, diplomats can help guide them toward the Slope of Enlightenment.

The Slope of Enlightenment and Plateau of Productivity are phases where science diplomats can leverage their anticipatory skills to promote widespread adoption of mature technologies. By understanding the practical applications and benefits that emerge during these phases, diplomats can craft policies and strategies that facilitate the integration of these technologies into broader societal and economic frameworks.

THE ROLE OF CRITICAL THINKING: THE INFERENCE DIMENSION

Critical thinking, as outlined by Facione's framework, is key in navigating the hype cycle. Each dimension of critical thinking plays a role in understanding and managing the phases of technological adoption. However, inference is crucial for generating hypotheses and anticipating future scenarios based on incomplete data (Boghossian, 2014).

Inference involves drawing logical conclusions from available evidence. In the context of science diplomacy, inference is essential for anticipating future scenarios and making decisions under uncertainty. Abductive reasoning, a key aspect of inference, allows science diplomats to generate hypotheses and explore innovative solutions based on incomplete data. This forward-thinking approach is essential for addressing the dynamic and evolving challenges posed by emerging technologies, and we will explain in detail.

Inference, AI, and Human Intelligence Development

In the field of human inference, knowledge text comprehension and understanding is the most researched field (Collins et al., 1977). In policymaking and diplomacy, text comprehension is the cornerstone of decision-making, but it may be attractive for automation through Large Language Models (LLMs) like ChatGPT. Due to the rapid development of LLMs, the inference field has expanded and gained importance. Large Language Models refer to advanced artificial intelligence (AI) systems that are capable of processing and generating human-like language. These models are typically trained on vast amounts of text data from the internet and other sources, allowing them to learn the nuances of language and context. ChatGPT is the most popular of these LLMs, and its impact on science diplomacy has been emerging as potential tool to aid decision makers, according to the recent European AI Act framework. Spector and Ma (2019) emphasize the importance of developing human intelligence alongside advancements in artificial intelligence. Their analysis concludes that more focus should be placed on human intelligence development with the support of artificial intelligence rather than prioritizing AI development at the expense of human cognitive skills. Conversely, the more we understand human inference skills, the more the LLM field can advance (Yiu et al., 2023). Hence, the importance of inference skill development to enhance critical thinking cognitive processes.

Predicting the future comes with its fair share of uncertainty even when based on scientific evidence. Data might be incomplete or misleading, and biases can skew our perceptions. Plus, unforeseen events can overturn even the most well-thought-out forecasts. Tackling these challenges requires an openness to new ideas, a willingness to learn from new developments, and the use of thorough systematic approaches to scenario planning. As a critical cognitive process that allows individuals to draw conclusions from available data, bridging the gap between known information and unknown phenomena, inference is a dimension to take attention to and to make sure inference awareness is present in decision makers.

The neuroscience of inferring

The cognitive process of inference is deeply rooted in principles of cognitive neuroscience. This discipline explores how brain functions underlie cognitive abilities, including how we draw conclusions from incomplete data. Multiple areas of the brain are involved in this inference elaboration, notably the prefrontal cortex, which is involved in decision-making, problem-solving, and applying logic (Miller & Cohen, 2001). This region coordinates with other areas like the hippocampus, crucial for memory formation, allowing individuals to apply previous knowledge to new situations. Furthermore, cognitive neuroscience emphasizes the role of neural networks in pattern recognition and prediction, essential components of inference (McClelland, McNaughton & O'Reilly, 1995). These neural circuits process sensory inputs and stored knowledge to generate predictions about future states, underpinning our ability to make informed decisions based on evidence.

Critical thinking can be done using either inductive, deductive, or abductive reasoning (Kennedy & Thornberg, 2018). There are three main types of inference: deductive, inductive, and abductive.

1. Deductive inference involves deriving specific conclusions from general principles or premises. It follows a top-down approach where conclusions logically follow from established facts or theories. For example, if all mammals are warm-blooded and a whale is a mammal, then we can deduce that whales are warm-blooded. Every premise must be true to be able to deduct a true statement always.
2. Inductive inference, on the other hand, involves drawing general conclusions from specific observations or examples. It operates on a bottom-up approach where patterns or trends observed in particular instances are generalized to broader principles. For instance, observing that every swan encountered is white may lead to the induction that all swans are white, though this conclusion can be disproven if a non-white swan is observed. The past is always used to explain future outcomes.

3. Abductive inference is a form of reasoning that involves making educated guesses or hypotheses to explain observed phenomena. Unlike deductive and inductive reasoning, which aim for certainty, abduction deals with the likelihood or plausibility of explanations. It is often described as "inference to the best explanation" (Harman, 1965). Abductive inference is particularly valuable in situations where there is incomplete information or uncertainty; the past alone cannot be used to fully conclude an outcome. It involves considering various possible explanations and selecting the one that best fits the available evidence, leading to explanations that may or may not be true.

Abductive inference is crucial, especially in the context of sustainability anticipation and emerging technologies, because it allows us to generate hypotheses and anticipate future scenarios based on limited data or unprecedented circumstances. In the face of new sustainability-related issues brought about by emerging technologies, past data or historical patterns may not always be applicable. Abductive reasoning enables decision-makers to explore innovative solutions and prepare for potential challenges by considering a range of possibilities beyond what has been previously observed or experienced. In essence, it empowers us to think creatively and adaptively in addressing complex sustainability issues shaped by rapidly evolving technological landscapes.

Forward-Thinking and the Role of Inferential Skills

Science anticipation is about forward thinking. It's a method of using what we know now to make educated guesses about what scientific and technological breakthroughs are coming, and how they might change our lives (Miller, 2018; Alvial-Palavicino, 2015). It's particularly important for emerging technologies like those described before. How does this converge with critical thinking? The skills of inferring and interpreting are especially important. This forward-thinking approach means that LATAM can plan for and shape its future rather than just reacting to changes as they happen. For example, understanding the potential for growth in clean energy allows for the creation of policies that support sustainable energy projects, while in biotechnology, it helps anticipate new treatments or agricultural methods that could benefit the region's people and its environment taking into consideration potential and probable risks. However, the task isn't straightforward, and it is not as easy at all given the multilateralism required due to the international nature of emerging technologies and sustainability global efforts.

Abductive inference is a highly present reasoning process used in science diplomacy, especially in the context of sustainability, anticipation, and emerging technologies, because it allows us to generate hypotheses and anticipate future scenarios based on limited data or unprecedented circumstances (Facione, 1990). In the face of new sustainability-related issues brought about by emerging technologies, past data or historical patterns may not always be applicable. Innovative solutions often are developed through abductive reasoning by considering a range of possibilities beyond what has been previously observed or experienced. In essence, it empowers us to think creatively and adaptively in addressing complex sustainability issues shaped by rapidly evolving technological landscapes. In figure 3, a description of the role of abductive inference in decision making is shown.

Figure 3. The Role of Abductive Inference in Forward-Thinking for Science Diplomacy

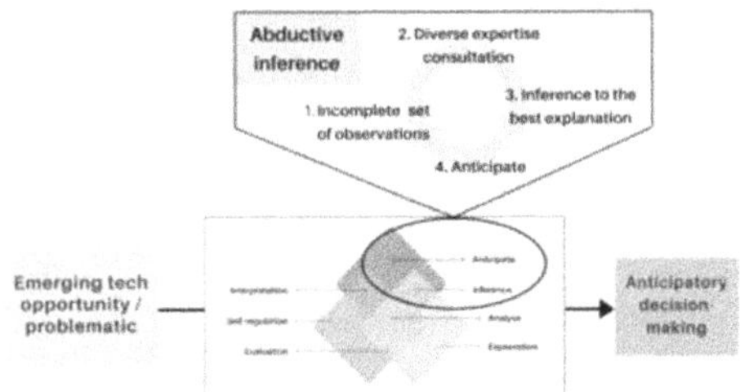

When we consider the process of abductive inference described in figure 3, it depicts a cyclical journey of understanding and decision-making, which starts with the acknowledgment of incomplete observations (1). This step is analogous to the scientific method's initial observations, which spur hypothesis generation. For emerging technologies, this means acknowledging the limitations of current data on their sustainability impacts. An incomplete picture might include unknowns about a technology's full environmental footprint, such as emissions during production, energy consumption during use, and waste at the end-of-life stage. Linton et al. (2007) emphasized the importance of a comprehensive life cycle assessment in determining the true environmental costs of a product or technology. Here, incomplete information necessitates a careful and cautious approach, relying on what is known while seeking further information to fill the knowledge gaps. Aiming for sustainability, this step is crucial in setting the stage for subsequent analysis and ensuring that no aspect of the technology's impact is overlooked due to insufficient data.In the realm of emerging technologies, this inquiry often grapples with sustainability questions: What are the environmental impacts of this new technology? Can it be produced, used, and disposed of in a way that aligns with our sustainability goals? Acknowledging that we start with partial information encourages humility and openness to learn and adapt as we gather more data.

Progressing to step (2) diverse expertise consultation, it expands the cognitive toolkit available for addressing these sustainability questions. It involves engaging with environmental scientists, engineers, ethicists, economists, and potentially affected communities. Each stakeholder group brings a unique perspective that, when integrated, provides a more holistic understanding of the technology's potential sustainability impact. This cross-pollination of knowledge is critical, as it challenges siloed thinking and uncovers interdependencies that may not be evident when viewed from a single perspective. Sustainability is inherently interdisciplinary, and each field brings valuable insights to the table. The consultation process benefits from the integration of perspectives from fields such as environmental science, economics, social science, and ethics. By consulting experts across a broad spectrum, a more holistic understanding of the sustainability of the technology in question can be achieved, and more effective strategies for its development and implementation can be formulated.

Inference to the best explanation (3) leverages the synthesized knowledge obtained from diverse experts. This abductive leap does not merely aim to fill gaps in understanding but strives to construct the most coherent and comprehensive narrative about the technology's sustainability trajectory. It seeks to answer questions such as: What is the most likely environmental footprint of this technology at scale? What sustainability challenges might we encounter, and how can we mitigate them? This reasoned conjecture is the driving force behind a pragmatic approach to sustainable technology development.Inference to the best explanation: When diverse experts have contributed their perspectives on an emerging technology's sustainability, the next logical step is to synthesize this knowledge into a coherent narrative—this is where 'inference to the best explanation' comes into play. This is not a simplistic filling of gaps; rather, it's a complex process of constructing the most plausible explanation for how the technology might interact with environmental, economic, and social systems.

This kind of inference requires a balancing act between competing explanations to identify which is most capable of comprehensively addressing the observations and insights at hand. It is at this juncture that the most pressing sustainability questions are addressed. Decision-makers engage in critical thinking to hypothesize about the technology's potential environmental footprint, its resource efficiency, or its energy demands at scale. Questions might include: How does this technology affect biodiversity? What are the implications for water and land use? Can waste be minimized or repurposed effectively? Moreover, it involves contemplating the indirect consequences that might not be immediately apparent, such as the potential for a new technology to exacerbate social inequalities or to displace existing sustainable practices. This inference acts as a crucial pivot, translating the broad-ranging insights into actionable strategies and policies aimed at sustainable technology development.

The robustness of this step cannot be overstated; it forms the bedrock upon which sustainable pathways for technological evolution are carved out.

Anticipation, the final step (4) extends the timeline of consideration beyond the immediate horizon. It's a forward-thinking exercise, an essential component of strategic sustainability. Anticipation is not merely about predicting the future but preparing for various plausible futures. For emerging technologies, this could mean considering the long-term availability of resources required for production, the technology's adaptability to regulatory changes, or its compatibility with circular economy principles. It also encompasses scenario planning for the unintended consequences that might arise from the technology's interaction with existing systems and societal structures. This stage goes beyond mere prediction; it is about preparing for a range of possible futures, each with its unique set of sustainability challenges and opportunities. In this context, anticipation might involve scenario planning, where different futures are imagined in detail, and strategies are developed for each. These scenarios could consider the effects of climate change, demographic shifts, policy changes, and economic fluctuations.

Anticipation is essentially about risk management and resilience building; it ensures that a technology's sustainability is future-proofed against a variety of potential developments. For example, how might a new biofuel source be influenced by shifts in agricultural policies or international trade agreements? Could advancements in energy storage technology respond to and accommodate the intermittent nature of renewable energy sources? In anticipating these various futures, the process acknowledges the dynamic nature of technology and society. This forward-thinking exercise ensures that when the time comes for the emerging technology to be fully integrated into society, it does so in a manner that enhances sustainability and aligns with long-term goals for environmental stewardship and social well-being. Through anticipation, decision-makers can develop flexible strategies that allow for adjustment as the technology matures and as new sustainability challenges emerge.

The process of abductive reasoning culminates in what is referred to as "anticipatory decision-making." This is a decision-making approach that, while grounded in the present, is dynamically oriented toward the future. It fuses the insights gathered from incomplete observations and expert consultations with the plausible narratives constructed during inference. The result is a series of actionable insights that inform strategies and policies. In the context of sustainability, this anticipatory approach is invaluable. Emerging technologies—from renewable energy systems to biodegradable materials—carry the potential to significantly alter our environmental footprint. Yet, they also carry risks and uncertainties. For example, a new technology might reduce emissions but increase water usage or demand for scarce materials. Anticipatory decision-making allows us to weigh such trade-offs proactively, integrating environmental, economic, and social considerations into the innovation pipeline.

The importance of this framework cannot be overstated in the face of global challenges. By being aware of the abductive inferences each decision-maker is making when evaluating the sustainability evaluation of emerging technologies, it can allow for a more self-aware bias understandment which, coupled with the understatement of the hype cycle for emerging technologies, it can improve a more realistic and holistic anticipation potentia. This process also emphasizes the dynamic nature of decision-making in the face of emerging technologies. It understands that decisions are not one-off events but part of a continuous process of adaptation and learning. As new information becomes available and the context evolves, so too must our inferences and anticipations. This iterative, reflexive approach is key to achieving sustainable outcomes in a world marked by rapid technological change and complex environmental challenges.

In summary, the abductive reasoning framework depicted in figure 3—through its steps of engaging with incomplete observations, consulting diverse expertise, making inferences, and anticipating impacts—can be a powerful approach to become self-aware of the realistic reasoning processes of fast-paced technological decisions. It aims to encourage a proactive, informed, and adaptive strategy for harnessing the power of emerging technologies to create a sustainable and desirable future when incorporated into a 6 dimensional critical thinking training. In emerging economies, the significance of abductive inference becomes amplified due to the pressing need for rapid adaptation and creative problem-solving in navigating dynamic technological landscapes. These economies, characterized by their agility and resilience, face unique challenges such as limited resources, infrastructure constraints, and socio-economic disparities.

The ability to have anticipatory thinking imagining probable future scenarios is crucial for leveraging emerging technologies to drive sustainable development. Abductive inference offers a valuable approach for decision-makers to address complex sustainability issues. Training this skill is important to be able to infer the most probable possibility and implement policies and frameworks that can take into consideration the needs of every region (Leo, 2008; Dewitz, 1987). Diverse cultural landscapes and entrepreneurial spirit provide fertile ground for innovative solutions in the region. However, traditional decision-making approaches, based more on deductive inferences, may fall short of addressing the intricacies of emerging technologies and sustainability challenges. By embracing ambiguity and uncertainty as a thinking exercise, decision-makers can generate novel hypotheses and strategies that capitalize on the region's strengths while mitigating its weaknesses, being open for dialogue, hence bringing all stakeholders, with first of importance experts of the current scientific knowledge, to the table. Currently, inferring, also colloquially referred to as guessing, is widely applied in policy-making, but evidence and science-based policymaking can be considered a new area (Likens, 2010).

Training inferential skills is crucial for science diplomacy, anticipatory science, and sustainability. Decision-makers must adopt a forward-thinking mindset and embrace a culture of innovation.

CONCLUSION

The intertwined relationship between emerging technologies and sustainability in Latin America and the Caribbean demands a multifaceted approach that incorporates critical thinking and forward-thinking strategies. As the region embraces quantum computing, biotechnology, clean energy, nanotechnology, artificial intelligence, and other emerging areas, it must do so with a keen awareness of the potential benefits and risks associated with these advancements. By investing in the critical thinking abilities of both current professionals and future generations, we can ensure a sustainable and prosperous future for Latin America and the Caribbean.

Facione's critical thinking framework provides a structured methodology for decision-makers to interpret, analyze, infer, explain, evaluate, and self-regulate their cognitive processes in crafting science-based policies and diplomatic strategies. Emphasizing abductive inference in this chapter underscores its crucial role in generating hypotheses and anticipating future scenarios based on incomplete data, which is essential for effective decision-making in the face of emerging technologies.

Inference awareness training enhances the ability of science diplomats to draw conclusions from evidence and explore innovative solutions, recognizing the preconceptions from past experiences, knowledge, and inherent biases. Tackling the awareness of our inferential processes is essential for achieving comprehensive critical thinking training, fostering a forward-thinking mindset crucial for strategic decision-making in science diplomacy.

The Hype Cycle, as presented in this chapter, provides valuable insights for decision-makers, helping them navigate the adoption and implementation of new technologies while avoiding the pitfalls of inflated expectations and leveraging the technology's benefits as it matures. Understanding the state of the art and its position in the hype cycle is crucial for viable and integral technology adoption strategies.

By comprehensively integrating Facione's critical thinking framework and emphasizing the role of abductive inference, policymakers can develop more effective strategies for addressing the multifaceted challenges of sustainable development. Through targeted educational interventions, practical applications, and a commitment to continuous learning, science diplomats can enhance their critical thinking skills and better navigate the complexities of the digital age. This holistic approach not only facilitates informed decision-making but also promotes the adoption of sustainable practices that align with the region's broader environmental and economic goals.

Science anticipation emerges as a complementary skill set, enabling decision-makers to proactively prepare for and shape future developments rather than merely reacting to them. By combining critical thinking with anticipation, stakeholders can navigate the uncertainties of technological progress more effectively and foster sustainable growth. This forward-thinking approach is essential for LATAM to harness the transformative potential of emerging technologies while ensuring that they align with the region's economic, social, and environmental goals.

Embracing science anticipation allows decision-makers to stay ahead of the curve, identifying emerging trends and potential challenges before they fully materialize. This proactive stance enables the formulation of strategic policies and interventions that promote innovation, resilience, and long-term sustainability. In essence, science anticipation complements critical thinking by providing decision-makers with the foresight and agility needed to navigate the rapidly evolving landscape of emerging technologies and steer LATAM towards a prosperous and sustainable future.

Bringing together scientists, policymakers, businesses, and community leaders to pool their expertise and perspectives is necessary for the best practices of science diplomacy. Through collective efforts, stakeholders will be able to enhance the reliability of science anticipation and make decisions that benefit everyone in the region both now and in the future. By embracing critical thinking and strategic foresight, Latin America and the Caribbean can pave the way for a sustainable and prosperous future driven by emerging technologies.

In summary, a multifaceted approach that integrates critical thinking and science anticipation is vital for navigating the complexities of emerging technologies and ensuring sustainable development in Latin America and the Caribbean. Investing in critical thinking education, promoting interdisciplinary collaboration, and fostering an anticipatory mindset will equip decision-makers with the tools they need to create a resilient and innovative future. This proactive and strategic approach will enable the region to harness the full potential of emerging technologies, drive sustainable growth, and improve the quality of life for all its inhabitants.

REFERENCES

Adler, J. E., & Rips, L. J. (2008). *Studies of Human Inference and Its Foundations*. Cambridge University Press. 10.1017/CBO9780511814273

Alavosius, M. P., & Houmanfar, R. A. (2020). Global warming: Analysis of behavior and organizational practices as climate impacts increase. In *Behavior science perspectives on culture and community* (pp. 221-256). Academic Press.

Alismail, H. A., & McGuire, P. (2015). 21st century standards and curriculum: Current research and practice. *Journal of Education and Practice*, 6(6), 150–154.

Alvial-Palavicino, C. (2015). The future as practice: A framework to understand anticipation in science and technology. *Tecnoscienza–Italian Journal of Science & Technology Studies*, 6(2), 135–172.

Babatunde, D. E., Denwigwe, I. H., Babatunde, O. M., Gbadamosi, S. L., Babalola, I. P., & Agboola, O. (2019). Environmental and societal impact of nanotechnology. *IEEE Access : Practical Innovations, Open Solutions*, 8, 4640–4667. 10.1109/ACCESS.2019.2961513

Boghossian, P. (2014). What is inference? *Philosophical Studies*, 169(1), 1–18. 10.1007/s11098-012-9903-x

Camacho Toro, R., Cumba Garcia, L. M., Galvis, L. A., Echeverría-King, L. F., Pantović, B., Alarcón-López, C., Suarez, V. R., Figueroa, P., Torres-Atencio, I., Widmaier, C., Fraga, T. R., & Benavides, S. (2024). The needed link between open science and science diplomacy -A Latin American perspective. [Frontiers.]. *Frontiers in Research Metrics and Analytics*, 9, 1355393. 10.3389/frma.2024.135539338903656

Dewitz, P., Carr, E. M., & Patberg, J. P. (1987). Effects of inference training on comprehension and comprehension monitoring. *Reading Research Quarterly*, 22(1), 99–121. 10.2307/747723

Doong, S. H., & Ho, S. C. (2012). The impact of ICT development on the global digital divide. *Electronic Commerce Research and Applications*, 11(5), 518–533. 10.1016/j.elerap.2012.02.002

Dwyer, C. P. (2017). Inference. In *Critical Thinking: Conceptual Perspectives and Practical Guidelines* (pp. 117–128). Cambridge University Press. 10.1017/9781316537411.009

Dwyer, C. P., Hogan, M. J., & Stewart, I. (2014). An integrated critical thinking framework for the 21st century. *Thinking Skills and Creativity*, 12, 43–52. 10.1016/j.tsc.2013.12.004

Economic Commission for Latin America and the Caribbean. (n.d.). *The Digital Economy for Latin America and Caribbean Initiative (DE4LAC)*. Retrieved from World Bank.

Ennis, R. H., Millman, J., & Tomko, T. N. (2005). *Cornell Critical Thinking Tests*. The Critical Thinking Co.

Ennis, R. H., & Weir, E. (1985). *The Ennis-Weir Critical Thinking Essay Test*. Critical Thinking Press and Software.

Facione, P. (1990). *Critical thinking: A statement of expert consensus for purposes of educational assessment and instruction (The Delphi Report)*. American Philosophical Association.

Facione, P., & Facione, N. (1992). *The California Critical Thinking Dispositions Inventory (CCTDI); and the CCTDI Test Manual*. The California Academic Press.

Harman, G. H. (1965). The inference to the best explanation. *The Philosophical Review*, 74(1), 88–95. 10.2307/2183532

Hersh, M. A. (1999). Sustainable decision making: The role of decision support systems. *IEEE Transactions on Systems, Man, and Cybernetics. Part C, Applications and Reviews*, 29(3), 395–408. 10.1109/5326.777075

Inter-American Development Bank. (2021). *IDB Report Highlights Future Growth Opportunities for Latin America and the Caribbean*. Retrieved from IDB.

Kennedy, B. L., & Thornberg, R. (2018). Deduction, induction, and abduction. In *The SAGE Handbook of Qualitative Data Collection* (pp. 49-64). 10.4135/9781526416070.n4

Leo, P. D., & Greene, A. J. (2008). Is awareness necessary for true inference? *Memory & Cognition*, 36(6), 1079–1086. 10.3758/MC.36.6.107918927026

Likens, G. E. (2010). The role of science in decision making: Does evidence-based science drive environmental policy? *Frontiers in Ecology and the Environment*, 8(6), e1–e9. 10.1890/090132

Mauduit, J. C., & Gual Soler, M. (2020, August). Building a science diplomacy curriculum. *Frontiers in Education*, 5, 138. 10.3389/feduc.2020.00138

McClelland, J. L., McNaughton, B. L., & O'Reilly, R. C. (1995). Why there are complementary learning systems in the hippocampus and neocortex: Insights from the successes and failures of connectionist models of learning and memory. *Psychological Review*, 102(3), 419–457. 10.1037/0033-295X.102.3.4197624455

Miller, E. K., & Cohen, J. D. (2001). An integrative theory of prefrontal cortex function. *Annual Review of Neuroscience*, 24(1), 167–202. 10.1146/annurev.neuro.24.1.16711283309

Miller, R. (2018). *Transforming the future: Anticipation in the 21st century*. Taylor & Francis.

Rotolo, D., Hicks, D., & Martin, B. R. (2015). What is an emerging technology? *Research Policy*, 44(10), 1827–1843. 10.1016/j.respol.2015.06.006

Soto, G. H. (2024). The effects of foreign direct investment on environmentally related technologies in Latin America. *Resources Policy*, 90, 104711. 10.1016/j.resourpol.2024.104711

Spector, J. M., & Ma, S. (2019). Inquiry and critical thinking skills for the next generation: From artificial intelligence back to human intelligence. *Smart Learning Environments*, 6(1), 1–11. 10.1186/s40561-019-0088-z

United Nations Economic Commission for Latin America and the Caribbean. (2021). *Emerging technologies and sustainability in Latin America and the Caribbean*. Retrieved from https://www.cepal.org

Yiu, E., Kosoy, E., & Gopnik, A. (2023). Imitation versus Innovation: What children can do that large language and language-and-vision models cannot (yet)? *arXiv preprint arXiv:2305.07666*. 10.31234/osf.io/kt9es

Chapter 2
A Taxonomy of Science Diplomacy From a Latin American and Caribbean Perspective

Luisa F. Echeverría King
https://orcid.org/0000-0002-5172-1487
Universidad Simón Bolívar, Colombia

Rafael Enrique Piñeros-Ayala
https://orcid.org/0000-0001-5539-9395
Universidad Externado de Colombia, Colombia

Branislav Pantović
https://orcid.org/0009-0005-5561-0476
Universidad Nacional de Río Negro, Argentina

Alejandra Fabiola Flores-Zamora
https://orcid.org/0000-0002-1783-615X
Universidad de Guadalajara, Mexico

Pedro Figueroa
DiploCientifica, Chile

ABSTRACT

The prospect for the solution of the most urgent challenges in Latin America, and also the projection of the countries and the region, is associated with the strength of the relation between science, technology, innovation, foreign policy, and decision making. This nexus requires a special, tailor-made science diplomacy that takes

DOI: 10.4018/979-8-3693-2746-3.ch002

into account the history, culture's characteristics, and particularity of governance. In other words, it is beneficial to start addressing issues from local perspectives, knowing its contexts and needs, and on the other hand, knowing how to calculate and predict diplomatic-political maneuvering and productivity. Therefore, an appropriate response to a challenge via science diplomacy should include the design of science diplomacy taxonomy, a tree of interconnected concepts and practices.

INTRODUCTION

While the studies of science diplomacy (SD) possess a historical lineage, a conspicuous emergence of the practice occurred during and after the Second World War. This pivotal period was catalyzed by the introduction of nuclear weapons of mass destruction which propelled technology into the forefront of the international arena (Vera & Echeverría King, 2022). Also, the significant surge in the conceptual application of science diplomacy since the formulations presented in 2010 by the Royal Society and the American Academy for the Advancement of Science has had a profound impact on theoretical discourse and practice worldwide. Its essentially transdisciplinary nature has contributed to fostering political relations that place science in a preponderant role. Likewise, it helps different actors beyond the State to find ways to use it and adapt it to the realities of the international agenda that best suit them, with the intention of favoring collaborative agreements at different levels of international governance.

SD nowadays is gaining global prominence from a political, technical, and practical perspective. Namely, the diverse nature of science diplomacy is evident through three main facets: bilateral and multilateral collaboration, policy and decision-making, and public engagement and international outreach. All this encompasses diplomatic interactions, policy shaping through scientific evidence, public engagement with non-state actors, etc.

Nevertheless, the majority of studies have been focused on SD practices in developed countries (Echeverría King et. al., 2023). This often obscures unique challenges and dynamics inherent in SD across diverse geopolitical contexts, thereby impeding a comprehensive understanding of the field. For instance, in the Global South, SD gives bigger importance to actors like universities or non-state entities. Moreover, it is noteworthy that discussions on SD in LAC are emerging. The SD landscape in the Latin American and Caribbean (LAC) region is distinguished by a diverse array of practices and frameworks, ranging from the formation of alliances to bolster Science, Technology, and Innovation (STI) systems to the integration of LAC nations into the knowledge economy. These actions often highlight concrete examples of national cases to illuminate the issues at hand (Vera & Echeverría King, 2022).

Despite the region's diverse historical, political, institutional, and socio-economic contexts leading to significant disparities across education, science, technology, and economic development, there has been a noted increase in research and development investment (Paz-Enrique et al., 2022). This trend is a characteristic driven by a discernible scientific revolution, marked by advancements in informatics, genetics, automation, and artificial intelligence, indicating a profound evolution in the region's technological landscape (Mejía Navarrete, 2020). However, it is crucial to acknowledge that investment in science, technology, and innovation (STI) remains relatively low overall compared to other regions of the world (Álvarez, 2023).

Furthermore, the evolving roles of academic actors in the development agendas of LAC nations is a characteristic that underscore the significance of understanding the interplay between academia and policymaking for effective science diplomacy strategies. For instance, Argentina's bid to host the 2027 World Expo emphasized nature and technology, particularly focusing on clean energy and human innovation. Notwithstanding the initiative's unsuccessful outcome, it demonstrated national unity and broad political support for promoting STI, with the shared goal of positioning Patagonia as a key player in clean energy production (Ministerio de Relaciones Exteriores, Comercio Internacional y Culto Argentina, 2023).

Despite the absence of a unified approach to SD in the LAC region, it is crucial for LAC countries to establish their own SD policies and promote instruments tailored to their specific needs and challenges. In this context governance frameworks are essential to facilitate dialogue among diverse actors, develop capacity-building strategies, and mobilize resources for STI. (Vera & Echeverría King, 2022). As an illustration, Panama and Peru have developed national frameworks for science diplomacy, managed by their Ministries of Foreign Affairs, while in Brazil, Colombia, Chile, and Mexico there is notable interest and progress in advancing the practice through initiatives such as the creation of university courses and government participation (Echeverria King et al., 2023; Gual Soler, 2021).

In this chapter we're targeting the design of SD taxonomy, a systematic classification and identification of various aspects, activities, actors and dimensions of SD. It is important to mention that our perspective is from the LAC region. By integrating this element into SD taxonomy, we desire to capture the specificities of it, thereby enhancing understanding, collaboration, and impact of this practice in the mentioned region.

We trust that this chapter will provide a framework for comprehending and analyzing SD that is both relevant and applicable to the unique context of the LAC region. Specifically, our aim is to demonstrate the significance of considering diverse approaches to executing and influencing with/by science diplomacy.

THEORETICAL FRAMEWORK

This theoretical framework aims to integrate the importance of taxonomic classification with the conceptual evolution of science diplomacy. Thus, the purpose of this section will be to generate a general exploratory framework that relates the relevance of taxonomic classification within the study of science diplomacy, with the intention of incorporating its most salient features, as well as the evolutionary process of the concept of science diplomacy.

Taxonomy and conceptual definition

The classification and organization of scientific knowledge has been a major concern throughout history. Observing the world around us with the intention of identifying perceptible characteristics, patterns of behavior, organizing them in a logical and coherent manner so that they have a useful meaning that can be generalized through concepts and theories, has been the source of inspiration not only for the development of the sciences in general, but at the same time for them to be used by society for the solution of problems of various kinds. Scientific knowledge, in other words, requires the understanding of patterns that have facilitated its birth, evolution and transformation and to project solutions to common problems.

This is why the evolution and specialization of scientific knowledge generates another important challenge, which consists in overcoming disciplinary boundaries. While disciplinary knowledge provides certain limits that facilitate the development of concepts, theories and methods, among others, it also favors the generation of controversies both within and across disciplines, whereby common and cross-cutting problems in contemporary multi-actor and multilevel governance involve making decisions ranging from global or regional agreements to the implementation of public policies at the smallest level of public administration. It also implies a growing variety of agendas, such as the climate crisis, food security, energy transition or the regulation of artificial intelligence.

Another effect of overcoming state-based decision making involves the inclusion of other actors. In this sense, the inclusion of private sector and civil society actors, such as universities, think tanks or NGOs, is considered a key element, as pointed out by Manassero-Mas & Vázquez-Alonso (2019), scientific knowledge is characterized by having a complex nature (interdisciplinary), dialectical experience (evolutionary); multifaceted expression (diverse perspectives) of analysis and controversial positions.

Thus, the creation of scientific knowledge through processes such as multidisciplinary, interdisciplinary or transdisciplinary[1], facilitates breaking the barriers that have traditionally been linked to a disciplinary structure, and is aimed at combining, superimposing, evolving or transgressing, static concepts and theories to generate new intellectual constructs (Klein, 2017).

The concept of science diplomacy is incorporated into the aforementioned transdisciplinary logic, insofar as it aspires to include diverse actors with the intention of promoting diverse purposes. Although the concepts behave in a differentiated manner according to the discipline of study, they face a constant struggle in terms of redefinition that the practical or theoretical evolution points out, with which, a good concept must be able to precisely define the phenomenon to be studied and establish certain indispensable properties or attributes -characteristics-, necessary for it to be considered relevant (Gerring, 1999). The above places us in a scenario in which, although the definitions provided on science diplomacy are recent, the relationship practices between political and diplomatic decision-making, on the one hand, and science, technology and innovation, on the other, have a much broader development in the history of mankind.

When referring to the relevance of taxonomic classification, it is understood as the need to include indispensable or necessary elements for theoretical or conceptual development. In the 1970s, the capacity of taxonomic classification was defined as the ability to specify the basic constructs through which the relevant domain will be defined and the definitional relationships between those constructs (Singer, 1971). In addition, in the words of Hudson, taxonomic classifications "identify a minimum number of properties that are sufficient to demarcate one concept from another, which is established from social practices in a variety of contexts" (Hodgson, 2019, p. 208). In that sense, taxonomic classification refers to categorizations useful for differentiating the meaning of one thing from another.

Although the birth of taxonomic classification took place in the natural sciences and that in the social sciences the definition of even the most basic concepts is not free of debate, discussion or interpretation, this does not mean that it is not a widely used technique. Taxonomic classifications are aimed at clarifying the essential characteristics of a concept in order to improve its analytical capacity, improve its proper use and contribute to its development depending on how it is being used in the practical world (Hodgson, 2019). In the words of Kales, taxonomic classifications are to minimize the controversy about the different types of objects under investigation (Klaes, 2004).

While the intention of taxonomic classification is not to delve into the specific characteristics that make up a concept, it does contribute to forging a general and broad consensus on the meaning of concepts, which is indispensable for the development of subsequent theories. In terms of its purpose, taxonomic classification establishes in a nominal and non-specific way the characteristics of a concept, provides certain shared conditions aimed at describing or explaining the scope that a scientific community has on something it is investigating or developing and, at the same time, are useful for finding novel ways of explaining phenomena that were previously not understandable, opening new research agendas that receive inputs that had previously not been contemplated (Hodgson, 2019; Harrison & Singer, 2006).

From the point of view of the ontological and methodological perspective, taxonomic classification contributes to understanding the importance of clarifying different types of codifications. That is why taxonomic classification facilitates the generation of certain objectivity, determination of characteristics, reproducibility on the occurrence of certain phenomena, differentiation from other definitions and certain acceptability in terms of knowledge production, by combining the observation of the real world, scientific and technical production with the world of practices exercised or executed by a community of experts, in a kind of methodological pluralism that favors the adaptation and combination of various methods for the production of knowledge (Scartozzi, 2018; Manassero-Mas & Vázquez-Alonso, 2019).

Reflection on the taxonomy of science diplomacy places the discussion on the relevance of questioning and eventually rethinking what has been defined as such, with the purpose of identifying practices, processes or relationships that develop in the real world and that contribute to the general evolution, and in particular in LAC on the very concept that is being worked on. Typologies and classifications are not neutral or static but reflect political choices aimed at including or excluding those particular activities contained within specific categories (Klein, 2017), which have the possibility of evolving, changing or modifying over time (Doty & Glick, 1994).

Thus, the interdisciplinary construction that has been generated on SD has had as its axis the integrative character, that is, the intentionality of aspiring to constitute a kind of structural generalizations that integrate sequences or interactions between actors, accompanied by a progressive conceptual transformation according to the uses and practices it acquires.

Likewise, in the words of Goertz, an adequate conceptual definition involves a theoretical and empirical part that has implications in the way the concept is used to characterize its object of study, whereby the concepts must provide explanations and causal mechanisms that are related to the empirical phenomenon they are analyzing (Goertz, 2006).

Finally, concepts must have certain attributes necessary to maximize their usefulness. In this sense, the concept must be familiar and acceptable to the scientific community that uses it, it must provide resonance in terms of the production it intends to contribute to develop, it must be clearly differentiable from other even similar concepts, and it must provide a certain degree of depth and both theoretical and empirical usefulness (Gerring, 1999).

Evolving Science Diplomacy

This section evidences the conceptual evolution of SD and its adaptation to different taxonomies, reflecting both its frontier nature, i.e. it inhabits different epistemic communities, as well as its transdisciplinary character, to the extent that it overlaps, integrates and transcends the traditional boundaries of disciplines with the aim of achieving a higher degree of precision adjustable to different actors of global governance.

When emphasizing SD approaches, it is necessary to bear in mind the particular context in which they are created and the scope and attributes they have been acquiring, to the extent that they contribute to a better understanding of the type of dynamic conceptual taxonomy that is emerging mainly in the 21st century.

As a starting point in terms of definitions, the one provided jointly by the Royal Society and the American Association for the Advancement of Science (AAAS) in 2010 will be used, as it continues to be a relevant meeting point that has also facilitated its subsequent evolution. Thus, SD is understood to be: (i) Science in diplomacy involves scientific knowledge influencing politics and policy in the international arena. (ii) Diplomacy for science facilitates and promotes scientific activities. (iii) Science for diplomacy enhances international relations by fostering scientific collaboration among various stakeholders. (The Royal Society, 2010).

SD can also be understood as a boundary object between different and consolidated disciplines, i.e., science, technology and innovation (STI) studies and International Relations. Boundary objects are characterized by linking different and in some cases antagonistic scientific communities, focused on the improvement, evolution or achievement of results of an idea that they share but observe and understand differently (Star, 2010; Fox, 2011). In addition, boundary objects are characterized by having a common object-idea, abstraction, representation or practice-that inhabits diverse social communities and that crosses their needs and interests in the pursuit of common purposes (Piñeros Ayala, Echeverría-King, & Andrade-Sastoque, 2022). This kind of conceptual adaptations from various disciplines contribute to the development of new frameworks of reference, joint but not exempt from contestation.

Under this perspective, SD emerges and develops as a world of encounter between scientific and technological advancement, on the one hand, with the governance of decision making in a multilevel and multiactor governance, on the other hand, and which, still does not develop a fluid and dynamic ecosystem between those, which contributes to the solution of common and transversal problems among different actors (Kaltofen & Acuto, 2018a). From this perspective, SD has a character of practical relationship between actors who aspire to the development and evolution of scientific knowledge, while at the same time a space for research, which focuses on the study of the phenomenon and its application.

The meaning and scope of the concept of SD has recently sought to move away from definitions whose central axis orbits around the decision making made by the State through its foreign policy or domestic regulations, with the intention of accounting for a broader perspective that incorporates the contributions of other actors, especially think tanks, civil society and even megacities that see how the local development of ITS ecosystems brings advantages for the promotion of interests at the local level. This situation may generate some conceptual ambiguity, because, for each actor, SD means a different scope and purpose.

However, what may initially be seen as a drawback opens space for the construction of new taxonomic categories capable of providing more precise explanations of a constantly evolving concept and practical phenomenon. For some, SD has acquired a holistic sense that must be perceived as an understanding differentiated magnitudes and impacts ranging from global decision making to local effects. This is why, "SD is understood as the concept and skills necessary for informed decision making and evolving from policy as a product to diplomacy as a process that includes STI" (Berkman, 2019, p. 72).

This approach implicitly welcomes the perspective of others who are inclined to see SD as scientific recommendations and who privilege scientific participation for decision-making (Gluckman, 2016). Though, noting that, this process should include at least three elements: the perspective of those who are involved or participate in decision-making, the evidence from scientific studies that generate a solid to guide informed decisions and, finally, the construction and strengthening of governance mechanisms -at different levels- for better implementation (Berkman, 2019).

Those orientations that have a nuanced approach to the definition of SD present varied interpretations, especially aimed at strengthening capacities and improving economic, social and scientific performance.

Cities plays an indispensable role in fostering horizontal relationships, aimed at strengthening or consolidating cross-cutting innovation and development ecosystems; perfecting mechanisms for interaction and communication among multiple actors to achieve the Sustainable Development Goals (SDGs), particularly Goal No. 11, which focused on creating a safer, more resilient, inclusive and sustainable cities. Finally, cities become a natural laboratory for the implementation of public policies with an ITS focus, capable of addressing the multilevel and cross-cutting challenges of the contemporary agenda (Gluckman, Turekian, Grimes, & Kishi, 2017; Abdullah & Roig, 2021).

The link between International Relations, as a discipline, and STI studies, as a field of experimentation and advancement of knowledge, also contribute to weave a perspective that relates the theoretical contributions of the former to the latter, as well as the disruptive, evolutionary and dynamic capacity of the latter in decision-making in the world of politics and international relations at least in three specific ways.

First, it is significant to note how the explanations provided on science and technology in International Relations have been linked to the way in which they affect a specific research agenda, i.e. climate crisis, international security, energy transition, but not necessarily to the evolutionary character they can produce within the international discipline. Science and technology are a central part of discussions in the international arena such as climate crisis, inputs and medicines to combat pandemics, nuclear security and disarmament or sustainable development (Kaltofen & Acuto, 2018b) in which States, companies and civil society, must find solutions to common and cross-cutting problems.

Thus, the core ideas constructed from the international perspective are imbued with an explanatory logic based on the major international paradigms, such as realism, liberalism or global governance. As for the first, it is inclined to link SD and technological development as a need to improve or expand the international and national security capabilities of the State (Wagner, 2002; Flink & Schreiterer, 2010); the second favors the increase of multilateral relations and the improvement of international regimes for the advancement of science, including the development of soft power as an element of scientific promotion (The Royal Society, 2010) and, finally, the third focuses on the added value that actors other than the State contribute to the advancement of scientific knowledge, such as epistemic communities (Bueger, 2015), while generating more democratic and participatory decision-making (Kaltofen & Acuto, 2018b).

Secondly, SD also contributes in various ways to strength the State Foreign Policy to other actors, as Intergovernmental Organizations, private sector or even NGO´s. With some elements previously exposed, for Flink & Schreiterer (2010), SD has a close relationship with foreign policy, to the extent that diplomatic action in relation to STI, should contribute to three important objectives: first, to expand the access of researchers and research centers to technology and relationships with other leading centers in their fields of study; second, STI contributes to promote national capabilities in international spaces and, third, collaborates in actions aimed at improving influence in international spaces.

Others who have the function of contributing to the development of both the relationship between STI and foreign policy through the development of national STI systems. In order to improve the performance of STI at the national and local levels, countries should implement foreign policy strategies that allow them to approach other countries with higher levels of development, with the intention of having access to founds to research, better training and capacity building programs, strengthening the relationship between actors in the STI ecosystem and, finally, improving co-production and collaboration between researchers, and improving the quality and effectiveness of their research (Kaltofen & Acuto, 2018a).

Thirdly, the relationship between science and scientific knowledge are a disruptive factor and intrinsically related to the international sphere, becoming that neither its study nor the practice it generates should be separated. For some, there are no international relations if there is no relationship with STI, and there is no STI if international relations are not cultivated or improved at different levels (Flink & Rüffin, 2019), whereby the transformational factor of science and technology on the international arena must be recognized.

In this sense, they have historically had a disruptive character to the extent that: first, it contributes to the ramification and specialization of international relations, based on the evolution of production in various global and national agendas; second, it becomes a player of change, promoting the advantages of some as opposed to the loss of competitiveness that others may have; thirdly, it causes situations of risk and instability derived from the need to reach new consensus within the community for the management of certain issues or agendas that do not have an adequate legal framework and, finally, scientific knowledge stimulates the development of new forms of production of goods and services, thus opening spaces for the renewal of foreign policy relations between states and between them and other international actors (Weiss, 2015).

As we have seen, SD adapts to particular contexts, so it is worth noting some preliminary approaches that have been developed in LAC. This region groups together several countries with profound differences, such as Haiti, the least developed country in the hemisphere, and Brazil or Mexico, which belong to the group of countries with larger economies such as the G-20, thus showing a profound heterogeneity in the meaning, available means, technical and financial resources, tools and scope of individual SD strategies. We point to individual strategies because at the regional level there is no clear or precise definition that is applicable.

Without expanding on the particular approaches that have been developed in recent years around SD in certain countries of the region, which would open up a space for analysis that goes beyond the scope of this chapter, certain noteworthy trends can be observed in the region. On the one hand, the development of training and capacity building programs in SD and ITS carried out by some regional intergovernmental organizations with a scientific vocation. On the other, the promotion of certain training programs within diplomatic academies and educational centers aimed at improving and strengthening national capabilities (Echeverría King, et al., 2023) and, finally, national strategies that adapt to the convenience of national capabilities and needs to solve internal needs or project external competitiveness (Gual Soler, 2021).

In short, science and diplomacy are interconnected on multiple levels, from international scientific cooperation to the promotion of good understanding between states to the resolution of global challenges. The international community sees science diplomacy today playing a significant role in building bridges between different cultures and communities and in finding joint solutions to some of the most pressing global challenges.

Approaches to the Taxonomy of Science Diplomacy

Science diplomacy is a constantly evolving field. The current literature has not yet agreed on a single, universally accepted taxonomy. However, there are several ways of classifying practices and actions within science diplomacy. Some of the main taxonomies of SD may include:

a) According to the number of stakeholders
   - Bilateral science diplomacy: cooperation between two countries, often those sharing a border. (See Rusieshvili, A. 2020).
   - Multilateral science diplomacy: cooperation involving three or more countries that are often united by a common interest. (See Bompongo Nkombe A. 2020).
   - Global science diplomacy: cooperation involving mainly international organizations of universal character (See Nair-Bedouelle, Sh. 2023).
   - Non-governmental science diplomacy: efforts by academic institutions, research centres, non-governmental organizations or private companies to promote international cooperation in research and development. (See Maass, A. et al., 2023).
   - Transregional science diplomacy: a tool with potential for the LAC region, which could address regional challenges and needs, and seek out new opportunities for integration into an increasingly dynamic world that also poses challenges in maintaining interactions under a North-South logic (Vera, et. al. 2023). It can function not only as a multi-regional bridge but also as a tool for LAC integration according to Gual Soler (2014).

b) According to the purpose.
   - Cross-border cooperation on science and technology (see Cormann, M. 2023).
   - Use of science and technology to promote foreign policy objectives (see López, D., and van Klaveren, A. 2023).
   - Cultural and educational diplomacy through science. (See Van der Berg, A., et al. 2023).
   - Using Science to Address Global Challenges (See Barrios, V., and Sancho, C. 2023).

c) According to the level of government involvement

- Intergovernmental science diplomacy: activities coordinated by a group of central governments of different states (Van Langenhove, L. 2016).
- National science diplomacy: activities coordinated by the central government of a given state (Gainza, C. 2023).
- Sub-national science diplomacy: initiatives carried out by sub-national entities such as municipalities, federated states, provinces, regional governments or partnerships between them (Ferraz da Silva, P. 2023).
- Non-governmental science diplomacy refers to science diplomacy that directly involves organized civil society. It ranges from international science outreach to public education on science and technology in the global arena (Pohl, N., and Luco, N. 2023).

d) According to the thematic area (Not restricted to):

Science diplomacy encompasses various fields within STI, showcasing its adaptability across agendas. For example, environmental science diplomacy addresses climate change, biodiversity, and resource management (Prieto, J., et al. 2024). On the other hand, health science diplomacy involves global collaboration in vaccine development, pandemic control, and medical research (Kalergis, A., et al. 2023). Likewise, space science diplomacy promotes cooperation in satellite technology and exploration (Barcons, X., and de Figueiredo Melo, C. 2022) and nuclear science diplomacy involves diplomatic efforts in nuclear technology and arms control, energy and medicine (Vera, 2021). These examples illustrate the complexity and versatile characteristics of science diplomacy, with each aspect deserving some further exploration.

In light of its literature, the taxonomies above are just some of the ways in which science diplomacy can be categorized. The exact taxonomy used might differ depending on the geopolitical context or the behavior of the stakeholders considered in a given analysis.

Science diplomacy in this chapter is viewed as a suitable practice for a challenging era of globalization. Nevertheless, there exists an insufficient amount of developed research concerning the governance of science diplomacy in emerging economies, notably in LAC countries (Echeverria King et. al., 2021). This gap can be costly, especially for the region, as the pressing challenges of today (such as global change, environmental issues, and regional health concerns) could be adequately addressed through the contributions of science diplomacy (Piñeros Ayala, et. al., 2023; Prieto et al. 2024).

It is noteworthy that science diplomacy, as a practice, can be viewed as a form of public diplomacy (Goldman 2022). This points to the fact that endeavors are directed towards a broad audience in international context, and by definition includes various actors from academia, primarily those involved in science, technological development and innovation. It is within this milieu that science diplomacy is presented as a viable diplomatic path that complements traditional diplomacy and official affairs of different sectors, aiming at both facilitating openness in the exercise of power on one hand, and achieving power on the other (Pantovic & Michelini, 2018). Therefore, we must emphasize that science diplomacy is the political arrangement that leads to the construction of a plurality of representations.

Diplomatic specializations emerge in the course of official performance, which seeks to be innovative due to certain difficulties and/or limitations of traditional diplomacy. In other words, science diplomacy is an instrument with the political ends of the actors exercising it. This can be a consequence of a quest to overcome challenges and consequently exceed problems associated with a disadvantaged position of the state, e.g., limited trade and economic diversification, lack of natural resources, geographical isolation, limited national security capacity, vulnerability to environmental risks and natural disasters, etc.

Therefore, the important question is not the resurgence of diplomacy, but rather its effectiveness today (Murray et al., 2011, p. 718). It's important to note that leveraging soft resources like science, technology, and innovation in international relations requires a certain mindset and proficiency. This entails training, as it's not necessarily common or straightforward to grasp and utilize these resources effectively. This is why, as highlighted by Echeverria-King et al. (2023), within the realm of science diplomacy, there's an emphasis on continuous education. This involves gaining a profound understanding of diplomatic activities alongside academic and scientific reasoning, and recognizing the influence and potential they hold. The focus lies in the capacity, opportunity, or potential to establish a conducive environment, thereby creating numerous pathways for desired interactions (Pantovic & Benavides, 2023).

So, the effectiveness of science diplomacy depends on the key factors conditioning its performance: national, regional and global challenges, diplomatic skills, and needs of actors. This highlights the changing and versatile work, observable in a contemporaneity where activities – both at the international and domestic levels – lie between the challenges and objectives on one hand, and resource management and communication on the other (Vickers 2004).

To evaluate the viability of the proposed concepts, it is imperative to analyze concrete cases. It is considered that the governments of LAC states should make greater use of science diplomacy in an appropriate manner: as a tool in their policies with specific purposes and by no means as an objective. Consequently, we will delve into an in-depth analysis of the LAC science diplomacy case, through the presentation of interview findings.

Criticisms about the birth and development of science diplomacy have been channeled to the imposition of developed countries on countries that have lower STI capabilities. This has contributed to some authors pointing out that SD becomes a form of colonization of scientific knowledge by those who have comparative advantages over less developed countries (Kaltofen & Acuto, 2018a).

Along the same lines, other authors called attention to the effort made by policy practitioners to overestimate the advantages that the instruments and practices implemented by SD, which are yet to be developed and implemented, have over the potential solution to global and cross-cutting problems. In the words of Ringius & Flink, the social construction of the SD concept is imbued with a narrative of crisis existence and the inability to find solutions to complex problems, thus requiring a type of immediate and rational action that can be executed collectively (Ringius & Flink, 2020) for the common good of global and local society.

This situation facilitates, in a certain way, the imposition of "common and collective", even neutral visions for the solution of transnational problems, assuming that the commitments, available resources and capacities and institutional arrangements can be assumed equally by a community of States especially differentiated by their capacities, magnitudes and influence.

In this sense, it appeals to a logic in which the SD considers applying by itself invisible hand skills for the solution of unresolved common and global problems, becomes an instrument of soft power without automatically taking into account the existing power relations and appeals to a myth of objectivity and rationality of public policy without taking into account the rationality of actors with divergent interests, eradicating the division, contestation and self-interest of international affairs, "from the impregnation of diplomacy by science, from an idealized and modernist understanding of science as an incorruptible and adjustable pacemaker to change and positive progress, which also promises the irrigation of the cooperative values of science in politics" (Ringius & Flink, 2020, p. 7).

Based on the above, interpretations of SD that incorporate and explain the behavior of countries in the Global South or with less developed capacities are emerging and developing with impetus. Based on a methodology that mixes the approach of SD objectives -scientific and non-scientific, political, economic or sustainable development, strategic drivers -intentions and principles that inspire action- and tools used, Ruffini & Krasnyak point out that for the countries of the Global South, the satisfaction of domestic needs based on the use of a heterogeneous framework for action, which combines the satisfaction of what they consider to be their national interest, facilitates the attainment of economic or development resources and contributes to the exercise of the influence of the State and international cooperation (Ruffini & Kransyak, 2023).

For countries in the global South, which usually have political, economic or scientific constraints that prevent them from efficiently implementing SD strategies, participation in intergovernmental science and technology events or forums contributes to achieving a heterogeneity of objectives. Forging agreements or participating in multilateral forums helps to mitigate internal national limitations, facilitates the sharing of resources and the exchange of experiences of more consolidated epistemic communities, participation can contribute to strategic positioning on global governance issues important to their local audience, and even helps to avoid international isolation (Rüland, et al., 2023).

In regard to the above, it is convenient to generate some preliminary ideas that both the study of scientific diplomacy and its exercise have been giving way to particular interpretations from Latin America. This sheds some light about the appropriation by countries of the Global South as adjusted to the needs of this region.

METHODOLOGY

This study aims to advance the discourse on science diplomacy for emerging economies by proposing a taxonomy tailored to the specific nuances and challenges of the LAC context. To achieve this, our methodology is designed to capture the perspectives and practices prevalent among practitioners and researchers in the region.

A qualitative approach was adopted, leveraging semi-structured interviews as the primary method of data collection. This approach was chosen for its flexibility, allowing for in-depth exploration of participants' insights and experiences (Gibbs, 2012). Through a series of open-ended questions, we sought to elucidate the dimensions of science diplomacy as understood and practiced in the LAC region. Interviews serve as an ideal method for understanding participants, their experiences and the perspectives of a specific object (Kvale, 2011). The interview protocol was crafted to probe into various aspects, including participants' conceptualizations of science diplomacy, their experiences with diplomatic endeavors involving scientific collaborations, and the strategies employed to navigate the complex interplay between science, diplomacy, and regional dynamics.

Data analysis followed a qualitative content analysis framework (Mayring, 2010). Transcripts of the interviews were carefully reviewed, with significant attention paid to recurring concepts, divergent perspectives, and emergent patterns. Through systematic coding and categorization, we aimed to distill the essence of participants' narratives, extracting meaningful insights that contribute to our understanding of science diplomacy within the LAC context, which can be also applied to other emerging economies.

A total of ten individuals were selected as participants for this study, representing a diverse range of backgrounds and expertise within the realm of science diplomacy, as shown in table 1. These individuals were chosen based on their active involvement in diplomatic initiatives, research endeavors, or policy formulation pertaining to scientific collaboration and international relations in LAC region. By engaging with this select group of practitioners and researchers, we sought to capture a comprehensive snapshot of the current landscape of science diplomacy in the region, thereby laying the groundwork for the development of a contextualized taxonomy that reflects the unique challenges and opportunities inherent in Latin America's scientific diplomacy endeavors.

Table 1. Participants

Name	Country	Role
Participant 1	Colombia	Career diplomat
Participant 2	Brazil	Researcher at a public university
Participant 3	Panama	Officer at a multilateral organization
Participant 4	Panama	Researcher at a public university and leader of a third sector organization dedicated to science and technology
Participant 5	Uruguay	Public officer working in international cooperation
Participant 6	Argentina	Researcher at a public university
Participant 7	Chile	Researcher at a public university and member of an organization dedicated to evidence-based public policy advice.
Participant 8	Argentina	Career diplomat
Participant 9	Argentina	Public officer working in international scientific cooperation
Participant 10	Colombia	Colombian doctoral student in diaspora and leader of international scientific collaboration networks.

RESULTS

Science diplomacy (SD) in the region has been configured in a particular way. Although some strategies or practices come from different contexts, there is a recognition of what has been done in the region and how each initiative impacted the regional identity and the prioritization of specific areas in which to collaborate. In this sense, a fundamental element to analyze how SD may be configured in LAC is the identification of certain qualities that, according to the interviews conducted, are the promotion of peace and horizontality in relations, this being a fundamental issue in the public agenda of the region.

To define the main properties or characteristics of SD in LAC, we propose a taxonomy of the initiatives developed in this field. Initially, it is recognized that SD has been seen as a strategy of "cooperation between countries, characterized by constant networking between different actors" (Participant 1), practitioners and researchers, who contribute to a global perspective of SD, its implications and challenges (Participant 6). Some interviews recognize that SD also contributes to improve communication and understanding skills among researchers, public officials and other actors involved in national STI ecosystem.

Thus, when asked about the main characteristics of SD in the region (Figure 1), the following stand out: cooperation, networking, collaboration, and development. These reflect a regional identity in which constant exchange is important, with countries in the region and abroad, always from a perspective of equality and understanding that mutual interests must be prioritized at all times.

Figure 1. Meaning of Science Diplomacy

Source: *Author's elaboration*

On the other hand, it is also relevant to consider what is understood by Science Diplomacy in LAC because, although different approaches were adopted, we are still a region characterized by being a recipient of initiatives or proposals. So, a pending task is to become more critical in terms of approaches from and towards SD (Participant 6).

Considering the above, SD is understood as "a type of inter-discipline between diplomacy and science but from a practical perspective" (Participant 7), which is not exclusive to a single sector but involves the participation of academia, industry, research centers, policymakers, among other actors that contribute to having a comprehensive perspective on the implications of SD in the region and how it is configured.

In other words, it is a "practice [that] must connect academia, government and include society. Without the connection of these actors [...] it will not be effective. It is not scientific dissemination; it is communication with society" (Participant 2). The aforementioned shows the relevance of thinking about science diplomacy from an interdisciplinary perspective since not only do actors or positions from a single discipline or area converge, but it also constitutes the ideal space for the exchange of knowledge and experiences that contribute to "building international and inter-institutional bridges so that there is knowledge, awareness and the promotion of its institutional framework to support this decision-making process at the international level" (Participant 3).

As mentioned above, there is no standardized definition of science diplomacy in the region, which is the result of the different approaches taken in this regard but also of the diversity of contexts and needs present in the region. However, it is recognized as an opportunity to build bridges and, through them, address the needs that arise in the region through cooperation, with a sense of social responsibility.

Thus, to classify the different regional strategies or initiatives, four categories are identified, specifically related to the activities that SD involves and the skills or qualities to which the science diplomacy contributes (Figure 2): foreign policy, governance, science and soft skills. These are not isolated but are linked to each other and imply a series of qualities that can support a better understanding of the approaches developed in the region.

Figure 2. Science Diplomacy Taxonomy

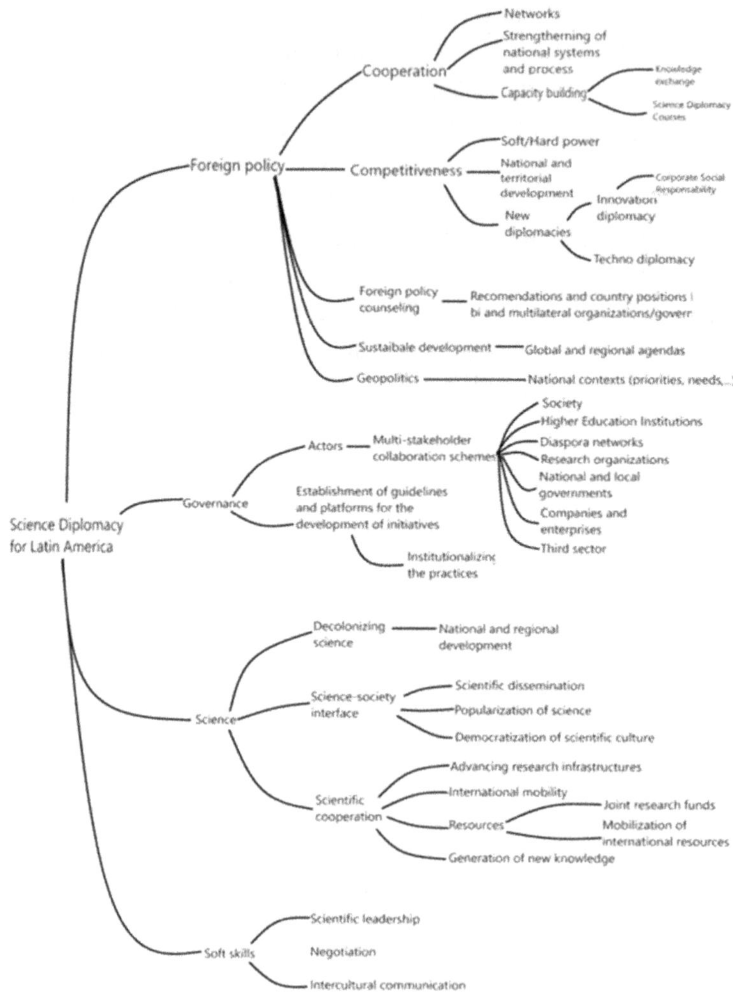

Source: *Author's elaboration*

Concerning foreign policies, we identified five subcategories: cooperation, competitiveness, counseling, sustainable development, and geopolitics. On the one hand, SD is considered a tool that fosters cooperation among different actors and institutions in the region, considering that this cooperation does not occur in only one direction or at only one level. But, as shown in the interviews, it also involves strengthening systems and processes, which in turn contributes to the institutionalization of the different strategies carried out within the framework of SD.

In this way, it also contributes to national and regional development through the identification and definition of common agendas, as proposed by the SDGs, but also prioritizing the development of capacities and competition between different sectors (Participant 6), understanding this as the incentive for the proposal of initiatives as well as the promotion of strategies framed in soft and hard power (Echeverria-King, et al. 2023; Vera, et. al. 2023). It gives rise to new disciplines related to SD, such as innovation diplomacy or techno-diplomacy, which are not mutually exclusive but complementary and interrelated. In addition, barriers between academia and politics are beginning to disappear, as well as the exclusivity of certain sectors in addressing particular issues (Jazbec 2010; Huijgh 2011).

On the other hand, linked to the idea of geopolitics, it cannot be ignored that power is an inherent component of any relation. As Participant 6 said: "The concept of power, which is something that is usually ignored, in definitions and works of scientific diplomacy but is very present in diplomacy and the sciences, [as well as the idea of] asymmetry, since power relations imply asymmetry". Thus, although the idea of horizontality frames the relationships established between actors and countries, we cannot ignore these components that condition how we link ourselves and assign roles for decision-making.

In regard to governance discussions, two fundamental factors are the actors and the development of guidelines and platforms for the implementation of initiatives that allow their institutionalization (Participant 10). As a result of the above, actors from different sectors participate in these initiatives, not only diplomats but also from society, higher education institutions, diaspora networks, research organizations, national and local governments, companies and enterprises and the third sector, to have a comprehensive overview of the characteristics and needs of the regional and local context, as well as how they are supported from different areas.

However, although the participation of different actors is recognized, how they are linked is not always based on the expected idea of horizontality. In addition, it is expected that the identified actors can work together since, as mentioned above, science diplomacy is a space for interdisciplinary collaboration between traditional and non-traditional sectors (Participant 2).

On the other hand, and from a regional perspective, it is necessary to see science diplomacy with a decolonial approach, in which, as mentioned above, regional and national development is also a priority through the identification of preferences, such as the promotion of a culture of peace through the different scenarios of scientific and political exchange in LAC region (Participant 6). It is essential to create a platform through which information on the proposed initiatives can be democratized, thus giving rise to scientific cooperation, which will have concrete objectives and, therefore, have a particular impact on different sectors and levels.

Finally, and although it is an element that is not exclusive to science diplomacy, the development of soft skills becomes a priority. On the one hand, being aware of the role that each actor has in the different relationships established at regional and global levels, as well as the ability to communicate and respect cultural diversity. In addition, scientific leadership and negotiation become fundamental qualities of the profile of a diplomat and scientist, given that, from a scientific perspective, what is proposed must also consider the context and how it affects them.

CONCLUSIONS AND RECOMMENDATIONS

Although the practice of linking diplomacy and decision-making on the one hand, with the advancement of scientific knowledge and the ITS ecosystem on the other, is quite old, attempts to find a sufficiently robust definition are rather young. However, since the commonly accepted Royal Society and AAAS definition was created in 2010, there has been an accelerated explosion of taxonomies linking SD as a constantly evolving concept, adaptable to multiple actors and contexts. Thus, its explanatory and practical capacity varies depending on who uses it, at what level of governance - global, regional, local or national - and for what particular purpose.

Based on this input, this chapter focused on developing several elements of relevance for the study of SD. On the one hand, an analysis was made of the importance of taxonomic classification for the conceptual and theoretical study in the social sciences, emphasizing its relevance for the construction of approaches that favor a transdisciplinary and interdisciplinary perspective, focused on the solution of practical problems or on the conceptual evolution from breaking disciplinary barriers to generate theoretical innovations. Thus, the concept of SD began with a State-centered perspective, but changed as other actors appropriated the explanatory capacity and practice demonstrated the need to generate innovations that account for what happens in the real world. SD generates great versatility so that it can be a suitable instrument in different scientific and practical environments. The theoretical perspective outlined in this chapter contributes both to the practical knowledge of SD in LAC states and to its discourse application.

From a practical perspective, four main lines of action were identified: first, foreign policy issues related to capacity building, improving economic competitiveness, recommendations for decision making, achieving the commitments derived from the 2030 agenda and, finally, as a geopolitical resource. Second, it was highlighted that SD has a capacity to strengthen ITS ecosystems based on the participation and contribution of actors of diverse nature and purpose and contributes to the generation of collaborative spaces for intergovernmental organizations, States and even sub-regional bodies such as cities.

Third, SD contributes to the generation and strengthening of its own autonomous scientific capacities, especially for developing countries such as those in LAC region. For decision-makers, scientists and other actors, it should facilitate better interaction between scientists and other actors, to identify issues, develop instruments and public policies that contribute to the solution of internal problems and the generation of scientific knowledge that enhances growth and sustainable development. Fourth, SD is also implemented to enhance soft skills by improving intercultural dialogue, negotiation among diverse actors, and scientific communication and dissemination skills.

Consequently, it was corroborated that the particular appropriations made of a concept, in this case SD, are relevant for the epistemic community and for the practices wherever it is used or implemented. In other words, situated knowledge contributes to the production in a region or country, for example, of conceptual innovations relevant to those who are using or intend to use the wide range of possibilities that SD has.

We need to emphasize that the conclusions drawn in this chapter primarily stem from qualitative analysis conducted through interviews. It is essential to recognize, however, that they do not represent the sole definitive taxonomy of science diplomacy. For instance, the authors underscore the significance of hard skills, such as the ability to interpret legal and political documents based on scientific evidence, draft recommendations through white papers, engage in policy debates, and possess negotiation skills for international events that aim to showcase scientific findings.

Notably, although these aspects were not explicitly mentioned by the interviewees, they remain critical components of effective science diplomacy. Additionally, another aspect that was highlighted in the theoretical work and in the interviews is the acquisition of communication skills. A better interaction between scientific knowledge, public policy recommendations and implementation of actions for the solution of public problems at different levels requires the development of certain skills that facilitate communication between scientists, decision-makers, public administrators and executive authorities.

As noted initially, this chapter intends to offer a preliminary perspective on a SD taxonomy for the LAC region, considering the scope, objectives and instruments traditionally used in that geographic space. As noted, this region as a whole, presents a profound heterogeneity among countries of the global South, with the result that their strategies are very dissimilar. In spite of this, there is an increasing intention to use SD in various ways, among them, those that the empirical work has shown us, which evidences an inclination towards the use of SD as an instrument to improve soft skills, such as scientific leadership, negotiation techniques or intercultural dialogue.

REFERENCES

Abdullah, H., & Roig, A. (2021). *City-led science diplomacy: building urban sustainability and resilience at the science-policy interface*. Cidob.

Álvarez, J. P. (2023). Esto invierten en ciencia los países de Latinoamérica: Brasil cuadriplica a México. Retrieved 1 March, 2024, from Bloomberg línea: https://www.bloomberglinea.com/2023/09/05/esto-invierten-en-ciencia-los-paises-de-latinoamerica-brasil-cuadriplica-a-mexico/

Barcons, X., & de Figueiredo Melo, C. (2022). Observatorio Europeo Austral (ESO): astronomía a través de la cooperación internacional, en Nuevos mapas para la diplomacia. Ciencia, Conocimiento, Tecnología e Innovación. Santiago de Chile: Academia diplomática de Chile.

Barrios, V., & Sancho, C. (2023). Ciberseguridad y Relaciones Internacionales: Interdependencia y Cooperación en la Era Digital. In Nuevos mapas para la diplomacia. Ciencia, conocimiento, tecnología e Innovación. Santiago de Chile: Academia diplomática de Chile.

Berkman, P. A. (2019). Evolution of science diplomacy and its local-global applications. *European Foreign Affairs Review*, 24(Special Issue), 63–80. 10.54648/EERR2019019

Bompongo Nkombe, A. (2020). Management of Congo Water Basin: Opportunities and Challenges for Science Diplomacy. In Research and Information System for Developing Countries (Coord.), Southern Perspectives on Science Diplomacy. Nueva Delhi: RIS.

Bueger, C. (2015). Making things known epistemic practices, the United Nations, and the translation of piracy. *International Political Sociology*, 9(1), 1–18. 10.1111/ips.12073

Cormann, M. (2023). Science, technology and innovation are changing the world: international cooperation to address global challenges. In Nuevos mapas para la diplomacia. Ciencia, conocimiento, tecnología e Innovación. Santiago de Chile: Academia diplomática de Chile.

da Silva, R., Ferreira, G., Onuki, J., & Nunes de Oliveira, J. N. (2021). The institutional building of science and innovation diplomacy in Latin America: Toward a comprehensive analytical typology. *Frontiers*, 49–62.33987524

Doty, D. H., & Glick, W. H. (1994). Typologies as a unique form of theory building: Toward improved understanding and modeling. *Academy of Management Review*, 19(2), 230–251. 10.2307/258704

Echeverría, L., Aquino, K., & Widmaier, C. (2020). Science Diplomacy and Sustainable Development Goals: A Latin American Perspective. *Science Diplomacy Review*, 2(1), 3–13.

Echeverría King, L. F., González, D. A., & Andrade-Sastoque, E. (2021). Science Diplomacy in Emerging Economies: A Phenomenological Analysis of the Colombian Case. *Frontiers in Research Metrics and Analytics*, 6(636538), 636538. Advance online publication. 10.3389/frma.2021.63653833997599

Echeverría King, L. F., Piñeros-Ayala, R., De Luque-Montaño, O., Flores-Zamora, A. F., Pantović, B., & Figueroa, P. (2023). Teaching Science Diplomacy at University Programs for Latin America. In Global Science 's Cooperation Opportunities, Challenges, and Good Practices. Academic Press.

Ferraz da Silva, P. (2023). Nuevos rumbos para la diplomacia científica. In Nuevos mapas para la diplomacia. Ciencia, conocimiento, tecnología e Innovación. Santiago de Chile: Academia diplomática de Chile.

Flink, T., & Rüffin, N. (2019). The current state of the art of science diplomacy. In Handbook on Science and Public Policy (pp. 104-121). Edgar Elgar Publishing. 10.4337/9781784715946.00015

Flink, T., & Schreiterer, U. (2010). Science diplomacy at the intersection of S&T policies and foreign affairs: Toward a typology of national approaches. *Science & Public Policy*, 37(9), 665–677. 10.3152/030234210X12778118264530

Fox, N. J. (2011). Boundary objects, social meanings and the success of new technologies. *Sociology*, 45(1), 70–85. 10.1177/0038038510387196

Gainza, C. (2023). Ciencia, tecnología, conocimiento e innovación en Chile: problemas complejos, miradas transdisciplinarias. In Nuevos mapas para la diplomacia. Ciencia, conocimiento, tecnología e Innovación. Santiago de Chile: Academia diplomática de Chile.

Gerring, J. (1999). What makes a concept good? A criterial framework for understanding concept formation in the social science. *Polity*, 31(3), 357–393. 10.2307/3235246

Gibbs, G. (2012). *El análisis de datos cualitativos en investigación cualitativa*. Ediciones Morata.

Gluckman, P. D. (2016). Science advice to governments: An emerging dimension of science diplomacy. *Science & Diplomacy*, 21–30.

Gluckman, P. D., Turekian, V. C., Grimes, R. W., & Kishi, T. (2017). Science diplomacy: A pragmatic perspective from the inside. *Science & Diplomacy*, 20–33.

Goertz, G. (2006). *Social science concepts*. Princeton university press. 10.1515/9781400842995

Goldman, I. (2022). El rol de la complementación entre ciencia y diplomacia en la Cuestión Malvinas. Revista de Investigación en Política Exterior Argentina, 2(4), 80-100.

Gual Soler, M. (2014). El Papel de las Redes Científicas Intergubernamentales en las Relaciones Regionales y la Integración de América Latina. *Science & Diplomacy*, 3(4).

Gual Soler, M. (2021). Science diplomacy in Latin America and the Caribbean: Current landscape, challenges, and future perspective. *Frontiers*, 1–9.

Harrison, N. E., & Singer, D. J. (2006). Complexity is more than systems theory. In Complexity in world politics. Concepts and methods of a new paradigm. State University of New York Press.

Hodgson, G. M. (2019). Taxonomic definitions in social science, with firms, markets and institutions as a case study. *Journal of Institutional Economics*, 15(2), 207–233. 10.1017/S1744137418000334

Kalergis, A., González, P., & Rivero, T. (2023). La paradiplomacia científica logra el acceso preferente, prioritario y temprano a vacunas contra el COVID-19. In Nuevos mapas para la diplomacia. Ciencia, conocimiento, tecnología e Innovación. Santiago de Chile: Academia diplomática de Chile.

Kaltofen, C., & Acuto, M. (2018a). Science diplomacy: Introduction to a boundary problem. *Global Policy*, 9(S3), 8–15. 10.1111/1758-5899.12621

Kaltofen, C., & Acuto, M. (2018b). Rebalancing the encounter between science diplomacy and international relations theory. *Global Policy*, 9(S3), 15–23. 10.1111/1758-5899.12620

Klaes, M. (2004). Evolutionary economics: In defence of vagueness. *Journal of Economic Methodology*, 11(3), 359–376. 10.1080/1350178042000252992

Klein, J. T. (2017). Typologies of interdisciplinarity. In Frodeman, R., Klein, J. T., & Pacheco, R. C. (Eds.), *The oxford handbook of interdisciplinarity (págs* (pp. 21–39). 10.1093/oxfordhb/9780198733522.013.3

Kvale, S. (2011). *Las entrevistas en la investigación cualitativa*. Ediciones Morata.

López, D., & van Klaveren, A. (2023). La diplomacia científica: una nueva dimensión de la Política Exterior. In Nuevos mapas para la diplomacia. Ciencia, conocimiento, tecnología e Innovación. Santiago de Chile: Academia diplomática de Chile.

Maass, A., Guerrero, G., San Martin, J., & Utreras, F. (2023). Un sistema robusto de infraestructuras críticas para la era de los datos: recurso humano para la investigación, capacidades de super cómputo y conexión internacional. In Nuevos mapas para la diplomacia. Ciencia, conocimiento, tecnología e Innovación. Santiago de Chile: Academia diplomática de Chile.

Manassero-Mas, M.-A., & Vázquez-Alonso, Á. (2019). Conceptualización y taxonomía para estructurar los conocimientos acerca de la ciencia. Revista sobre enseñanza y divulgación de la ciencia, 3104-3116.

Mayring, P. (2010). *Qualitative Inhaltsanalyse*. Grundlagen und Techniken. Weinheim Basel: Beltz Verlag.

Mejía Navarrete, J. (2020). Epistemología de las políticas de ciencia y tecnología en América Latina. Cinta De Moebio. *Revista De Epistemología De Ciencias Sociales*, (67), 14–25.

Ministerio de Relaciones, E., Comercio, I., & Argentina, C. (2023). Candidatura de Bariloche para ser sede de la Exposición Especializada 2027. https://www.cancilleria.gob.ar/es/destacados/candidatura-de-bariloche-para-ser-sede-de-la-exposicion-especializada-2027

Murray, S., Sharp, G., Wiseman, D., Criekemans, D., & Melissen, J. (2011). Criekmans, y J. Melissen. 2011. The Present and Future of Diplomacy and Diplomatic Studies. *International Studies Review*, 13(4), 709–728. 10.1111/j.1468-2486.2011.01079.x

Nair-Bedouelle, Sh. (2023). Science diplomacy is part of UNESCO's DNA. In Nuevos mapas para la diplomacia. Ciencia, conocimiento, tecnología e Innovación. Santiago de Chile: Academia diplomática de Chile.

Pantović, B., & Benavides, S. (2023, August 29). La red de diplomacia científica en Latinoamérica y El Caribe. Retrieved March 1, 2024, from Las habilidades para la diplomacia científica en los futuros investigadores e investigadoras: Reflexiones desde Latinoamérica: https://diplomaciacientifica.org/las-habilidades-para-la-diplomacia-cientifica-en-los-futuros-investigadores-e-investigadoras/

Pantović, B., & Michelini, G. (2018). Ciencia y Cultura de la Memoria en la Diplomacia Serbia / The Science and Culture of Memory in Serbian Diplomacy. *Revista CIDOB D'Afers Internacionals*, (120), 241–258. 10.24241/rcai.2018.120.3.241

Paz-Enrique, L. E., Núñez-Jover, J. R., & Hernández-Alfonso, E. A. (2022). Pensamiento latinoamericano en ciencia, tecnología e innovación: políticas, determinantes y prácticas. Desde el Sur, 14(1), e0008. 10.21142/DES-1401-2022-0008

Piñeros Ayala, R., Echeverría-King, L. F., & Andrade-Sastoque, E. (2022). Política exterior y de CTI en Colombia: la diplomacia científica como frontera. Desafíos, 34(Especial), 1-36.

Piñeros Ayala, R., Echeverría-King, L. F., & Pantović, B. (2024). La diplomacia científica frente a los retos globales: Acciones para mitigar el cambio climático en la región de América Latina. In *La Nueva Agenda de la globalización* (pp. 257–277). Universidad de Xalapa.

Pohl, N., & Luco, N. (2023). Encontrarnos desde el conocimiento: comunicación para la diplomacia en ciencia. In Nuevos mapas para la diplomacia. Ciencia, conocimiento, tecnología e Innovación. Santiago de Chile: Academia diplomática de Chile.

Prieto, J., Cumba García, L., Echeverría-King, L., Benavides, S., & Alarcón, C. (2024). Recomendaciones para una estrategia de Diplomacia Científica para la salvaguarda de la Amazonía. DOI:)10.5281/zenodo.10695010

Ramos, G. (2023). UNESCO's Recommendation on the Ethics of Artificial Intelligence as a reflection of global cooperation in diplomacy, science, technology and policy. In Nuevos mapas para la diplomacia. Ciencia, conocimiento, tecnología e Innovación. Santiago de Chile: Academia diplomática de Chile.

Ruffini, P.-B., & Kransyak, O. (2023). Science diplomacy from a nation-state's perspective: A general framing and its application to Global South countries. *Science & Public Policy*, 50(4), 771–781. 10.1093/scipol/scad023

Rüland, A.-L., Rüffin, N., Cramer, K., Ngabonziza, P., Saxena, M., & Skupien, S. (2023). Science diplomacy from the Global South: The case of intergovernmental science organizations. *Science & Public Policy*, 50(4), 782–793. 10.1093/scipol/scad024

Rungius, C., & Flink, T. (2020). Romancing science diplomacy for global solutions: On narratives and interpretative schemas of science diplomacy. *Humanities & Social Sciences Communications*, 1–10.

Rusieshvili, A. (2020). Role of Science Diplomacy in Russian-Georgian Relations. In Research and Information System for Developing Countries (Coord.), Southern Perspectives on Science Diplomacy. Nueva Delhi: RIS.

Scartozzi, C. (2018). A new taxonomy for international relations: Rethinking the international system as a complex adaptative system. Journal of Policy and Complex Systems, 4(1), 109-132.

Singer, J. D. (1971). A general systems taxonomy for political science. General Learning.

Star, S. L. (2010). This is not a boundary object: Reflections on the origin of a concept. *Science, Technology & Human Values*, 35(5), 601–617. 10.1177/0162243910377624

The Royal Society. (2010). *New frontiers in science diplomacy*. The Royal Society.

Van der Berg, A., Latorre, P., Diaz, R., Vargas, N., Leiva, S., & Olmos, F. (2023). Internacionalización en universidades chilenas: un proceso en construcción. In Nuevos mapas para la diplomacia. Ciencia, conocimiento, tecnología e Innovación. Santiago de Chile: Academia diplomática de Chile.

Van Langenhove, L. (2016). Multilateral EU Action through Science Diplomacy. In Ujvari, B. (Ed.), *The EU Global Strategy: going beyond effective multilateralism?* European Policy Centre.

Vera, N. (2021). *Ciencia, tecnología y Relaciones Internacionales: visibilizar la simbiosis. Desarrollo y políticas de ciencia, tecnología e innovación en un mundo en transformación. Reflexiones sobre la Argentina contemporánea*. CEIPIL-UNICEN.

Vera, N., & Echeverría King, L. (2022) Hacia una diplomacia científica para América Latina. En Oregioni (comp.) Derribando estructuras, construyendo puentes, tejiendo Futuro. Aportes para pensar la internacionalización desde una perspectiva situada en el Sur. Editorial-Cooperativa El Zócalo, Buenos Aires, Argentina. https://ceipil.fch.unicen.edu.ar/wp-content/uploads/2023/02/Libro-PICT-final.pdf

Vera, N., Salvati, A., de Angelis, I., & Pantović, B. (2023). Diplomacia Científica Transregional y prioridades en la agenda de cooperación científico-tecnológica de la Comunidad de Estados Latinoamericanos y caribeños (2013 - 2023). Ciencia, tecnología y cooperación internacional en clave de desarrollo: reflexiones sobre Argentina y América Latina, de María Paz López. Tandil: Centro de Estudios Interdisciplinarios en Problemáticas Internacionales y Locales (CEIPIL).

Vickers, R. (2004). The New Public Diplomacy: Britain and Canada Compared. *British Journal of Politics and International Relations*, 6(2), 182–194. 10.1111/j.1467-856X.2004.00133.x

Wagner, C. S. (2002). The elusive partnership: Science and foreign policy. *Science & Public Policy*, 29(6), 409–417. 10.3152/147154302781780741

Weiss, C. (2015). How do science and technology affect international affairs? Minerva, 411-430.

ENDNOTE

1 For the purposes of this chapter, multidisciplinary refers to the explanations of the same social phenomenon offered by the disciplines of knowledge; interdisciplinary refers to the construction of scientific knowledge by applying concepts, theories or approaches from different disciplines; transdisciplinary refers to overcoming the traditional limits of the disciplines to face problems with a greater variety of methodological, epistemological and practical approaches, aimed at overcoming specific problems (Klein, 2017).

Chapter 3
Innovation Diplomacy in the LAC Region and InnSciD Case Study From Brazil

Almas Taj Awan
University of Sao Paulo, Brazil

Humberto Filipe de Andrade Januário Bettini
University of Sao Paulo, Brazil

Gabriela Gomes Coelho Ferreira
https://orcid.org/0000-0002-3548-395X
University of Sao Paulo, Brazil

Janina Onuki
https://orcid.org/0000-0001-9033-1863
University of Sao Paulo, Brazil

Amâncio Jorge Silva Nunes de Oliveira
University of Sao Paulo, Brazil

ABSTRACT

With rapid globalization, efforts focusing on science diplomacy in the Global South have proven essential elements of foreign policy. Such efforts mainly include transnational research partnerships and guiding foreign policy agendas with scientific advice. Nevertheless, rapidly trending innovation ecosystems in different parts of the world have brought forward a new perspective of science diplomacy, more inclined towards collaborative innovation. This aspect of science diplomacy, known

DOI: 10.4018/979-8-3693-2746-3.ch003

as innovation diplomacy focuses on financial/commercial growth and shared mutual benefits that may improve international relations between countries. In this chapter, the authors discuss why the Global South, especially the LAC region, could benefit from focusing on Innovation diplomacy. The authors believe that there is a growing interest of stakeholders from the LAC region for getting familiar with this subject. To illustrate this, they share some quantitative analyses of "The Sao Paulo School of Advanced Sciences on Science and Innovation Diplomacy" (InnSciD-SP).

1. INTRODUCTION

Globalization has led societies all over the world, to understand the importance of science and technology as essential components of foreign policy (Flink, 2022). In the past after Cold War, science and technology have been used as an important tool for strengthening international relationships (The Royal Society, 2010) primarily via international research partnerships and the use of scientific advice to inform foreign policy goals. Two major objectives of such international scientific cooperation were the advancement of knowledge and/or the advancement of national interests (Gluckman et al., 2017). However, quickly emerging innovation ecosystems (Granstrand & Holgersson 2020) around the globe have led to growing interest of public and private sectors for investing in innovative technologies. At the national and foreign policy level, it has changed the objectives of science diplomacy by adding greater emphasis on collaborative or joint innovation with competitiveness and economic gains (Leijten, 2017). This perspective of science diplomacy commonly known as "Innovation Diplomacy", emphasizes on economic and commercial expansion as well as the sharing of benefits that strengthen international ties between nations. Such tendency requires reshaping government policies that favour the development of a suitable environment where the foundations of innovation-led smart and sustainable future can be laid (Mazzucato & Penna, 2016).

Innovation diplomacy is a dimension of public diplomacy consolidated in several countries of the Global North. Some major examples worth mentioning include US programs aimed at guiding and organizing innovation missions and delegations; UK programs that promote science and business around the world being part of the 'GREAT Britain' campaign (Department for Science, 2024) such as the UK's Science and Innovation Network (SIN 2024) jointly funded by the Department for Business, Energy and Industrial Strategy (BEIS) and the Offices of Science and Innovation (OSI) of Sweden based in six embassies around the world (Brasília, New Delhi, Beijing, Seoul, Washington and Tokyo (OSI 2024)).

In the past, Science Diplomacy programs in the Global South were more of a rarity than the norm. However, the rapidly changing world politics and the economic situation of emerging economies are gradually changing this scenario, shifting the focus from western centered programs to the ones that focus on multi-centered approaches (Ruffini & Krasnyak, 2023). For understanding such changes and to navigate global science diplomacy agendas, it is imperative that the rise in popularity of science diplomacy elements in the foreign policy agenda among developing nations from the Global South are taken into consideration (Ruffini & Krasnyak, 2023). In this context, the case of Brazil is atypical. Despite not having a science diplomacy policy at national level, Brazil adopted an Innovation Diplomacy Program (PDI) in 2017, promoted by the Ministry of Foreign Affairs (MRE). The program was created to promote the formation and intensification of strategic partnerships in the area of Science, Technology and Innovation (ST&I), attracting investments to the ST&I sector and attracting new talents in science and innovation (Brazilian Ministry of Foreign Affairs, 2023). In 6 years, the number of initiatives implemented under the program jumped from a shy 6 in 2017 to a more pronounced figure of 180 in 2023, with the participation of 56 Consulates and Embassies that started carrying out actions under the PDI program (Brazilian Ministry of Foreign Affairs, 2023). Till date in 2024, 234 initiatives have been approved for 50 Brazilian Consulates and Embassies around the world (PDI, 2024). From the budgetary point of view, in this period, the budget for this program jumped from just over 300 thousand to more than 2.8 million reals in 2024 (PDI, 2024)[1].

In the Latin American region, the narrative and practice of science diplomacy have been strengthening, but not that of innovation diplomacy, except for the case of Brazil. It is valuable to understand the impacts of Brazil's context and the imminent results it has brought for its competitiveness. The 2023 report from the Brazilian Ministry of Foreign Affairs reports the impacts of the PDI as internationalization of agents related to Brazilian innovation ecosystems, promoting Brazil's image within innovation ecosystems, increased international partnerships and networks, attracting investments, increasing exports and attracting talent (Brazilian Ministry of Foreign Affairs 2023).

In this chapter, we will discuss how innovation diplomacy can enhance other forms of diplomacy such as science diplomacy, economic diplomacy, public diplomacy etc. We believe that there is a strong interest of people from LAC region to understand and apply concepts of Science and Innovation Diplomacy. "The Sao Paulo School of Advanced Sciences on Science and Innovation Diplomacy" (InnSciD-SP, 2023) provides 2 weeks of thorough training sessions to the audience interested in the topic. In this chapter, we will share some analyses on the participation of stakeholders from LAC region in the Science and Innovation Diplomacy (S&ID)

training by InnSciD-SP and the importance of building social networks of national and international stakeholders by such training.

2. INNOVATION DIPLOMACY COMPARED TO SCIENCE DIPLOMACY

Complex interdependence (Keohane & Nye, 1977), marked by digital technological development and the increase in cross-border flows beyond the reach and control of the State – with special attention to the mobility of capital (Hay, 2020), and the mobility of information through digital means – changed significantly foreign policy in scope, processes, and actors.

The intersection between science, public policy and the market has garnered attention from both scholarly works, governance and governmental bodies in recent years. The pivotal role of scientific knowledge in decision-making stems from the very essence of strategic actions as public policy decisions. In its turn, public policy decisions can be characterized as actions that strategically mobilize resources to a certain end, or the diversity of actions executed, mobilizing human, financial, and technological resources by governments to promote citizenship.

While science diplomacy can be described as the intersection between science and diplomacy, it can also be described as an instructive example of the intersection between science and a specific category of public policy: foreign policy. Since talking about foreign policy is also to talk about power in the transnational arena, it is important to understand the double nature of science as soft and hard power. According to the "Madrid Declaration on Scientific Diplomacy" (2018), "*scientific diplomacy is understood as a series of practices at the intersection of science, technology and foreign policy*" to respond to challenges "*where a greater scientific voice could both add value to bilateral and multilateral discussions and decisions on shared global concerns*" (S4D4C 2019).

From a theoretical perspective from the international relations field regarding realists and institutionalists, it can also be framed as competitive or cooperative. As a soft power instrument, science can, on the one hand, increase national (or shared) interests between countries by improving political-scientific international relations between them and by enhancing countries' socioeconomic development. On the other hand, as hard power, it can be an instrument to militarily or economically coerce others.

When inserted in the economic realm, with the economic value generation process, science and its developments are framed as innovation. Innovation can be understood as the products or production processes, developed with new knowledge or an innovative combination of pre-existing knowledge, introduced into markets and social life (OECD/Eurostat, 2018). Currently, the concept of innovation is encapsulated in a nonlinear model that mirrors an evolutionary vision. In this dynamic framework, interactions among science, technology, and the market are not confined to linear trajectories but are instead fluid, adaptable, and frequently unpredictable. This model recognizes the diverse array of actors contributing to innovation, beyond traditional boundaries and involving various segments of society.

Innovation diplomacy, therefore, encompasses a larger myriad of actors than those initially mobilized by science diplomacy, while bringing the market competition to the negotiating table. Even though we have a common core to the concept, the innovation diplomacy narrative has been developed differently by Global South and Global North countries. Global North authors tend to emphasize the economic and competitive elements, as highlighted by scholars such as Jos Leijten (Leijten, 2017) and Kristen Bound (Bound, 2016), to underscore the multifaceted nature of innovation diplomacy. Simultaneously, drawing from Ademar Seabra (da Cruz Junior, 2010), Global South countries, such as Brazil, tend to leverage their diplomatic resources to align with their productive-technological profiles. This strategic alignment empowers them to redefine their international position within the dynamic landscape of the global division of labor, moving towards sustainable development goals. In this sense, innovation diplomacy serves as a crucial bridge between science, sustainable development, and economic perspectives, particularly when considering the unique context of Global South countries. Hence, the major milestone "Sao Paulo Framework of Innovation Diplomacy", defines Innovation Diplomacy (ID) as the collection of concepts, practices and strategies put forth by national and subnational players, using diplomatic channels to boost creative innovative capacities and paying particular attention to sustainable development goals (InnScid, 2019).

Innovation diplomacy appears as a concept that, at the same time, narrows and enlargers the perspectives related to the interplay between science and foreign policy: while the concept of innovation diplomacy is related to the interplay between foreign policy and scientific knowledge applied to generate socioeconomic development – what brings competition to the cooperative conceptual framework, it also enlarges it by bringing inherently a myriad of actors part of the innovation ecosystems that are commonly not encompassed by the idea of science diplomacy.

By intertwining scientific and economic strategies with diplomatic initiatives, countries can foster sustainable development, catalyze technological advancements, and fortify their roles on the world stage. This tension between the competitive and the collaborative approach, coupled with the active involvement of the private sector, reinforces the potential for shared prosperity and global progress.

Consequently, science and innovation diplomacy (S&ID) perform as conceptual frames to describe and analyze a set of actions that occur in the intersection of science and a specific set of public policies, and can be used as important tools for understanding, intervening and practicing. Consequently, while science diplomacy (SD) tends to bring a more cooperative approach, innovation diplomacy (ID) performs a competitive one (Leijten, 2017). The next section will discuss context of innovation in countries from the Global South, especially the LAC region.

3. GLOBAL SOUTH AND INNOVATION

Gone are the days when countries from the Global South (South American, African, and South Asian regions) were viewed as passive users of technology developed by the Global North (Srinivasan, 2019). In recent years, countries from the Global South have proven their innovative worth by creating affordable solutions to the local problems meanwhile facing all the economic constraints. A remarkable degree of optimism has existed in the communities across the Global South with the concepts of jugaad innovation, reverse innovation and frugal innovation (Marcelle, 2016). Moreover, when it comes to building technologies, they are developed not just to support the financial gains but also the moral and ethical values that are guided by the local society, while making use of the natural resources and taking care of the local ecosystem.

While analyzing the potential of Latin America and the Caribbean (LAC) region in the Global context, its vast biodiversity feature, endless sources of renewable energy, minerals, the Amazon Forest, huge amounts of natural drinking water, strong agriculture, political and religious harmony among the neighbours, and neutral geopolitical stance makes it an ideal place for exploring, developing and spreading sustainable global practices (Beltran, 2024). Provided that the public and private sectors play their crucial roles by investing and catering to the economic and systemic challenges, the LAC region is capable for creating innovative solutions related to the green energy revolution, preservation of biodiversity and food security (Beltran, 2024).

To understand this further, below are a few examples of local innovations from the LAC region:

a. **Rhizomatica:** In 2009, from the Oaxaca region of Mexico, the largest autonomous cellular network in the world was constructed by Rhizomatica organization and it is totally community-owned. They avail sustainable energy resources and deploy projects related to bio-economy. This idea is now taken to other communities across the world. Alternative telecommunication networks are important for connecting with people who are living in a place that is considered extremely poor, or too far and isolated to establish a connection, or in case of any environmental threat of natural disaster, or people who are unable to connect with the outer world due to oppressive government regimes (Rhizomatica, 2024).

b. **Nilus:** To tackle food insecurity, this startup has developed a technology for reducing the cost of living for low-income communities and currently operates in Mexico City (Mexico), Lima (Peru) and Buenos Aires (Argentina). It has developed a technology that connects directly to farmers and local producers for securing food items and groceries. Later community buying networks distribute those items to the public who live with food insecurity in areas with limited access to food supplies. The best part of Nilus is that its products are nutritious and affordable, at 70-75% cheap rates compared to the market price (Nilus, 2024).

c. **Amazonia Emprende**: It is a pioneering 'Escuela Bosque' (forest school) whose main goals are to educate and train local communities, students and local businesses on the development, replication and benefits of nature-based solutions. By connecting and working together with local communities and the business sector, Amazonia Emprende believes that by 2030, it will be able to conserve and restore up to 100,000 hectares of the Amazon Forest (Amazonía Emprende 2024).

The above examples show that the concept of innovation is much broader when it comes to resource-scarce regions (Qadri, 2021). Many technology-based innovation models from Silicon Valley have proven to be unsustainable for communities from the Global South due to less developed rural / urban structures (Qadri, 2021). The proposed failed solutions by the foreign innovators mostly get fixed when they embrace practices that had been around in the current location for decades. The human factor of the innovative process is what makes the whole solution trickier and more complicated. Human factor means designing solutions, systems, products, and equipment around humans. It involves use of psychological and physiological measures to cater to the mental and physical limitations of human beings, thus making innovation more effective and useable (Thales, 2024).

It is also noticeable that most of the innovations from the Global North usually come at the cost of irreversible side effects such as e-waste with hazardous chemicals, scrutiny, invasion of privacy, censorship controversies, or algorithmic gatekeeping from the big tech companies (Srinivasan, 2019). Thus, innovation is not just about novelty, neither it is the synonym for R&D (research and development). Usefulness and multiple source knowledge integration are equally important as profitability (Marcelle, 2016). Also, traditional ethics of living in a human society that creates social harmony must be taken into account for maintaining or creating better societies in the coming years.

Data shows that around 2.4% of unicorn companies have their origin and establishment in the LAC region, which makes this region stand at the second position among developing regions after Asia (López-Calva, 2022). "Unicorn" is a private venture capital start-up / industry with a market value of more than US 1 billion dollars. Further regional analyses show that the unicorn landscape from LAC is rather uneven with Brazil and Mexico having more than three-quarters of the unicorns. Both of these countries are not just the top regional performers accounting number of unicorns but also ranked high on the Global scale. In June 2022, Brazil was ranked 9th (with 19 unicorns) while Mexico was ranked 15th (with 8 unicorns) in the world (López-Calva, 2022). In the LAC region, such innovation patterns were clearly catalyzed by the local needs and investments during the crises of the COVID-19 pandemic. Although it is clear that the distribution patterns of such unicorns are uneven among the countries of the region, the state-level policies are required to establish transparency in strong local business and regulatory environments, so that countries from the LAC region with low economic conditions may find ways to attract foreign investment. This will push sustainable industrialization with local resources, hence creating affordable products while the flourishing job sector that fuels the overall economic growth of a country (Sustainability Times, 2023).

In the next 20 years, the global energy transition from fossil fuels to renewable power and electric vehicles is going to be drastic and lithium demand is expected to grow 40-fold due to its use in the batteries. Latin America can be a global player in energy transition as it is the owner of around 60% of identified lithium reserves: found in Bolivia, Argentina and Chile, these countries form the 'lithium triangle'. At the moment Latin American countries are at the bottom of the supply chain as raw material providers, while the Asian countries own the lead manufacturers. Thus, bigger investments are needed for transforming this material into an end-user product which will provide higher profits and will help grow the market share of Latin American industries (Ellerbeck, 2023). This is an example of one of the major areas in which innovation diplomacy may play a vital role, fostering regional cooperation for the solution of common challenges.

The Brazilian Innovation Diplomacy program (PDI) is an effort to portray Brazil as an innovative nation while broadening its perception beyond that of a supplier of raw materials. The PDI brings together technological promotion activities developed by Ministry of Foreign Affairs posts (currently at 56 Consulates and Embassies) abroad, which are guided by three institutional missions: to promote Brazil's image as an innovative nation; to foster connections between Brazilian and foreign partners; and to engage agents from Brazilian innovation ecosystems in internationalization processes. The main activities in 2023 included: the mapping of Innovation-Promoting

Environments, 28 face-to-face seminars and conferences, 21 webinars and 18 start-up missions. Other activities include diaspora mobilization, technological fairs, innovation days, researchers´ missions, media campaigns, and startup incubation among others. For the most part, technological promotion activities within the PDI are multisectoral. For example, in 2023, 73% of the activities were multisectoral, in 2022, such activities corresponded to 67%, in 2021, 44% and in 2020, 64%. The top three categories aimed at specific sectors included artificial intelligence sectors, agritechs and energy (Brazilian Ministry of Foreign Affairs, 2023).

4. INNOVATION DIPLOMACY AS A TOOL FOR REGIONAL ECONOMIC GROWTH (SOUTH-SOUTH COOPERATION)

One of the most perennial questions in Economics relates to the sources of economic growth, development, and the wealth of nations in a broad approach. Authors as important as Adam Smith, very frequently referred to as *The Father of Economics*, and Nobel Laureates Robert Solow and Paul Romer have dealt with such a question in different times in the History of Economic Thought, by means of a vast intellectual production. Another feature uniting these authors – and thousands more of disciples each one of them have gathered – is that, although in different extents and under different perspectives, both stock and dynamics of human resources are understood to be one of the very key variables explaining differences in countries' fortunes and performances along years, decades or centuries. These authors have suggested that human resources are always under the spotlight while discussing about means for improving growth and development trajectories of nations, either while being dealt with in a way closer to absolute terms as the total count of population, or while being dealt with in a fashion closer to more relative, sophisticated terms, as the educational profile of (segments of) the population.

A pattern exists in the evolution of Economic Thought though: in a progressive manner, human resources-based explanations for economic development of nations departed from more absolute perspectives and became increasingly rooted at more relative grounds. As a matter of fact, it is currently widely accepted that the level of economic development of a given country is positively correlated with its base of high-skilled labor force, either on technological or scientific functions, and then further correlated to how the scientific and technological sectors of the given country interact with others of the sectors and stakeholders that compose – and promote – the productive and innovation ecosystems. All the praise towards education, science, and technology has direct affiliation with the prior belief that these elements are functional for societies overcoming natural forces and circumventing material scarcity. Key multilateral organizations are quite vocal on this subject (Barrett et al. 2021;

OECD, 2020) and empirical evidence is widespread across a number of countries such as Brazil (FGV, 2023), Mexico (Gálvez-Soriano, 2020), and Turkiye (Beskaya et al., 2010). Indeed, world recognized experts such as Eric Hanushek, Hirofumi Uzawa, and Robert Lucas have devoted their careers to the concrete understanding of the mechanisms through which public, private expenditure in education, science and technology ultimately impact wages, productivity, innovation, product increase and so forth.

One striking fact amidst this discussion on human resources and economic development is that world is globalized, meaning that borders for circulation of capital, goods and people are pretty much porous. In this sense, and in parallel with the discussion on the quantitative and qualitative profile of *local* human resources, the fact economies are pretty much open requires that the role of *global* human resources become investigated and understood. Indeed, and herein refraining discussion to human populations, it is a fact that nations see themselves in both the positions of sending human resources abroad, and of receiving such resources from foreign countries. Open-economy countries with lower economic and / or development levels constantly send lower-skilled students and workers for acquiring better grades at foreign universities and companies, while also constantly watching to a flow of higher-skilled students and workers migrating – temporarily or permanently – for foreign countries. Conversely, any given country of higher economic level also constantly receives such dual flow of foreign students and workers – both unskilled looking for improvement of their skills by means of national resources such as attendance to local educational institutions, and skilled ones looking for migration, and the final balance regarding the net benefits of such a dual flow of dual quality human resources remain as an empirical subject, and it is in the ambit of such a discussion that literature (and also the deliberate political positionings) on topics such as brain drain (or gain) exists. Indeed, although the most common thesis on this topic assumes that such outflux of qualified human resources hurts the sending (usually lower income) country and benefits receiving (usually higher income) country (and the human resource him or herself), some researches could dispute such a common belief.

It is not in our interest to discuss such a topic. At least, not at a purely national level. Our interest lies on a regional (multi-national) level, and it departs from a theoretical approach according to which some groups of nations can more easily benefit from international cooperation than other groups. In fact, one very strong hypothesis guides us as scholars approaching such a topic in international economics: countries with closer cultural, social profiles should be better off on the task of retaining the gains from such international exchange of human resources. Holding all other factors constant, and according to what theories such as the absorptive capacity propose (the ability for retaining information in learning processes vary according to the prior circumstances of the agent), countries with more similar socio-economic structures should perform better in such an endeavor of cooperating for the betterment of the skills of students and professionals that both emigrate and immigrate, in an argument that favors initiatives such as South-South cooperation and boosts the scope of Innovation Diplomacy. Theoretical pieces of argument can be found on authors as diverse as Alfred Marshall (pioneer on topic of knowledge spillover), Douglass North, and Paul David. More specifically, studies such as Cohen & Levinthal (Cohen & Levinthal, 1990) and, more recently and relating to topics closer to culture and cultural studies, Straubhaar (Straubhaar, 2007) show that the level of (institutional, cognitive, religious, or cultural) proximity is a strong determinant for the success of human collaboration on social, economic grounds. Empirical evidence seems to support such hypothesis, as can be seen in Garcia et al (Garcia et al., 2018).

Hence, it seems to us appropriate to embrace the hypothesis that collaboration on scientific and educational levels among countries that share common cultural, historical and institutional aspects (what we call "symmetrical countries") can be more productive than collaborations between countries that do not share such aspects ("asymmetrical countries" by contrast): if by one side knowledge gap is generally smaller, by the other features such as the absorptive capacity of migrating human resources can be higher[2].

If this belief stands correct, then there is a strong case for initiatives that reduce intercontinental technological dependence on the traditional basis of North-South collaborations, and their replacement with South-South collaborations. In this sense, identifying new partnerships within regions where one can locate countries that share common cultural, historical and institutional aspects, such as the majority of the countries that are part of Latin America and Caribbean region, become a subject high in the agenda of Federal Governments. Furthermore, the agenda according to which national innovation ecosystems should be encouraged can gain a second coating: fostering such ecosystems in world regions such as Latin America and the Caribbean region becomes a valid option for the national (local) and also international (regional) contributions that can be generated. Indeed, initiatives such as Start-Up Chile and Start-out Brazil can be understood as very compelling and functional in the task of training not only national citizens and ventures, but also those from regional (the LAC region in our example and as focus of this chapter) origin. Indeed, ecosystems supporting business models, internationalization of local startups can be put at service of entire regions, and speech should be calibrated in order to argument in favour of their competitiveness, just in the terms we expressed in footnote 2: even trainings of lower intrinsic depth can result in higher impact if absorptive capacity differs and matters.

In short, Innovation Diplomacy, by means of initiatives such as the opening of national innovation ecosystems for citizens (individuals such as students and entrepreneurs) originating from countries that share common cultural, historical and institutional aspects with the host country, should be seen as a tool for regional economic growth. In ultimate analysis, initiatives such as these should help enhancing R&D efforts as they act as supportive policies related to technological innovation in the R&D sectors within industries, and contributes to the economic development of a country by means of attracting investments and fostering collaboration between home country and foreign technology innovation areas.

5. INNOVATION DIPLOMACY AS A TOOL FOR PUBLIC DIPLOMACY

The term public diplomacy changed its form and definition multiple times over decades and due to lack of definitional consensus, there is still no single definition (CPD 2024; Parliament of Australia, 2024). Public diplomacy supports the constructing of a secure international environment by investing in suitable mechanisms that seed trust and mutual understanding among nations. Public diplomacy is a multidisciplinary concept which links several knowledge areas, such as mass communication, world history, media studies, international relations, regional studies, and public relations (CPD, 2024).

Projecting country´s image that showcases its ability to innovate, generate knowledge and provide services related to R&D builds its modern and technological worth internationally. Most of the countries from the LAC regions have an image related to suppliers of raw materials within global production chains with very few countries invested in productive structures (OECD et al. 2021). Investing in suitable innovation strategies can change this perspective of raw material providers to technological diversification, what may lead to opening new horizons in the diplomatic arena ultimately building trust for cooperation on technological frontiers.

As mentioned previously in Section 3, Brazilian Ministry of Foreign Affairs started *Programa de Diplomacia da Inovação* (PDI, or Innovation Diplomacy Program in English) with the aim to propagate Brazil´s technological potential abroad, in accordance with its Policy of National Innovation (MRE, 2024). Since then, innumerous activities have been held in 56 Brazilian Consulates and Embassies around the world. On one hand, such activities have the potential to impact researchers and entrepreneurs not just in the home country but also the host one, while on the other hand, they help in diminishing insecurities and building new relationships that helps in strengthening policies related to public diplomacy.

When we look at data from 2023, activities developed under PDI were mainly concentrated in Europe (43% of activities, or 79 activities in total). Activities held in Asia accounted for 23.8% of total number, and that was followed by Latin America and Caribbean (16.6% of activities, or 30 activities in total) and North America (12.2% of activities, or 22 activities in total). Three remaining world regions held a much lesser number of activities: Australia held 1.6% of activities, while Africa held 1.1% and the Middle East held 0.56% of them (1 sole activity) (Brazilian Ministry of Foreign Affairs, 2023).

Although skewed towards Europe and China, it is worth noticing that top 10 embassies / consulates in terms of number of activities include Buenos Aires (Argentina), Ciudad de Mexico (Mexico) and Santiago (Chile), all these cities and countries belonging to LAC region, plus Barcelona (Spain) and Lisbon (Portugal), countries that share linguistic similarity (former) and historical ties (later).

Activities from PDI reflect the current state of the hierarchy of knowledge and money transfer on world level, at least according to how The Brazilian Foreign Ministry interprets it and tries actively promoting Brazil on official terms. However, how can we change such a scenario, especially if we are to believe that both countries and individuals are heterogeneous and differences in absorptive capacity exist? As the concept of democratic quality of co-creation suggests that for partnerships across borders, empowering the less skilled actors may result in balanced joint agreements with novel public value solutions by creative problem-solving (Ansell et al. 2024). In the next section we present an initiative that can help building bridges that may enable more and better cooperation initiatives on South-South level by means of both, exploring existing and promoting new elements of common grounds between more symmetric countries.

6. SAO PAULO SCHOOL OF INNOVATION AND SCIENCE DIPLOMACY

Training highly qualified human capital from various disciplines from the triple helix in Innovation and Science diplomacy is of utmost importance. To cater for this need, while focusing on the Global South, the University of Sao Paulo organized the first Sao Paulo Innovation and Science Diplomacy School (InnSciD SP), a 2-week-long training workshop in 2019. Its success motivated the organization of next editions of the school in the consecutive years. To date, five editions have been held, where more than 500+ high-profile participants were trained.

In the next few sections, participants' data from the LAC region who were trained during these five editions is briefly discussed. Such data has extreme importance as it helps in mapping skills and future collaborative initiatives that may connect the existing gap areas within the LAC region.

6.1. Background

InnSciD started as Sao Paulo School of Advanced Science (SPSAS), funded by The Sao Paulo Research Foundation (FAPESP). This line of funding provides monetary resources for organizing short-term courses focused on advanced research across various fields in the State of Sao Paulo. SPSAS aims to offer two-week courses covering advanced topics in science and technology to enhance the education of young participants. Through this initiative, FAPESP aims to establish a globally competitive center for talented researchers within Sao Paulo. SPSAS instructors are highly qualified scientists renowned in their respective research areas, including invited experts from abroad.

The InnSciD school was designed to fuse, on the one side, seminars and panels merging a strong theoretical foundation with cutting edge themes, and on the other, practical activities in which the participants could mobilize their own knowledge from different areas and the themes discussed throughout the school. In this regard, the selection process played a significant role by gathering determinants as gender, country of origin, nationality, area of knowledge and professional background. The objective laid on guaranteeing the participants' diversity, one of the key features related to innovation processes and outputs (Lungeanu & Contractor 2015; Nieto & Santamaría, 2007a).

6.2. Focus Areas of School

The first edition of InnSciD took place in 2019, when more than 100 high level scientists, diplomats and practitioners were brought to Sao Paulo to discuss Science and Innovation Diplomacy as a concept in development, with an important Global South perspective. In the same edition, the Sao Paulo Innovation Diplomacy Framework (InnScid, 2019) was launched, as part of the negotiation sessions that involved all the participants. The in-person model allowed for two technical visits: the first one took participants to Embraer, a Brazilian aerospace company renowned for its contributions to science and innovation to commercial, military, executive, and agricultural aircrafts (Embraer, 2024) while the second one to Sirius, Brazil's cutting-edge synchrotron light source. This state-of-the-art facility employs particle accelerators to generate synchrotron light, a specialized form of illumination crucial for probing the composition and structure of matter across a wide spectrum of forms. With its unparalleled capabilities, Sirius serves as a vital tool for scientific exploration, offering invaluable insights across diverse fields of knowledge (SIRIUS, 2024).

In 2020, forced by the physical mobility limitations imposed by the COVID-19 pandemic, InnSciD was conducted online. Taking advantage from digital platforms, the school also organized an Open Lecture week, aiming at sharing knowledge with a broader audience to raise attention to the new obstacles to be tackled with the help of Science and Innovation Diplomacy. Another important feature of the online mode was the creation of practical activities, as simulations and problem-solving discussions. Therefore, the second edition had over 130 students from all regions in the world and included lectures and panels on International Collaboration in Life Sciences, Internationalization of Technology Startups, Amazon 4.0 and the Global Innovation Index, among many others. The program commenced with an exploration of the fundamental concepts of science diplomacy, setting the stage for subsequent sessions that delved into the multifaceted dimensions of the field. At the end of the school, participants were asked to prepare Roadmaps on Innovation Diplomacy for their respective regions, which was published as an e-book. It is meaningful to stress the UNESCO support to this edition, also encouraged by the continued interest of researchers from different institutions.

The third edition, in 2021, received support from The World Academy of Sciences (TWAS) and American Association for the Advancement of Science (AAAS). InnSciD 2021 convened a dynamic program exploring a diverse array of topics, from the foundational principles of science diplomacy to the practical applications of international collaboration in scientific endeavors. Themes such as the role of multinational scientific infrastructures, brain circulation, and the internationalization of science, technology, and innovation policies took the center stage, offering attendees a comprehensive understanding of the global landscape of scientific cooperation and its diplomatic implications.

The same structure followed the 2022 online edition, which aimed at presenting an overview of subjects in which science and innovation diplomacy are key instruments. As standard, the discussions also explored emerging trends in science diplomacy, stemming from geopolitical implications of technology to the role of venture capital in fostering innovation ecosystems, the objective was to capture the potential areas of impact. InnSciD aimed at fostering innovative approaches to addressing pressing global challenges and advancing the agenda of science and innovation diplomacy in the 21st century.

In 2023, InnSciD received a new funding from Sao Paulo School of Advanced Science (SPSAS). With this line of funding, the InnSciD took place in person, at the University of Sao Paulo. InnSciD 2023 analysed the success and challenges of the field, and specifically delved into diplomatic dimensions associated with the international exchanges of scientific data, exploring emergent policy issues, especially in terms of the relationship between power and data. The fifth edition was designed in partnership with Neworld@A - Negotiating World Research Data, a research project funded by the European Research Council (ERC). Among the specific topics, the school provided a deep dive on global dissemination of research data, open science, open innovation, and data economy.

As an in-person school, InnSciD 2023 had two technical visits: one at Sirius, as in 2019, and other at the recently renewed Ipiranga Museum. Ipiranga is a university museum, officially part of the University of Sao Paulo, and known as Museu Paulista. It is one of the most important historical museums in Brazil, located in Sao Paulo (Ipiranga, 2024). During its restoration, the Ipiranga Museum incorporated technological solutions with the goal of enhance the visitor experience and provide a more interactive and immersive exploration of Brazilian and Paulista (gentilic for Sao Paulo) history and culture. Through the combination of new technologies such as augmented reality, virtual reality, interactive displays, and digital storytelling, visitors can engage with the museum's exhibits in innovative ways, while learning from Brazilian culture through an up-to-date perspective regarding historical facts.

Next sections will analyse the profile of InnSciD-SP school participants according to a number of criteria. For the topic that interests us the most, Section 5.5 is the most important.

6.3. Gender Distribution and Age Groups

Table 1 shows the gender distribution of participants from the LAC region in 5 editions of InnSciD-SP school. The female versus male ratio was almost equal except in the year 2020 when a higher proportion of female participants was observed. That was the time of COVID-19 lockdown and InnSciD-SP training was organized virtually. During times of lockdown, some studies report a decline in the productivity of female scientists in the research area leading to fewer science publications as first author (Philippe Vincent-Lamarre et al., 2020). However, when it came to participation in the virtual conferences, the data shows that female participation increased (Skiles et al., 2022). A similar tendency was found in the year 2020 when InnSciD-SP organizers received more highly qualified female registrations compared to male registrations in that specific year, that made it to the final list of selected candidates. Research shows that virtual events make women empowered, by not travelling far for attending events, they have the freedom to take care of both family and work, without missing out or making a choice between them (Morressier, 2022). After the 2021 edition, candidates were given a wide choice for declaring their gender variation in case of preference and hence, on receiving a suitable candidate profile, participation was made possible.

Table 1. Percentage Distribution of Training Candidates from the LAC Region According to their Gender

GENDER	YEARS					
	2019	**2020**	**2021**	**2022**	**2023**	**Grand Total**
Female	48.00%	61.54%	45.87%	49.15%	52.00%	50.75%
Gender variant/ Non-conforming	0.00%	0.00%	0.00%	1.69%	0.00%	0.30%
Male	52.00%	38.46%	54.13%	49.15%	48.00%	48.95%
Grand Total	**100.00%**	**100.00%**	**100.00%**	**100.00%**	**100.00%**	**100.00%**

Figure 1 shows the bar chart of the age range related to gender. A 10-year range is represented by each bar. A higher proportion of attendees was in the age range of 31-40, followed by 21-30 both for male and female participants.

Figure 1. Age Range and Gender Distribution

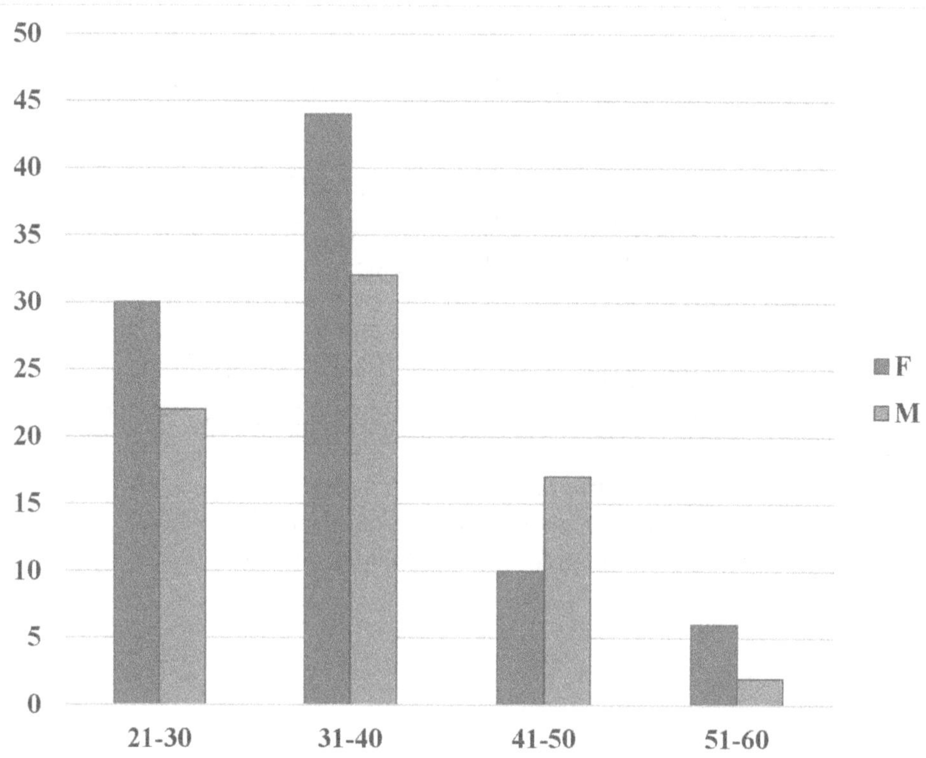

6.4. Academic Area of Knowledge

Figure 2 shows the area of knowledge based on the academic profiles of participants in the form of a pie chart. Coordination for the Improvement of Higher Education Personnel (CAPES) from the Ministry of Education Brazil provides a list of areas of knowledge that was considered while categorizing the subject areas into 9 different categories ('CAPES' 2023). Each segment of the pie chart shows the percentage of an individual category with a different colour. The top three categories were Humanities, Applied Social Sciences, and Exact and Earth Sciences.

Figure 2. Pie Chart Showing Academic Profiles of Participants from the LAC Region

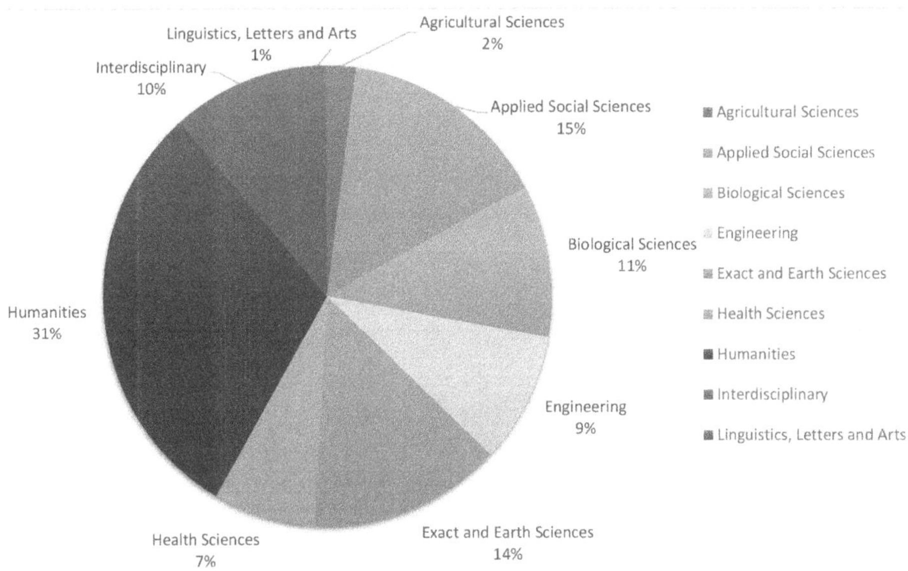

For alignment of science and innovation with global challenges, interdisciplinarity is a must. Establishing bridges between different forms of knowledge areas, broadly the hard/ natural sciences and the social sciences and humanities is a prerequisite for innovation (Royal Society of Great Britain, 2010). Commonly, scientists are expected to deliver results of their particular research area in short timeframe, that on one hand confines them to their research discipline boundaries, while on the other, prohibits them from thinking out of the box for solving bigger challenges faced by humanity. InnSciD training workshops provided interdisciplinary engagements where participants from various knowledge backgrounds worked together on well-defined regional and global challenges and sorted out roadmaps for possible solutions. Such collaborative interdisciplinary approach for innovation is important for the technological growth of the LAC region.

6.5. Stakeholders From the LAC Region

Due to the global nature of all the InnSciD school editions, participants were from different corners of the world. The date shows that out of 516 participants, 331 were from the LAC region, with 178 participants just from Brazil. By the program design, it can be deduced that InnSciD gives preference to Brazil-originating participants as there are half reserved seats for them in each edition, due to ease in logistics and the desired need of training the local citizens. If we take out Brazilian nationals from the LAC region, LAC participants minus Brazil, still accounts for almost 45% of the overall participants compared to other regions of the worlds.

Table 2 shows participants from the LAC region and their continent-wise distribution in relation to residence and nationality. Comparing to the overall number of participants, the LAC region dominated in terms of diaspora studies. Out of 331 LAC participants, 252 were residing in the LAC region while the remaining 79 had different country of current residence or nationality.

Table 2. Continent-wise Allocation of Participants from the LAC region, Current Residence Versus Nationality

NATIONALITY (columns); **RESIDENCE** (rows)

Count of Full Name	**Column Labels**					
Row Labels	**AF**	**EU**	**LAC**	**NA**	**SA**	**Grand Total**
AF			4			4
EA			9			9
EU			25			25
LAC	5	7	*252*	1	2	267
NA			23			23
PA			1			1
SA			1			1
SEA			1			1
Grand Total	**5**	**7**	**316**	**1**	**2**	**331**

*AF = Africa, EA = East Asia, EU = Europe, LAC = Latin America and the Caribbean, NA = North America, PA = Pacific, SA = South Asia, SEA = South East Asia.

For diaspora studies, it is clear from the Table 2 that nationals from the LAC region have a preference to reside in North America (NA) and Europe (EU). On one hand this is considered as brain drain while on the other hand, the concept of brain circulation opens doors to future collaborations and healthy interactions between the home and the host countries (Echeverría-King et al. 2022). Also, out of 252 people from LAC, 21 are diaspora within LAC, which means they have different countries of origin and nationalities within the LAC region. Some public and private initiatives/organization support the development of activities involving diaspora that establish connections between the home and host countries (CD4D2 2023; GDC 2023).

7. CONCLUSION

Innovation Diplomacy approaches on one side may accelerate economic growth, while on the other, sharing mutual benefits may help in improving international relations between countries of the LAC region. For bringing such approach to life requires adjustments to governmental policies to support the creation of an environment that is conducive to the development of the framework for building an innovative, intelligent, and sustainable future. According to what we referred to while discussing the topic of absorptive capacity of people, firms and nations when having possibility of knowledge acquisition, prioritize learning from nations with closer ground realities to their own. From the results of InnSciD-SP participant data, it could be seen that most of the registered candidates for InnSciD-SP are indeed originating from the LAC region. This data seems to testify that people appreciate and privilege sources of knowledge and experiences that are culturally closer to them. This fact and assumption exposed in footnote 2, reinforces the importance of InnSciD-SP as a pioneer in the Global South for science and innovation diplomacy training that also puts Brazil in the forefront of other nations belonging to the Global South concerning this initiative.

The ultimate effect of the network ties that were established during these trainings among stakeholders from diverse knowledge areas, sectors and nationality, will help in the future development and evolution of ST&I practices over time. Furthermore, InnSciD-SP trainings will be useful for future developments and long-term impact studies on this topic, enabling the promotion of innovation diplomacy agendas in different countries both in academic and non-academic environments.

REFERENCES

Ansell, C., Sørensen, E., & Torfing, J. (2024). The democratic quality of co-creation: A theoretical exploration. *Public Policy and Administration*, 39(2), 149–170. 10.1177/09520767231170715

Barrett, P., Hansen, N.-J., Natal, J.-M., & Noureldin, D. (2021). Why Basic Science Matters for Economic Growth. *IMF Blog*. Retrieved March 1, 2024, from https://www.imf.org/en/Blogs/Articles/2021/10/06/blog-ch3-weo-why-basic-science-matters-for-economic-growth

Beltran, L. (2024). Latin America is moving towards a sustainable economy. These are the challenges and opportunities' *World Economic Forum Annual Meeting*. Retrieved April 4, 2024, from https://www.weforum.org/agenda/2024/01/latin-america-is-moving-towards-a-sustainable-economy-these-are-the-challenges-and-opportunities/

Beskaya, A., Savas, B., & Samiloglu, F. (2010). The impact of education on economic growth in Turkey. *The Journal of Faculty of Economics and Administrative Sciences*, 15(3), 43–62.

Bound, K. (2016). *'Innovating Together? The Age of Innovation Diplomacy'. Global Innovation Index 2016* (Vol. 1). World Intellectual Property Organization.

Brazilian Ministry of Foreign Affairs. (2023). *Relatório Programa de Diplomacia da Inovação*. Retrieved April 5, 2024, from https://www.gov.br/mre/pt-br/assuntos/ciencia-tecnologia-e-inovacao/Relatrio2023ProgramadeDiplomaciadaInovao.pdf

CD4D2. (2023). Through the transfer of knowledge and expertise, CD4D2 engages diaspora to support the development of their countries of origin. Retrieved December 28, 2023, from https://connectingdiaspora.org/about-cd4d2/

Cohen, W. M., & Levinthal, D. A. (1990). Absorptive Capacity: A New Perspective on Learning and Innovation. *Administrative Science Quarterly*, 35(1), 128. 10.2307/2393553

CPD. (2024). USC Center on Public Diplomacy. Retrieved January 14, 2024, from https://uscpublicdiplomacy.org/page/what-is-pd

da Cruz, A. S., Jr. (2010). Diplomacia da inovação. *Parcerias Estratégicas*, 15, 297–308.

Department for Science. (2024). *The UK Science and Technology Framework*. Retrieved July 13, 2024, from https://www.gov.uk/government/publications/uk-science-and-technology-framework/the-uk-science-and-technology-framework

Echeverría-King, L. F., Camacho Toro, R., Figueroa, P., Galvis, L. A., González, A., Suárez, V. R., & Torres Atencio, I. (2022). Organized Scientific Diaspora and Its Contributions to Science Diplomacy in Emerging Economies: The Case of Latin America and the Caribbean. *Frontiers in Research Metrics and Analytics*, 7. 10.3389/frma.2022.893593

Ellerbeck, S. (2023). Lithium: Here's why Latin America is key to the global energy transition. *World Economic Forum*. Retrieved April 4, 2024, from https://www.weforum.org/agenda/2023/01/lithium-latin-america-energy-transition/

Embraer. (2024). History of EMBRAER. Retrieved March 22, 2024, from https://historicalcenter.embraer.com/global/en/history

Emprende, A. (2024). Amazonía Emprende-Escuela Bosque. Retrieved April 4, 2024, from https://uplink.weforum.org/uplink/s/uplink-contribution/a012o00001pTVciAAG/A%20knowledge%20and%20incubation%20hub%20for%20nature-based%20solutions%E2%9C%AA

FGV. (2023). Higher quality education associated with higher growth rates, study finds. Retrieved March 1, 2024, from https://portal.fgv.br/en/news/higher-quality-education-associated-higher-growth-rates-study-finds

Flink, T. (2022). *Taking the pulse of science diplomacy and developing practices of valuation. Science and Public Policy, 49(2), 191–200*. Oxford University Press. 10.1093/scipol/scab074

Gálvez-Soriano, O. de J. (2020). Could Education Increase the Economic Growth of Mexico? *Análisis económico- SciELO Analytics*.

Garcia, R., Araújo, V., Mascarini, S., dos Santos, E. G., & Costa, A. R. (2018). An analysis of the relation between geographical and cognitive proximity in university-industry linkages. *Anais do XLIV Encontro Nacional de Economia*, 1–20.

GDC. (2023). Welcome to the Global Diaspora Confederation. GDC.

Gluckman, P. D., Turekian, V., Grimes, R. W., & Kishi, T. (2017). Science Diplomacy: A Pragmatic Perspective from the Inside. *Science & Diplomacy*, 6(4).

Granstrand, O., & Holgersson, M. (2020). Innovation ecosystems: A conceptual review and a new definition. *Technovation*, 90–91, 102098. 10.1016/j.technovation.2019.102098

Hay, C. (2020). *International Relations Theory and Globalization* (Dunne, T., Kurki, M., & Smith, S., Eds.; 5th ed.). International Relations Theories. 10.1093/hepl/9780198814443.003.0015

InnScid. (2019). *São Paulo Framework of Innovation Diplomacy*. Retrieved September 5, 2021, from https://innscidsp.com/framework.pdf

Ipiranga. (2024). Museu do Ipiranga-USP. Retrieved March 22, 2024, from <https://museudoipiranga.org.br/en/>

Keohane, R. O., & Nye, J. S.Jr. (1977). Power and interdependence. *Survival*, 15(4), 158–165. 10.1080/00396337308441409

Leijten, J. (2017). Exploring the future of innovation diplomacy. *European Journal of Futures Research*, 5(1), 1–13. 10.1007/s40309-017-0122-8

López-Calva, L. F. (2022). Finding Unicorns: Investing in innovation-led growth in LAC. *United Nations Development Programme, Latin America and the Caribbean*. Retrieved April 4, 2024, from <https://www.undp.org/latin-america/blog/graph-for-thought/finding-unicorns-investing-innovation-led-growth-lac>

Lungeanu, A., & Contractor, N. S. (2015). The Effects of Diversity and Network Ties on Innovations: The Emergence of a New Scientific Field. *American Behavioral Scientist*, 59(5), 548–64. 10.1177/0002764214556804

Marcelle, G. M. (2016). *Redefining innovation in the Global South: Critical imperatives*. Retrieved April 3, 2024, from https://unctad.org/system/files/non-official-document/CSTD_2015_ppt15_Marcelle_en.pdf

Mazzucato, M., & Penna, C. (2016). *The Brazilian Innovation System: A Mission-Oriented Policy Proposal.*

Morressier. (2022). Women's participation rates are 253% higher at virtual conferences. Here's why. Morressier *Research Team*. Retrieved December 28, 2023, from https://www.morressier.com/post/womens-participation-rates-are-253-higher-at-virtual-conferences-heres-why

MRE. (2024). Innovation Diplomacy Program. Retrieved January 14, 2024, from https://www.gov.br/mre/en/subjects/science-technology-and-innovation/innovation-diplomacy-program

Nieto, M. J., & Santamaría, L. (2007). The importance of diverse collaborative networks for the novelty of product innovation. *Technovation*, 27(6–7), 367–377. 10.1016/j.technovation.2006.10.001

Nilus. (2024). La pobreza no debería ser más cara. Retrieved April 4, 2024, from https://www.nilus.co/

OECD. (2020). Science, Technology and Innovation in the New Economy. Retrieved March 1, 2024, from https://www.oecd.org/science/inno/1918259.pdf

OECD/Eurostat. (2018). Oslo Manual 2018: Guidelines for Collecting, Reporting and Using Data on Innovation, 4th Edition. *The Measurement of Scientific, Technological and Innovation Activities*. OECD Publishing: Paris/Eurostat, Luxembourg. 10.1787/9789264304604-en

OSI. (2024). Office of Science and Innovation, Embassy of Sweden. Retrieved July 13, 2024, from https://www.swedenabroad.se/en/embassies/japan-tokyo/about-us/the-office-of-science-and-innovation/

Parliament of Australia. (2024). Defining "Public Diplomacy". Retrieved January 14, 2024, from https://www.aph.gov.au/Parliamentary_Business/Committees/Senate/Foreign_Affairs_Defence_and_Trade/Completed_inquiries/2004-07/public_diplomacy/report/c02

PDI. (2024). Programa Diplomacia da Inovação. *Ministério das Relações Exteriores*. Retrieved July 13, 2024, from https://www.gov.br/mre/pt-br/assuntos/ciencia-tecnologia-e-inovacao/programa-de-diplomacia-da-inovacao

Qadri, R. (2021). Global South startups can do better than Silicon Valley. *Reporting Global Tech Stories*. Retrieved April 2, 2024, from https://restofworld.org/2021/grab-indonesia-silicon-valley/

Royal Society of Great Britain. (2010). *The scientific century : securing our future prosperity*. Royal Society.

Ruffini, P. B., & Krasnyak, O. (2023). *Science diplomacy from a nation-state's perspective: A general framing and its application to Global South countries. Science and Public Policy, 50/4: 771–81*. Oxford University Press., 10.1093/scipol/scad023

SIN. (2024). UK Science and Innovation Network. Retrieved July 13, 2024, from <https://www.gov.uk/world/organisations/uk-science-and-innovation-network>

SIRIUS. (2024). SIRIUS: Accelerating The Future Of Brazilian Science. Retrieved March 22, 2024, from https://lnls.cnpem.br/sirius-en/

Skiles, M., Yang, E., Reshef, O., Muñoz, D. R., Cintron, D., Lind, M. L., & Rush, A. (2022). *'Conference demographics and footprint changed by virtual platforms'. Nature Sustainability*. Nature Research. 10.1038/s41893-021-00823-2

Srinivasan, R. (2019). Opinion: The Global South is redefining Tech Innovation. *Wired*. Retrieved April 2, 2024, from https://www.wired.com/story/opinion-the-global-south-is-redefining-tech-innovation/

Straubhaar, J. (2007). *World Television: From Global to Local*. SAGE Publications, Inc. 10.4135/9781452204147

Sustainability Times. (2023). Global South needs locally driven innovation to turn tide on UN's SDG agenda. *Sustainability Times*. Retrieved April 2, 2024, from https://www.sustainability-times.com/in-depth/global-south-needs-locally-driven-innovation-to-turn-tide-on-uns-sdg-agenda/

Thales. (2024). What is Human Factors? Retrieved April 3, 2024, from <https://www.thalesgroup.com/en/human-factors-design#:~:text=Human%20Factors%20is%20about%20designing,things%20better%20and%20more%20effective>

The Royal Society. (2010). *New frontiers in science diplomacy*. Retrieved July 14, 2024, from https://www.aaas.org/sites/default/files/New_Frontiers.pdf

ENDNOTES

1 As of July 15th, 2024, exchange rate was R$ 5.40 to US$ 1.00.

2 Imagine individual A presents a knowledge level of 100 and can (temporarily or permanently) migrate to country B (symmetrical country to him's or her's) or country C (asymmetrical country to him's or her's). Imagine country B enables him or her access to a knowledge level of 140, while country C enables him or her access to a knowledge level of 180. Imagine further absorptive capacity of such individual is 80% when exposed to educational system from country B, and 30% when exposed to educational system from country C. Such a situation yields new knowledge levels of 132 and 124 respectively. All else constant, such a human resource is better off after having access to the educational system of the lower-level country, and such a result is explained because of differences in the absorptive capacity of the individual.

Chapter 4
Synergies Between Science Communication and Science Diplomacy to Address Global Challenges:
Lessons From the LAC Region

Branislav Pantović
https://orcid.org/0009-0005-5561-0476
Universidad Nacional de Río Negro, Argentina

Luz M. Cumba Garcia
DiploCientifica, Argentina

María Inés Carabajal
Inter-American Institute for Global Change Research, Argentina

ABSTRACT

This chapter explores the role of science communication in advancing science diplomacy in the Latin America and the Caribbean (LAC) region. It examines communication initiatives and collaborative approaches among LAC institutions through science diplomacy amidst challenges posed by the COVID-19 pandemic and global environmental change (GEC). Case studies featuring organizations such as the National Secretariat of Science, Technology, and Innovation from Panama (SENACYT), the Science Diplomacy Network in Latin America and the Caribbean (DiploCientifica), the Inter-American Institute for Global Change Research (IAI),

DOI: 10.4018/979-8-3693-2746-3.ch004

Puerto Rico Science Policy Action Network (PRSPAN), and Ciencia Puerto Rico (CienciaPR) highlight their efforts in disseminating science, technology, and innovation (STI). These organizations serve as intermediaries, translating scientific insights for decision-makers and facilitating connections between scientists, diplomats, policymakers, and society. By analyzing the organizations' missions, visions, and contributions, the chapter aims to identify key factors for effective scientific communication in science diplomacy and essential skills.

PURPOSE OF THE CHAPTER

This chapter explores the integral role of effective science communication in advancing the practice of science diplomacy within the Latin America and the Caribbean (LAC) region. Through an exploration of communication initiatives and collaborative efforts during pivotal events such as the COVID-19 pandemic and efforts to tackle environmental change, the chapter aims to underscore the critical intersection between science, technology, and innovation (STI) with diplomacy and policy. By analyzing the contributions of various organizations in the LAC region, including the National Secretariat of Science, Technology and Innovation from Panama (SENACYT), the Science Diplomacy Network in Latin America and the Caribbean (DiploCientifica), the Inter-American Institute for Global Change Research (IAI), the Puerto Rico Science Policy Action Network (PRSPAN), and Ciencia Puerto Rico (CienciaPR), the chapter highlights their significant roles in disseminating STI information and advocating for informed decision-making. These entities serve as intermediaries, translating complex scientific insights into accessible information for decision-makers and fostering connections among stakeholders during critical periods. Through this examination, the chapter aims to identify key factors for effective science communication in the sphere of science diplomacy and highlight the essential skills required for tailored communication relevant to the unique context of the LAC region. Ultimately, the chapter seeks to offer insights and recommendations that can contribute to addressing global and regional challenges by leveraging the synergies between communication, STI, and diplomacy for the benefit of societies in the LAC region.

METHODOLOGY

This study utilizes a qualitative approach to investigate the critical role of effective science communication in advancing both theoretical and practical science diplomacy frameworks. Case studies involving interviews with representatives from intergovernmental, governmental, nonprofit, and third-sector organizations engaged in science communication and science diplomacy across the LAC region were conducted (Table 1). These organizations include Panama's SENACYT, DiploCientifica, IAI, PRSPAN, and CienciaPR. The interviewees were selected purposely to ensure diverse representation across different types of organizations and provide rich insights into the diverse approaches, challenges, and successes in science communication and science diplomacy efforts within the region. While there is no conflict of interest, it is important to note that all interviewees are known to the authors of this chapter through previous collaborations in science diplomacy and science communication in the LAC region with their respective organizations.

Three questions were posed to the interviewees to enable the identification of main priorities, strategies, and challenges faced by their institutions in communicating scientific knowledge to diverse audiences. These questions focused on communication initiatives and strategies implemented during the COVID-19 pandemic and efforts to address environmental change, underscoring the significance of science diplomacy in tackling shared global challenges in the LAC region. These questions were:

1. How do you perceive the relationship between science diplomacy and science communication? In your experiences, do you find them to be closely linked?
2. What initiatives did your institution and its members undertake to contribute to addressing regional challenges like COVID-19 or/and global environmental change? How did science communication and dissemination play a role in these efforts?
3. Which skills were deemed essential, and which were considered secondary when addressing these regional challenges?

In addition to conducting interviews, a detailed analysis of the selected organizations was carried out. This analysis involved examining their visions, missions, objectives, past and ongoing initiatives and collaborations among institutions, communication strategies, and roles in science diplomacy within the region. Information gathered from organizational websites, publications, and other sources contributed to contextualizing the findings and understanding the landscape of science communication in the LAC region. A non-exhaustive literature review was also conducted to explore the relationship between science communication and science diplomacy. Thematic analysis was used to identify recurring patterns, key themes, and critical factors influencing effective science communication within the context of science diplomacy among the studied organizations. By integrating insights from interviews, literature review, organizational analysis, and case studies, this research aims to provide a comprehensive understanding of science communication practices and their role in advancing science diplomacy agendas in the LAC region.

Table 1. Case Study Participants

Participant name and role within the organization	Organization	Participant country of origin
Dr. Eduardo Ortega-Barría, National Secretary of Science, Technology, and Innovation	National Secretariat of Science, Technology, and Innovation from Panama (SENACYT)	Panama
Pedro Figueroa, President	Science Diplomacy Network in Latin America and the Caribbean (DiploCientifica)	Chile
Dr. Luisa Fernanda Echeverría-King, Executive Director	Science Diplomacy Network in Latin America and the Caribbean (DiploCientifica)	Colombia
Dr. Anne-Teresa Birthwright, IAI Science, Technology, and Policy (STeP) Alumni	Inter-American Institute for Global Change Research (IAI)	Jamaica
Dr. Anna Stewart-Ibarra, Executive Director	Inter-American Institute for Global Change Research (IAI)	Uruguay
Dr. Luis Alexis Rodríguez-Cruz, Former Co-Chair	Puerto Rico Science Policy Action Network (PRSPAN)	Puerto Rico
Dr. Mónica Feliú-Mójer, Director of Communications and Science Outreach	Ciencia Puerto Rico (CienciaPR)	Puerto Rico

CONTEXTUALIZING GLOBAL AND REGIONAL CHALLENGES

STI plays a central role in government investments and strategies, particularly in high-income nations, by enhancing decision-making at the science-policy interface. Effective science-policy interfaces are characterized by inclusivity, transparency, accountability, legitimacy, relevance, and adaptability. These attributes foster collaboration and ensure that scientific knowledge informs policymaking processes in response to national and international needs (van den Hove, 2007). Additionally, STI is indispensable for tackling global environmental change (GEC) challenges, such as climate change, biodiversity conservation, ecosystem services, and ensuring water and food security. Given the transboundary nature of these issues, international collaboration is imperative as no single country can address them in isolation.

The significance of international collaboration and science policy in addressing these challenges is underscored by significant milestones achieved in various United Nations (UN) conferences and forums specifically convened to address biodiversity matters. Examples of these milestones include the establishment of the Intergovernmental Panel on Climate Change (IPCC) by the UN in 1988[1], the adoption of Agenda 21 for sustainable development and economic growth during the Earth Summit in Brazil in 1992[2], and the adoption of the Paris Agreement in 2015 at the UN Climate Change Conference (COP)[3]. These events exemplify the critical link between governance and global change on a global scale, providing platforms not only for the exchange of knowledge, expertise, and scientific findings but also for policymaking and reaching mutual agreements (Piñeros Ayala et al., 2023).

In the LAC region, interest in science diplomacy is increasing among countries, leading to numerous efforts to engage governments, diplomats and scientists in science diplomacy. These efforts include capacity-building initiatives, the development of national strategies, and the formulation of international agreements. However, these mechanisms remain fragmented or insufficient in comprehensively addressing the multifaceted challenges at hand (Camacho Toro, et al., 2024) (Echeverria King, et al., 2023). Therefore, we assert that strengthening science communication and science diplomacy is crucial to bolstering regional capacities and fostering regional collaboration within the LAC region.

It is crucial to recognize the evolving complexity of contemporary challenges, which often transcend isolated issues and involve interconnected factors. Disruptions to one element can lead to cascading effects, akin to the toppling of dominos. These challenges extend to addressing problems at the intersection of climate, environment, and health (CEH), exemplified by the COVID-19 pandemic. The pandemic has highlighted the critical need for collaborative science diplomacy in developing, sharing, and implementing strategies to combat such complex health and social crises (Widmaier, et al., 2020). This interconnectedness underscores the need for comprehensive and systemic approaches to address the multifaceted nature of these kinds of challenges successfully.

Effective communication among scientists, policymakers, and the public within this framework is essential for coordinated pandemic responses and preparedness for future challenges that may emerge from human-nature interactions. Strengthening these areas can enhance readiness and promote a more cohesive global and regional approach to impending health and environmental crises. COVID-19 and climate change magnify existing vulnerabilities, making collaboration and networking increasingly imperative in the region to effectively address historical and emerging challenges (Cumba García and Carabajal, 2024; Khojasteh et al., 2022).

THE EVOLUTION OF SCIENCE DIPLOMACY AND ITS IMPACT

Diplomacy involves communication and representation by government actors to manage relationships and achieve outcomes (Gregory, 2016). Considering the asymmetrical and growing power of citizens and civil society actors, it is essential to emphasize that contemporary diplomacy should prioritize dialogue and foster collaboration among a diverse array of actors. This approach, known as public diplomacy, directs its activities towards a broader international audience, contrasting with traditional diplomacy focused on state or multilateral relationships (Melissen, 2005).

The evolving structure of international relations, especially since September 11, 2001, has underscored the necessity for a more nuanced diplomatic approach to address global issues like terrorism, migration, and financial instability, necessitating collaboration between governments, non-governmental organizations (NGOs), and individuals (Gonesh and Melissen, 2005). In both bilateral and multilateral contexts, public diplomacy has shifted from rigid official interactions to more open and dynamic engagements with the public, signaling a departure from conventional practices and embracing diverse actors to achieve objectives in innovative ways. This evolution has also led to the adoption of inter/transdisciplinary and intersectoral approaches within diplomacy, making public diplomacy a comprehensive term encompassing a variety of activities and interactions conducted by (inter)government representatives.

The increasing use of adjectives in diplomacy could be partly attributed to Mark Leonard's 2002 book "Public Diplomacy", which introduced new terms such as co-operative diplomacy, competitive diplomacy, NGO diplomacy, diaspora diplomacy, and corporate diplomacy, among others (Brown, 2017). This trend reflects a shift towards more inclusive and multifaceted approaches to international relations, reflecting a world where power is diffuse and global challenges require collective action.

In the context of this chapter, the practical application of science diplomacy can be interpreted as a manifestation of public diplomacy (Goldman, 2022). The Vienna Convention of 1961, in Article 3[4], outlines the core functions of a diplomatic mission, which include promoting and enhancing cultural, economic, and scientific relations, alongside other types of relations, within the framework of international law, while safeguarding the interests of the accrediting state and its nationals. This marks perhaps the earliest formal acknowledgement of science within diplomatic practice.

Ruffini suggests that science diplomacy seeks to shape international relations by harnessing the universal values of science to tackle a myriad of critical issues (Ruffini, 2020). However, defining science diplomacy poses a challenge due to the lack of universal consensus among stakeholders involved. It is widely understood that science diplomacy encompasses a range of actions and concepts that bridge the gap between politics, policy, and STI within an international framework. These actions include providing international science advice and advocacy, securing international financial support for technological advancements, negotiating international agreements and treaties, and navigating scientific and technological competition among nations, among others. Several pivotal historical instances have shaped the emergence and development of science diplomacy. These include the development of nuclear technology during World War II (Vera and Echeverría King, 2022), a series of United Nations conferences and treaties addressing global environmental challenges (Piñeros Ayala et al., 2023), the refinement of the science diplomacy conceptual framework in 2010 by the Royal Society and the American Association for the Advancement of Science (AAAS) (2010), and, more recently, the COVID-19 pandemic, which highlighted the importance of a science-based approach and international collaboration to address shared challenges (Pérez Valerino, 2021). These events highlight the continuous significance and dynamic evolution of science diplomacy across different historical periods, showcasing its adaptability as scientific focuses and applications change. They emphasize the need for a comprehensive understanding and analysis of science diplomacy in effectively addressing global challenges.

SCIENCE DIPLOMACY IN LAC: CHALLENGES & OPPORTUNITIES

The LAC context of science diplomacy is characterized by the region's diverse socio-economic, political, and environmental factors. This complexity presents challenges in the implementation and development of science diplomacy initiatives, often tied to issues such as funding constraints, resource limitations, inequality, political and economic instability, and institutional fragmentation. However, in sectors related to environmental sciences, public health, agriculture, and renewable energy, the significance of science diplomacy is particularly evident. (Piñeros et al., 2023). These areas require international collaboration to effectively address shared challenges and meet mutual needs. As a result, there is a clear demand and overwhelming interest in science diplomacy for specific topics within the LAC region.

Emerging progress in the implementation of science diplomacy and public policies within the LAC region is evident, despite existing disparities in capacity among countries (López-Vergès, et al., 2021). For instance, Panama has taken a pioneering step by defining its national strategy for STI diplomacy. Similarly, Colombia is in the process of formulating a national science diplomacy strategy, Brazil is actively promoting an Innovation Diplomacy Program through its Department of Science and Technology within the Ministry of Foreign Affairs, and Cuba has established itself as a leader in medical diplomacy (Gual Soler, 2021). These examples illustrate a growing culture of collaboration, training, international knowledge exchange, and the development of scientific policies, highlighting the increasing importance of science diplomacy in the region.

It is crucial to acknowledge that the development of science diplomacy in the LAC region is closely tied to capacity building initiatives. Currently, there are only a few science diplomacy courses in the region with ongoing offerings, such as the São Paulo Innovation and Science Diplomacy School (InnSciD SP)[5], the Critical

Introduction to Science Diplomacy course at the Faculty of Medicine of Universidad Nacional Autónoma de México since 2024[6], and virtual outreach courses provided by institutions such as Universidad de Chile[7], and Universidad Externado de Colombia[8]. This highlights the pressing need for expanded implementation of capacity building programs, particularly within universities, to address this gap (Echeverria King et al., 2023; Hernández-Mondragón, 2022). Such training initiatives, though scarce, not only enhance scientific and diplomatic competencies but also empower countries to assume more proactive roles in global and regional scientific networks and negotiations. Moreover, these training programs foster a culture of innovation and offer support to startups in the science and technology sectors. Consequently, this could attract foreign investment in technology, create job opportunities, and stimulate economic growth in the region.

Given the multifaceted nature of science diplomacy, its objectives can span across various sectors, as exemplified by Argentina. This country has historically used science to build diplomatic relations and mitigate tensions in other areas (Gual Soler, 2021), particularly through its significant commitment to nuclear and space diplomacy. This commitment is evident in the inclusion of the Directorate of International Security, Nuclear Affairs, and Space (DIGAN) within the Ministry of Foreign Affairs. Through this commitment, Argentina not only advances technological capabilities and economic opportunities but also fosters strategic collaboration within the region. A notable demonstration of this is Argentina's leadership in establishing the Latin American and Caribbean Space Agency (ALCE) (Vera and Guglielminotti, 2018). Furthermore, science diplomacy serves to safeguard Argentina's sovereignty and national integrity, particularly concerning the Malvinas Islands, a key objective of Argentine foreign policy (D'agrosa Okita and Pantovic, 2023). The Pampa Azul interministerial initiative, which promotes marine research in the southern part of the country and was established as law by the Argentine National Congress in 2015, exemplifies this approach (Goldman, 2022).

Each example from the LAC region exhibits unique characteristics, yet they share a commonality in the involvement of multiple and diverse actors, which is acknowledged as an important ingredient in science diplomacy. This highlights the importance of collaboration between scientific institutions, government agencies, and diplomatic corps, aligning with development goals and global/regional agendas for success. On the regional level, this becomes paramount as it enables countries to harness their collective expertise and resources, thereby enhancing their capacity to effectively address shared issues such as those concerning biodiversity in the Amazonian region (Prieto et al., 2024) or energy transformation, which is recognized in many countries of the LAC region as a shared problem (Piñeros et al., 2023). This collaborative approach not only enhances regional development but also contributes to global cooperation and sustainability efforts in the LAC region.

In all these instances, elements of science diplomacy such as intersectoral collaboration, strategic networking, and science communication play crucial roles. This chapter focuses specifically on effective science communication, exploring its role, impact, and connection with science diplomacy in addressing global challenges such as GEC and public health through advocacy and awareness-raising initiatives.

ENGAGING POLICYMAKERS THROUGH SCIENCE COMMUNICATION

We define science communication as any exchange of information, including scientific content, aimed at engaging audiences in learning, discussions, or activities related to science, technology, engineering, mathematics, and medicine (STEMM) in a clear, accessible, and compelling manner (DiCenzo et al., 2021). Its main goal is to bridge the gap between the scientific community and diverse stakeholders, including the public, journalists, policymakers, and diplomats, by promoting understanding and appreciation of scientific concepts, discoveries, and their societal implications. Notably, the COVID-19 pandemic and current GEC challenges have underscored the importance of effective science communication in combating the spread of misinformation (Awandare et al., 2020), thus making scientific dissemination not only a right but also a duty for scientists (Lopez-Goñi, 2020; Cumba García, 2021). Some scientists argue that the complexity of this task necessitates having specialists working full-time on scientific dissemination (Galvão et al., 2021).

Science communication is critical in advancing foreign policy objectives through science diplomacy (Moronta-Barrios et al., 2021), particularly during complex and challenging circumstances, by facilitating decision-making and negotiation for international agreements and treaties. We recognize the importance of disseminating scientific knowledge beyond academia to address global and regional challenges and to contribute to evidence-informed policymaking. To accomplish this, scientists need to possess effective communication skills tailored to engage with diplomats and policymakers. Some key strategies for effective and ethical science communication with policymakers and diplomats include: 1) understanding their perspective, priorities, political context and timing, and policy goals to frame scientific content in a relevant and understandable way; 2) highlighting the relevance and impact of scientific findings to policy issues using real-world applications and concrete examples; 3) providing clear, accurate, and concise scientific information, avoiding technical jargon and using visual aids; 4) building relationships and trust through ongoing dialogue, listening to their perspectives, and demonstrating openness and respect; 5) tailoring the message to their specific knowledge, interests, concerns, and decision-making process; 6) being proactive and responsive by anticipating their needs and questions and providing feedback in a timely manner; and 7) maintaining objectivity to prevent manipulation of scientific evidence for political or personal gain (Suhay et al., 2019). By adopting these strategies, scientists can effectively and ethically communicate with diplomats and policymakers, facilitating the integration of scientific evidence into regional and foreign policymaking processes and contributing to evidence-informed decision-making on critical societal issues.

Along these lines, ongoing efforts to improve and strengthen science communication are oriented to co-produce knowledge with different social and institutional actors. This approach of knowledge co-production serves to democratize science, foster community engagement, and establish a unified framework to address common challenges and promote sustainable development. By engaging various actors in the process, including scientists, diplomats, policymakers, and local communities, knowledge co-production ensures that outcomes are more relevant, legitimate, and salient, thus improving its social appropriation and their societal relevance (Carabajal and Hidalgo, 2021; Cash et al., 2003). Furthermore, this approach fosters the exploration of diverse alternatives, informs policy formulation, and addresses real-world challenges to facilitate informed decision-making (Dilling & Lemos, 2011; Cash et al., 2003).

The co-production and dissemination of scientifically accurate information is crucial in fostering trust and collaboration across diverse communities worldwide. During the COVID-19 pandemic, efforts were directed not only at combating misinformation and disinformation but also at nurturing trust and catalyzing collective action, highlighting the indispensable role of communication in addressing global challenges (Cumba García, 2020; Cumba García, 2021). Thus, science communication emerges as an indispensable tool for effectively addressing transboundary issues. This holds particularly true in the field of science diplomacy, where effective science communication includes: 1) facilitating trust and collaboration between nations to address shared challenges; 2) informing policymaking and encouraging international cooperation by enabling decision-makers and diplomats grasp complex scientific concepts; 3) facilitating efforts in addressing global challenges that require coordinated, cross-border responses; and 4) promoting cultural exchange and engagement between nations through collaborative scientific projects.

Effective science communication combines essential soft skills, such as interpersonal communication, ability to listen and convey complex concepts, and empathy with hard skills, such as technical writing, data interpretation, and multimedia production. This fusion forms a critical element for an inclusive and impactful science diplomacy strategy, underlining the integration of STI elements into diplomatic initiatives and facilitating knowledge exchange on the international stage.

SCIENCE COMMUNICATION IN LAC: CHALLENGES AND OPPORTUNITIES

Within the vast and exceptionally diverse Latin American region, science communication has a rich history spanning at least two centuries. It began with the emergence of science stories in newspapers and the establishment of the first science museums and botanical gardens (Vessuri, 1994; Massarani, 2015). Various entities, including research institutes, universities, governmental and intergovernmental organizations, NGOs, and third-sector organizations, have all contributed significantly to generating scientific knowledge, offering expertise, raising public awareness about critical issues, and fostering dialogue among nations within the LAC region. In recent decades, there has been significant growth in science communication efforts across the LAC region. The rise of the internet and social media platforms has played a crucial role in facilitating the dissemination of scientific information to broader audiences (Pulido-Salgado et. al., 2021; Massarani, 2018).

Despite the progress made, there are still significant opportunities for enhancing science communication practices across the region and for co-designing and co-creating inclusive spaces among different stakeholders. In this sense, several challenges persist, including the need to actively incorporate the voices, perspectives, and insights from local and indigenous communities already impacted by the effects of GEC. Building inclusive solutions requires new approaches, such as transdisciplinarity (TD) or co-production of knowledge (Carabajal, 2022). Both approaches are challenging as they focus on building trust relationships that take time to develop a common ground (Sandroni et al., 2023). This could encompass not only language but also the imperative to adopt culturally relevant science communication and TD methods tailored to the unique lives and experiences of specific and marginalized communities. This is crucial because individuals value, comprehend, and engage with science through their own cultural perspectives (Feliú-Mojer, 2022). Moreover, inadequate funding and resources hinder science communication initiatives, while insufficient engagement with indigenous knowledge and perspectives persists.

In contexts of complex crisis and uncertainty, like the COVID-19 pandemic and climate change, the proliferation of misinformation and disinformation presents additional hurdles that require a holistic and transdisciplinary approach that can engage all the stakeholders affected directly or indirectly by these issues. Efforts must transcend effective communication and involve concerted, collaborative approaches to overcome barriers and sustain interaction and dialogue.

DESCRIPTION OF CASE STUDY ORGANIZATIONS AND INITIATIVES ADDRESSING THE COVID-19 PANDEMIC AND GEC

1. Inter-American Institute for Global Change Research (IAI)

The five organizations highlighted as case studies in this chapter have been instrumental in addressing science communication challenges in the LAC region during times of emergency and uncertainty. The Inter-American Institute for Global Change Research (IAI)[9] stands out as a regional intergovernmental organization established in 1992, following the Declaration of Montevideo, and comprising 19 countries across the Americas. It coordinates scientific research on global change in the Americas, fostering transdisciplinary research and capacity building to inform decision-making and promote science-based policies. The IAI's initiatives extend to funding, developing, and launching multinational grants through partnerships like the Belmont Forum, aimed at addressing regional challenges and member countries' needs through transdisciplinary approaches. Science communication is integral throughout their processes, from training transdisciplinary teams to translating science into policy.

A significant milestone has been the creation of the Science Diplomacy Center (SDC), an innovative platform aiming at fostering science policy dialogue and collaboration among countries in the region. The SDC operates on three main pillars: capacity building, knowledge hub, and addressing emerging and evolving issues. Through these pillars, the SDC aims to strengthen the connection between science and policy by facilitating regional collaboration across the Americas. The IAI SDC has established the foundation for enhancing both human and institutional capacities to address pressing global environmental issues, mobilizing science diplomacy efforts. These issues encompass the interconnected fields of climate change, environmental and public health, energy, food security, water resources, and biodiversity conservation, with a particular focus on the LAC region.

Additionally, the creation of the IAI Science, Technology, and Policy (STeP)[10] Program has also been a pivotal initiative to address GEC challenges in the region. The program focuses on training the next generation of science and policy advisors through innovative capacity building initiatives that enhance science communication, policy advice, and negotiation skills, among others. The IAI Directorate, its staff, and the STeP Fellows have been actively involved in publishing reports, white papers, policy briefs, opinion articles, and columns to raise awareness of these global issues and engage the public, scientific community, and diplomatic circles (Prieto et al., 2024; Cumba Garcia and Carabajal, 2024).

2. Panama's National Secretariat of Science, Technology, and Innovation (SENACYT)

The National Secretariat of Science, Technology, and Innovation (SENACYT)[11] is an autonomous governmental body dedicated to utilizing science and technology for Panama's sustainable development. SENACYT has developed a communication strategy aimed at disseminating the results of its supported projects, aligning with its mission to harness science and technology for achieving sustainable development goals. Through workshops, symposiums, and international meetings, SENACYT facilitates international collaboration in science, technology, and innovation ecosystems, fostering connections and strengthening networks among scientists, national institutions, and academia (Gittens et al., 2021). SENACYT's International Office has organized workshops in recent years to promote science diplomacy and international scientific collaboration in addressing regional challenges, thereby positioning Panama as a key player in the LAC region's science diplomacy efforts. Notably, Panama stands out as the first Latin American country to have a National Strategic Plan for Science, Technology, and Innovation Diplomacy[12]. This plan, developed by SENACYT and Panama's Ministry of Foreign Affairs, aims to reduce inequality and promote equitable progress by advancing science, technology, and innovation diplomacy across various sectors.

Moreover, SENACYT conducts targeted public relations activities with the media and designs advertising campaigns to promote science awareness among the public, utilizing various communication channels such as the magazine IMAGiNA, the radio program IMAGiNA Radio, and the YouTube channel IMAGiNA TV. During the COVID-19 pandemic, SENACYT spearheaded impactful initiatives to communicate scientific breakthroughs and promote science diplomacy. One notable effort was the Rapid Response Call for COVID-19, providing a platform for researchers to address pandemic-related challenges. This initiative yielded diverse projects, including epidemiological modeling, innovative disinfection methods, and robotics for remote sanitization, showcased in the "Panamanian Minds vs. COVID-19" advertising campaign. By highlighting local scientists' proactive role, the campaign boosted public confidence in scientific expertise and underscored research's societal benefits. Another campaign that exemplified the intersection of science diplomacy and public health communication was "Voices of Science in the Face of Misinformation". By featuring well-known Panamanian researchers from diverse institutions, such as the Gorgas Memorial Institute of Health Studies and the University of Panama, the campaign aimed to combat misinformation surrounding COVID-19 vaccines and promote evidence-informed decision-making. Through targeted messaging on social media platforms, this initiative fostered community trust and encouraged responsible information consumption, thereby mitigating the spread of false information and raising awareness of public health guidelines. Panama's collaborative approach, guided by a strategy grounded in science diplomacy since 2018 (SENACYT, 2018), resulted in the country conducting the highest number of tests per capita among LAC countries for many months following the detection of SARS-CoV-2 (López-Vergès et al., 2021).

3. Ciencia Puerto Rico (CienciaPR)

Ciencia Puerto Rico (CienciaPR)[13] is a nonprofit organization comprising over 15,000 scientists, students, educators, and allies who recognize the transformative power of science in enhancing lives and society. Through various programs, CienciaPR fosters civic engagement within its diverse community to democratize science, reform science education and professional development, and harness the potential of science for the betterment of Puerto Rico and the LAC region. CienciaPR and its members played a pioneering role in COVID-19-related science communication and advocacy efforts in Spanish, publishing over 90 opinion columns and articles in Puerto Rico's leading newspaper "El Nuevo Día" ("A New Day"), and participating in more than 60 radio and TV interviews (Cumba García, 2020; Cumba García 2021; Feliú-Mójer, 2020; Shortwave Podcast, 2020). Another milestone was the launch of the award-winning multimedia collection of educational materials and community engagement program "Aquí nos Cuidamos" ("Here we take care of each other"). The "Aquí nos Cuidamos" multimedia collection comprises five public service campaigns about prevention and vaccination against COVID-19 and other infectious diseases, as well as addressing mental health and other topics. With 250 multimedia assets available in Spanish and sign language, the collection offers extensive resources to support public health initiatives and educational efforts across more than 100 countries in the LAC region (Feliú-Mójer et. al., 2024).

Additionally, CienciaPR collaborated on a rumor-tracking program concerning COVID-19 and vaccines with Mercy Corps PR and the Puerto Rico Public Health Trust (PRPHT), titled "Infórmate, Protégete, Vacúnate" ("Get informed, Protect yourself, Get vaccinated") (Mercy Corps, 2021). They also participated in a public service campaign with PRPHT and Mercy Corps PR called "Cuídate, Cuídame" ("Take care of yourself, take care of me"). These efforts, spearheaded by CienciaPR's Director of Communications and Science Outreach, Dr. Mónica Feliú-Mójer, amplified the voices of scientists and addressed the need for COVID-19 information in Spanish-speaking and marginalized communities in Puerto Rico and across the LAC region.

All these initiatives have focused on community engagement activities, organizing community forums and discussions involving leaders and representatives to ensure communication strategies resonated with marginalized and vulnerable communities. They also implement a diversity of communication strategies to convey scientific information effectively and empower individuals to take actionable steps based on scientific knowledge.

4. Science Diplomacy Network in Latin America and the Caribbean (DiploCientifica)

Third-sector organizations like the Science Diplomacy Network in Latin America and the Caribbean (DiploCientifica)[14] and the Puerto Rico Science Policy Action Network (PRSPAN)[15] are also highlighted as case studies for their science communication efforts in the LAC region during the COVID-19 pandemic and beyond. Founded in 2020, DiploCientifica aims to advance science diplomacy in the LAC region by fostering dialogue among various stakeholders, including the scientific and diplomatic communities, policymakers, civil society, and the private sector. It seeks to create conducive environments for engagement, capacity building, and knowledge exchange to contribute to the region's sustainable development. The network envisions becoming a central hub for science diplomacy in the region within this decade, advocating for active participation and leadership from the LAC region. DiploCientifica curates a repository of scientific and outreach materials showcasing the intersection between science and diplomacy. The repository features peer-reviewed scientific articles, white papers, audiovisual materials, and science communication activities carried out by its members.

Notably, DiploCientifica recently collaborated with the IAI on a white paper proposing policy recommendations derived from a science diplomacy strategy for protecting the Amazon (Prieto et al., 2024). Additionally, members of DiploCientifica have actively contributed to addressing regional challenges such as COVID-19, environmental change, and international scientific cooperation. These contributions include over 50 blog posts, opinion articles, reports, interviews with key figures in science diplomacy within the region, and materials aimed at enhancing regional collaboration and understanding in these critical areas.

5. Puerto Rico Science Policy Action Network (PRSPAN)

PRSPAN, founded in 2018 as an initiative of CienciaPR, aims to activate the Puerto Rican scientific community to inform the development and implementation of policies based on scientific evidence. PRSPAN seeks to enhance the capacity of Puerto Rican scientists and graduate students in science communication, science policy, advocacy, and outreach skills. Members of PRSPAN have authored and participated in over 30 newspaper articles, opinion columns, blog posts, and television and radio interventions in both English and Spanish, shedding light on critical issues affecting marginalized and underserved communities in Puerto Rico and the broader LAC region. These efforts were particularly pronounced during the pandemic, with close collaboration with CienciaPR. In 2020, PRSPAN strived to highlight the lack of science-based decision-making and policymaking in Puerto Rico's governmental response to the pandemic. To achieve this, they collaborated on joint organizational petitions and letters to government elected officials, urging improved policy decisions (Feliú-Mójer, 2020). Members of PRSPAN also advocate for the inclusion of Puerto Rico, as a U.S. Territory, in national data sets concerning the pandemic response in the United States. PRSPAN leveraged its network in the Committee of Natural Resources in the House of Representatives of the United States, resulting in a bipartisan letter to the U.S. Department of Health and Human Services demanding equal treatment for Puerto Rico in the COVID-19 response. Members also partnered with the International Network For Government Science Advice (INGSA) to compare COVID-19 policy decisions around the world with the ones made in Puerto Rico and analyze their effectiveness. Additionally, during the 2020 Puerto Rican election, PRSPAN developed the "Visibilidad y conCiencia en Puerto Rico de cara a las elecciones 2020" ("Visibility and science in Puerto Rico facing the 2020 elections") project, spearheaded by co-founders Dr. Luis Alexis Rodríguez-Cruz and Dr. Bianca Valdés-Fernández (Rodríguez Cruz, 2020). This initiative aimed to increase awareness of overlooked issues in public policy, such as gender and racial inequity in Puerto Rico, food security, misinformation, and community-led emergency management, among other important issues, during the election cycle.

These organizations might differ in their priorities, responsibilities, and approaches based on their vision, mission and roles. However, they share a collective goal of making science relevant to society through science communication and science diplomacy. Together, these entities strive to overcome barriers and harness the transformative power of science communication and science diplomacy to improve the well-being of communities across the LAC region. Amidst the challenges posed by the COVID-19 pandemic and GEC, they have been instrumental in disseminating and co-creating actionable information for civil society, particularly underserved and marginalized populations. Moreover, they advocate for evidence-informed policymaking in the LAC region. Their commitment and accomplishments serve as a successful demonstration of the impact of collaboration and innovation in driving positive societal change.

PERSPECTIVES ON SCIENCE COMMUNICATION & DIPLOMACY SKILLS

Through interviews, online resources, and established literature, we observe a symbiotic relationship between science diplomacy and science communication, characterized by their mutual reinforcement in advancing scientific understanding, promoting international collaboration, and addressing global challenges. Is not surprising that key themes emerging from expert interviews align closely with the mission and vision of their respective institutions, as well as with those of other institutions, emphasizing interconnectedness and mutual support. The symbiosis is vital; effective science communication transparently communicates findings to policymakers and the public, enhancing the outcomes of scientific collaborations, while science diplomacy provides platforms for direct scientist-stakeholder engagement, highlighting science's pivotal role in addressing global challenges.

Both science diplomacy and science communication, although distinct concepts, are deeply intertwined, each serving complementary roles in advancing global scientific cooperation and understanding. Interviews underscore the imperative of clear and transparent communication, essential for collaboration, trust-building among nations, informed decision-making, and fostering public trust in science. Effective communication skills are paramount for both fields, encompassing intercultural communication, translating scientific concepts into accessible language, and navigating diverse networks with varied backgrounds and perspectives. Ultimately, science diplomacy and science communication promote international collaboration by bridging divides, fostering meaningful partnerships, and highlighting shared challenges. Leveraging these concepts enables countries to mobilize cooperation and address complex global issues such as environmental change, public health, pandemic preparedness, and biodiversity loss. In addition to these shared concerns, each organization formulates its unique strategies and cultivates specific skills to strengthen the intersection of science diplomacy and science communication.

Given the diversity in their visions and missions, we explored the key skills that each organization deemed essential for addressing regional and global challenges as well as their individual perspectives on the symbiosis between science communication and science diplomacy.

1. DiploCientifica Perspectives on Science Communication and Diplomacy Skills

Interviews with the President and Executive Director from DiploCientifica provided insights into the complex dynamics of science diplomacy and science communication, particularly within the Latin American context. They concurred that both fields rely on a spectrum of skills essential for navigating the complexities of regional challenges and effectively disseminating scientific knowledge to diverse audiences. In the field of science diplomacy, key skills include intercultural and assertive communication, skillful storytelling, ability to summarize and present scientific findings, proficient design of presentations, translation of knowledge for broader accessibility, and proficiency in academic writing in English. These competencies facilitate meaningful engagement with stakeholders across borders, fostering collaboration and negotiation to address common challenges. Furthermore, a profound understanding of the scientific culture within the region, coupled with networking, leadership, and collaboration abilities, are paramount for successful science diplomacy initiatives. In parallel, effective science communication requires the skill to translate intricate information into accessible narratives, utilization of audiovisual materials, and tailoring messages to suit diverse audiences. Additionally, problem-solving capacity, adaptability, and the encouragement of innovation and creativity are vital for tackling dynamic regional challenges such as those posed by COVID-19, climate change, and environmental conservation. Through a diverse array of activities including publications, interviews, policy briefs, and webinars, DiploCientifica exemplifies the integration of these skills to advance science diplomacy and science communication in Latin America and the Caribbean.

2. CienciaPR and PRSPAN Perspectives on Science Communication and Diplomacy Skills

CienciaPR's Director of Communications and Science Outreach and PRSPAN's Former Co-Chair highlighted the essential nature of a diverse skill set, including cultural competence, empathy, and adaptability, for effective science communication and science diplomacy. Given the diverse and dynamic landscape of Puerto Rico's communities, effective science communication necessitates a profound understanding of cultural nuances and contextual sensitivities. Cultural relevance serves as a cornerstone, demanding communicators tailor messaging and outreach strategies to resonate with various audiences. Moreover, fostering humility and empathy enables engagement with individuals amidst uncertainty and skepticism, crucial during crises such as the COVID-19 pandemic. Additionally, science diplomacy, inherent in connecting scientific expertise with policymaking and community needs, calls for proficient interpersonal skills and collaborative approaches. Embracing diverse communication modalities, from multimedia resources to community forums, ensures accessibility and inclusivity across different populations. Furthermore, navigating the intersectionality of science and public policy demands astute negotiation skills and strategic partnerships to advocate for equitable representation and address overlooked issues. Ultimately, the efficacy of science communication and diplomacy hinges on an interdisciplinary skill set that bridges scientific expertise with cultural fluency, empathy, and strategic collaboration to foster meaningful engagement and drive impactful change.

3. Panama's SENACYT Perspectives on Science Communication and Diplomacy Skills

SENACYT's National Secretary of Science, Technology, and Innovation discussed Panama's extensive history in the fields of science diplomacy and science communication, demonstrating a proactive approach in formulating robust strategies amidst the COVID-19 pandemic. Through targeted initiatives, the organization has exhibited a dedication to enhancing accessibility and clarity in disseminating scientific information. They have skillfully tailored messages to suit various stakeholders, ensuring effective communication across diverse audiences, including policymakers, the media, and the general public. SENACYT's efforts have underscored the significance of strategic messaging, utilizing appropriate channels to convey essential information effectively. Additionally, the organization has fostered interdisciplinary collaborations and partnerships, both domestically and internationally, highlighting its commitment to cross-cultural communication and negotiation. Through prioritizing these strategies, SENACYT Panama has made significant contributions to global endeavors in addressing public health crises, serving as a beacon of excellence in science communication and diplomacy.

4. IAI Perspectives on Science Communication and Diplomacy Skills

Interviews with IAI's Executive Director and STeP alumni underscored that advancing science communication and diplomacy requires a holistic approach, emphasizing the development of technical expertise alongside soft skills. Key competencies include effective communication, cultural awareness, collaboration, and the ability to bridge disciplinary divides. Given the intricate nature of global challenges such as climate change and health, it is imperative for scientists and policymakers to communicate clearly and engage in constructive dialogue to formulate evidence-informed policies. Moreover, interpersonal skills such as humility, active listening, and the capacity to cultivate trust are vital for successful science diplomacy endeavors. These proficiencies empower stakeholders to navigate diverse viewpoints, negotiate effectively, and cultivate enduring collaborations across borders and disciplines. Additionally, leadership models that prioritize collaboration and inclusivity play a pivotal role in advancing science diplomacy objectives within the IAI. Collaborative leadership styles, characterized by traits such as attentiveness, connectivity, and introspection, foster an environment of openness and innovation essential for addressing pressing global issues. As science and diplomacy intersect in addressing challenges like climate change, nurturing leaders who embody these attributes becomes increasingly important. By prioritizing the cultivation of such skills through training initiatives, mentorship programs, and interdisciplinary partnerships, the IAI can enhance its capacity to facilitate meaningful scientific discourse, support effective policy making, and foster positive transformations within the LAC region.

The preceding paragraphs illustrate how each organization utilizes science communication effectively to tackle significant contemporary challenges, effectively linking science with policy. Additionally, our analysis reveals commonalities among these organizations, emphasizing their role as intermediaries bridging gaps between policy and science in the LAC region. Currently, collaborative efforts among these entities are being reinforced. For instance, under the IAI Science Diplomacy Center, the IAI and SENACYT Panama have conducted two training programs on science diplomacy for policymakers and scientists across the LAC region. These workshops aim to empower various social and institutional actors to support national and multilateral science policy processes in addressing regional and global challenges effectively. The workshops mark the beginning of a series of training sessions in science diplomacy, science communication, and scientific advice to foster dialogue and collaboration among countries. DiploCientifica experts have contributed as speakers and co-organizers of these workshops between the IAI and SENACYT, following the Memorandum of Understanding (MoU) signed by IAI and DiploCientifica in early 2024. Moreover, this MoU led DiploCientifica members to publish a white paper on recommendations to protect the Amazon (Prieto et al., 2024) and engage in extensive dissemination efforts across the LAC region with IAI support. Another collaborative example is CienciaPR and PRSPAN, where PRSPAN was created to address a need identified by CienciaPR to link Puerto Rican scientists with science diplomacy and evidence-informed policymaking. These examples demonstrate effective collaboration among the organizations highlighted as case studies to collectively address pressing transboundary challenges of our time.

Some scholars classify the endeavors of these organizations within the framework of "boundary organizations". A boundary organization is an entity that operates at the intersection of scientific research and public policy, serving as a channel for communication and collaboration between these two domains. These organizations are proficient at translating complex scientific findings into actionable policy recommendations while also ensuring that policy decisions are informed by the latest scientific insights. They often facilitate dialogue, foster partnerships, and promote knowledge exchange between scientists, policymakers, and other stakeholders. Through their unique position, boundary organizations play a crucial role in facilitating evidence-informed decision-making, promoting interdisciplinary approaches, and addressing complex societal challenges (Guston, 2001; Gustafsson et al., 2018).

CONCLUSION

The conclusions and reflections drawn in this chapter extend beyond the conventional theoretical-practical approach that typically emphasizes the role of politics and academia in science diplomacy. Instead, they underscore the need to acknowledge the evolving landscape of international relations and the diverse range of actors involved in shaping contemporary diplomacy. This includes recognizing the importance of reinventing initiatives, fostering partnerships and collaborations among various stakeholders, and adapting approaches based on the specific regional context. Our findings support that science diplomacy should not be viewed as a modern replacement for traditional diplomacy but rather as a complementary and essential endeavor to address shared challenges in a globalized and interconnected world. By embracing this perspective, practitioners can better navigate these complexities through collaborative and inter/transdisciplinary efforts.

Recognizing the significance of science communication as a fundamental component of science diplomacy is crucial. The interview findings support this idea, highlighting the importance of integrating science communication into diplomatic endeavors. Moreover, the interviews underscore how the convergence of science communication and diplomacy strengthens official initiatives and magnifies scientific engagement's broader influence and relevance on a global scale. From these interviews, three fundamental lessons emerge, emphasizing the necessity for: capacity building for scientists, policymakers, and diplomats to tackle regional and global challenges from an intersectoral angle; the development of a customized interface between science communication and science diplomacy strategies that are culturally relevant and inclusive; and the promotion of effective and ethical science communication practices to support evidence-informed policymaking and facilitate international cooperation.

All the organizations discussed in this chapter employed various approaches and diverse strategies to overcome obstacles and harness the power of science communication and science diplomacy to improve the well-being of communities across the LAC region, addressing the hurdles presented by the COVID-19 pandemic and GEC. Despite their unique approaches, these institutions serve as vital "boundary organizations" within the LAC region. They act as intermediaries and facilitators, bridging three distinct areas: science, policy, and society. Operating at the intersection of these different social spheres, they contribute to ensure comprehensive representation and effective communication.

In the LAC region, the concept of boundary organizations holds significant relevance due to the region's diverse socio-political landscapes and complex scientific endeavors. These organizations play a pivotal role in fostering collaboration and knowledge exchange by facilitating interactions among scientists, stakeholders and society. By promoting inter/transdisciplinary dialogue, they enhance the integration of scientific findings into policy making processes for the benefit of society. Moreover, their role in co-producing knowledge with different sectors of society aligns with the region's emphasis on inclusive decision-making and participatory governance.

Through the generation of different communication strategies and outreach efforts, these organizations contribute to developing shared understandings and effective solutions for pressing challenges, such as environmental conservation, public health, and sustainable development. As a result, they serve as a vital channel for navigating the complex interplay between science, politics, and societal needs in the LAC region. Ultimately, they foster innovation and advance regional development agendas.

During pandemic times and amidst widespread fear and skepticism, effective communication has been essential in disseminating accurate information, countering misinformation, and building public trust. The interviews underscore how the uncertainties surrounding the COVID-19 pandemic highlighted the importance of humility and transparency as scientific understanding rapidly evolved. Additionally, cultural sensitivity became crucial in addressing the unique needs and concerns of diverse communities, recognizing that one-size-fits-all approaches may not be adequate. In this context, institutions acknowledge that science diplomacy emerged as a critical framework for coordinating regional and international efforts, facilitating data sharing, and fostering collaboration among researchers and policymakers across borders.

Diplomatic endeavors have focused on negotiating access to essential resources such as vaccines and medical supplies, highlighting the interconnectedness of global health and the need for collective action. Furthermore, science communication during the pandemic involved disseminating vital information and countering misinformation that proliferated online. Leveraging their background and expertise, each institution has implemented innovative approaches, including multimedia educational resources such as interviews, podcasts, advertising campaigns, and magazines, to effectively reach various social and institutional actors.

The urgency of the pandemic and the environmental change context underscored the importance of science communication in shaping public perceptions, influencing policy decisions, and disseminating scientific information in critical times. In this sense, science communication has been instrumental in mobilizing societal responses and guiding humanity toward a more resilient and equitable future by fostering an environment of openness, dialogue, and evidence-informed decision-making. All the organizations emphasize the importance of tailoring communication strategies to ensure accuracy and cultural sensitivity. Moreover, they highlight the significance of humility and respect, along with acknowledging the limitations of scientific understanding and valuing alternative perspectives. Engaging and co-creating plural spaces to produce relevant, salient, and robust inter- and transdisciplinary science is key to overcoming barriers and making science useful to society.

As scholars dedicated to bridging the gap between science and policy and involved in many initiatives that these organizations are planning and organizing, we have observed firsthand how science communication can inspire action, empowering individuals, institutions, and countries to make informed decisions. By fostering an environment of openness, dialogue, and collaboration, science communication plays a pivotal role in addressing regional and global challenges.

Despite the fragmented approaches to science diplomacy in the LAC region, the efforts of these organizations demonstrate the effectiveness of tailored science diplomacy policies and communication strategies in tackling local challenges and addressing specific needs. They also underscore the potential for regional and international partnerships. As demonstrated in this chapter, these collaborative institutional efforts emphasize the significance of regional perspectives and context-specific priorities. By laying a solid foundation for inter- and transdisciplinary action, such initiatives are instrumental in addressing shared environmental challenges and fostering meaningful progress in the LAC region and beyond.

ACKNOWLEDGMENT

The authors extend their gratitude to Dr. Anna Stewart Ibarra and Dr. Anne-Teresa Birthwright from the IAI, Dr. Luisa Echeverría-King and Pedro Figueroa from DiploCientifica, Dr. Mónica Feliú-Mójer from CienciaPR, Dr. Luis Alexis Rodríguez-Cruz from PRSPAN, and Dr. Eduardo Ortega Barría from SENACYT for their valuable contributions to the case studies. Their efforts have highlighted the remarkable work undertaken by their respective organizations in tackling global and regional challenges by harnessing the synergies between science communication and science diplomacy. The authors also express their appreciation to Claudia Soto from DiploCientifica for her invaluable assistance in transcribing an interview for the case study.

REFERENCES

Awandare, G., André, E., Corrales-Aguilar, E., Chen, C., Mostajo-Radji, M. A., Jancoriene, L., & Nemer, M. (2020). Science advisers around the world on 2020. *Nature*, 588(7839), 586–588. 10.1038/d41586-020-03557-x33340028

Brown, J. (2016, March 23). Diplomacies, from public to public. *Huffington Post*. Retrieved February 20, 2024, from https://www.huffingtonpost.com/john-brown/diplomacies-from-public-to-public_b_9530412.html

Camacho Toro, R., Cumba García, L. M., Galvis, L. A., Echeverría-King, L. F., Pantović, B., Alarcón-López, C., Suarez, V. R., Figueroa, P., Torres-Atencio, I., Widmaier, C., Fraga, T. R., & Benavides, S. (2024). The needed link between open science and science diplomacy—A Latin American perspective. *Frontiers in Research Metrics and Analytics*, 9, 1355393. Advance online publication. 10.3389/frma.2024.135539338903656

Carabajal, M. I. (2022). Coproducción de conocimiento y redes de colaboración en el Sur de América del Sur. *Revista Redes - Revista de Estudios Sociales de la Ciencia y la Tecnología, 53*. https://doi.org/10.48160/18517072re54

Carabajal, M. I., & Hidalgo, C. (2021). Making sense of climate science: From climate knowledge to decision making. In Sillitoe, P. (Ed.), *The Anthroposcene of weather and climate: Ethnographic contributions to the climate change debate* (pp. 120–140). Berghahn Books., https://www.berghahnbooks.com/title/SillitoeAnthroposcene10.2307/j.ctv29sfz69.12

Cash, D. W., Clark, W. C., Alcock, F., Dickson, N. M., Eckley, N., Guston, D. H., & Mitchell, R. B. (2003). Knowledge systems for sustainable development. *Proceedings of the National Academy of Sciences of the United States of America*, 100(14), 8086–8091. 10.1073/pnas.1231332100012777623

Cumba García, L., & Carbajal, M. I. (2024, February 22). “One Health” as a pillar of the Loss and Damage Fund. *Latinoamerica21*. Retrieved March 2, 2024, from https://latinoamerica21.com/en/one-health-as-a-pillar-of-the-loss-and-damage-fund/

Cumba García, L. M. (2020, September 15). Reinventing myself in a pandemic: From immunologist to science communicator. *Sister Stem*. Retrieved March 1, 2024, from https://sisterstem.org/2020/09/15/reinventing-myself-in-a-pandemic-from-immunologist-to-science-communicator/

Cumba García, L. M. (2021, October 19). La importancia de la divulgación científica en tiempos de pandemia: Un llamado. *Editorial Politics*. Retrieved from https://editorialpoliticscom.wordpress.com/2021/10/19/la-importancia-de-la-divulgacion-cientifica-en-tiempos-de-pandemia-un-llamado/

D’agrosa Okita, S., & Pantovic, B. (2023, October 10). La investigación marítima para la diplomacia científica en Argentina: La Iniciativa Pampa Azul. *La red de diplomacia científica en Latinoamérica y El Caribe*. Retrieved March 1, 2024, from https://diplomaciacientifica.org/la-investigacion-maritima-para-la-diplomacia-cientifica-en-argentina-la-iniciativa-pampa-azul/

DiCenzo, C., Menezes, S., Smith, H., Murray-Johnson, K., Azizi, M., & McDuffie, K. (2021). *Inclusive science communication starter kit*. Metcalf Institute, University of Rhode Island., Retrieved from https://inclusivescicomm.org/wp-content/uploads/sites/1568/ISC-Starter-Kit_FINAL.pdf

Dilling, L., & Lemos, M. C. (2011). Creating usable science: Opportunities and constraints for climate knowledge use and their implications for science policy. *Global Environmental Change*, 21(2), 680–689. 10.1016/j.gloenvcha.2010.11.006

Echeverría King, L. F., Piñeros-Ayala, R., De Luque-Montaño, O., Flores-Zamora, A. F., Pantović, B., & Figueroa, P. (2023). Teaching science diplomacy at university programs for Latin America. In M. Moussaoui, *Global science’s cooperation opportunities, challenges, and good practices* (pp. 165-189). IGI Global.

Feliu-Mojer, M. (2022). Advancing inclusion through culturally relevant science communication: A perspective from Puerto Rico. *JCOM, Journal of Science Communication*, 21(07), C04. 10.22323/2.21070304

Feliú-Mójer, M. I. (2020, March 19). The Boricua science community responds to Covid-19. *Cienciapr*. Retrieved March 21, 2024, from https://www.cienciapr.org/en/blogs/equipo-informa/boricua-science-community-responds-covid-19

Feliú-Mójer, M. I. (2020, May 26). Scientific organizations ask governor to prioritize purchase of molecular tests for COVID-19. *Cienciapr*. Retrieved from https://www.cienciapr.org/en/blogs/pr-span/scientific-organizations-ask-governor-prioritize-purchase-molecular-tests-covid-19

Feliú Mójer, M. I., Ayala Rosado, E., Ortiz Torres, X., González Otero, R., & López, A. I. (2024). Creating risk know-how amongst marginalized communities in Puerto Rico during Covid-19. *Risk know how*. Retrieved March 2, 2024, from https://riskknowhow.org/risk-know-how-around-the-world/stories/aqui-nos-cuidamos/

Galvão, T., Tavares, K. A., Fernandes, M., Souza, J. C. M., & Noll, M. (2021). Case report: Scientific dissemination as a mediator of the research consolidation process: Proposals for institutionalization [version 1; peer review: awaiting peer review]. *F1000 Research*, 10, 886. 10.12688/f1000research.55655.1

Gittens, R. A., Lopez-Verges, S., Collado, T., Pimentel, J., Vazquez, A., Pulido-Salgado, M., & Torres-Atencio, I. (2021). Science Diplomacy as an Umbrella Term for Science Advisory in Public and Foreign Relations in Small Developing Countries: The Case of Panama. *Frontiers in Research Metrics and Analytics*, 6, 655335. Advance online publication. 10.3389/frma.2021.65533533912787

Goldman, I. (2022). El rol de la complementación entre ciencia y diplomacia en la Cuestión Malvinas. *Revista de Investigación en Política Exterior Argentina, 2*(4), 80-100. Retrieved from https://politicaexteriorargentina.org/wp-content/uploads/2022/12/Goldman-I.-El-rol-de-la-complementacion-entre-ciencia-y-diplomacia-en-la-Cuestion-Malvinas.pdf

Gonesh, A., & Melissen, J. (2005). *Public diplomacy: Improving practice*. Netherlands Institute of International Relations Clingendael.

Gregory, B. (2016). Mapping boundaries in diplomacy's public dimension. *The Hague Journal of Diplomacy*, 11(1), 1–25. 10.1163/1871191X-12341317

Gual Soler, M. (2021). Science diplomacy in Latin America and the Caribbean: Current landscape, challenges, and future perspective. *Frontiers in Research Metrics and Analytics*, 6, 670001. Advance online publication. 10.3389/frma.2021.67000134222772

Gustafsson, K. M., & Lidskog, R. (2018). Boundary organizations and environmental governance: Performance, institutional design, and conceptual development. *Climate Risk Management*, 19, 1–11. 10.1016/j.crm.2017.11.001

Guston, D. H. (2001). Boundary organizations in environmental policy and science: An introduction. *Science, Technology & Human Values*, 26(4), 399–408. 10.1177/016224390102600401

Hernández-Mondragón, A. C. (2022). From lab to science policy advisor. *Nature Human Behaviour*, 6(4), 477. 10.1038/s41562-022-01314-w35177810

Khojasteh, D., Davani, E., Shamsipour, A., Haghani, M., & Glamore, W. (2022). Climate change and COVID-19: Interdisciplinary perspectives from two global crises. *The Science of the Total Environment*, 844, 157142. Advance online publication. 10.1016/j.scitotenv.2022.15714235798107

Lopez-Goñi, I. (2020). *Preparados para la Próxima Pandemia*. Editorial Planeta.

López-Vergès, S., Collado, T., Pulido-Salgado, M., Pimentel, J., Vásquez Fábrega, A., & Torres Atencio, I. (2021, February 11). The role of science diplomacy in Panama's response to COVID-19. *Science & Diplomacy*. Retrieved from https://www.sciencediplomacy.org/perspective/2021/role-science-diplomacy-in-panamas-response-covid-19#note6

López-Vergès, S., Macías-Navarro, L., Hernández-Mondragón, A. C., Corrales-Aguilar, E., Soler, M. G., & Guerra, M. (2021). Closing the gap between emerging initiatives and integrated strategies to strengthen science diplomacy in Latin America. *Frontiers in Research Metrics and Analytics*, 6, 664880. 10.3389/frma.2021.66488033912788

Massarani, L. (2018). Estado del arte de la divulgación de la ciencia en América Latina. *Journal of Science Communication in Latin America*, 1(01), A01. 10.22323/3.01010201

Massarani, L., Aguirre, C. C., Reynoso, E., & Lindergaard, L. (2015). RedPOP: 25 años de red en comunicación de la ciencia en América Latina. *Journal of Science Communication*, 14(03), Y06. 10.22323/2.14030406

Melissen, J. (2005). Wielding soft power: The new public diplomacy. Netherlands Institute of International Relations Clingendael. The Hague: *Clingendael Diplomacy Papers No. 2*.

Mercy Corps. (2021). Rumour tracker programme: A community based approach to address information gaps and misinformation on COVID-19. *World Health Organization*. Retrieved March 11, 2024, from https://cdn.who.int/media/docs/default-source/science-translation/case-studies-1/cs13_rumourtracking.pdf?sfvrsn=829a4b42_4

Moronta-Barrios, F., Vargas-Domínguez, S., Nuesch-Germano, M., Torres, V., Selvaggi, K., Di Prinzio, C., O'Brien, E., Hernandez, V., & Monteiro, M. (2021). Latin American Network for Scientific Culture (RedLCC): A regional science communication initiative. *Frontiers in Research Metrics and Analytics*, 6, 654022. 10.3389/frma.2021.65402233870070

Pérez Valerino, D. R. (2021, July 2). Vacunas y diplomacia en tiempos de pandemia. *La red de diplomacia científica en Latinoamérica y El Caribe*. Retrieved from https://diplomaciacientifica.org/vacunas-y-diplomacia-en-tiempos-de-pandemia/

Piñeros Ayala, R., Echeverría King, L., & Pantovic, B. (2023). La diplomacia científica frente a los retos globales: Acciones para mitigar el cambio climático en la región de América Latina. In *M. Lascurain Fernández, & A. M. Chípul Castillo, La nueva agenda de la globalización* (pp. 257–277). Universidad de Xalapa.

Prieto, J., Cumba García, L. M., Echeverría-King, L. F., Alarcón López, C., & Benavides, S. (2024). Recommendations for a science diplomacy strategy for the safeguarding of the Amazon. *Inter-American Institute for Global Change Research & DiploCientífica*. https://zenodo.org/doi/10.5281/zenodo.10695010

Pulido-Salgado, M., & Castaneda Mena, F. A. (2021). Bringing policymakers to science through communication: A perspective from Latin America. *Frontiers in Research Metrics and Analytics*, 6, 654191. 10.3389/frma.2021.65419133981947

Rodríguez Cruz, L. A. (2020, October 3). PR-SPAN's science rising initiative to bring visibility and awareness in Puerto Rico's election cycle. *Cienciapr*. Retrieved from https://www.cienciapr.org/en/blogs/pr-span/pr-spans-science-rising-initiative-bring-visibility-and-awareness-puerto-ricos

Ruffini, P. (2020). Collaboration and competition: The twofold logic of science diplomacy. *The Hague Journal of Diplomacy*, 15(3), 371–382. 10.1163/1871191X-BJA10028

Sandroni, L., Ramos Quispe, L., House-Peters, L., Alonso-Yanez, G., Carabajal, M. I., Valentine, M., Schweizer, S., Murisa, M., Roy, N., & de Vernal, A. (2023). Transdisciplinarity 101: Short-term training in knowledge co-production to face global environmental change. *Journal of Science Policy & Governance*. 10.38126/JSPG220207

SENACYT. (2019). Lineamientos de la estrategia de diplomacia científica. Panama City: *Ministerio de Relaciones Exteriores de Panamá y la Secretaría Nacional de Ciencia, Tecnología e Innovación*. Retrieved from https://www.senacyt.gob.pa/publicaciones/?p=21846

Shortwave Podcast. (2020, April 7). Puerto Ricans are at risk from the coronavirus and a lack of information. *NPR*. Retrieved from https://www.npr.org/2020/04/06/828439886/puerto-ricans-are-at-risk-from-the-coronavirus-and-a-lack-of-information

Suhay, E., Cloyd, E., Heath, E., & Nash, E. (2019). Recommended practices for science communication with policymakers. Retrieved from https://www.american.edu/spa/scicomm/upload/recommended-practices-booklet_v17-digital.pdf

The Royal Society. (2010). *New frontiers in science diplomacy*. Retrieved from https://royalsociety.org/-/media/policy/publications/2010/4294969468.pdf

van den Hove, S. (2007). A rationale for science–policy interfaces. *Futures*, 39(7), 807–826. 10.1016/j.futures.2006.12.004

Vera, N., & Echeverría King, L. (2022). Hacia una diplomacia científica para América Latina. In Oregioni, S. (Ed.), *Derribando estructuras, construyendo puentes, tejiendo futuro: Aportes para pensar la internacionalización desde una perspectiva situada en el Sur* (pp. 213–236). Editorial-Cooperativa El Zócalo.

Vera, N., & Guglielminotti, C. (2018). Potencialidades de la agenda de cooperación sur-sur en tecnología espacial: Entre las proyecciones y las posibilidades para la Argentina (2003-2015). *Saber y Tiempo*, 77-94. https://ri.conicet.gov.ar/bitstream/handle/11336/112594/CONICET_Digital_Nro.48-53-PB%20%281%29.pdf?sequence=5&isAllowed=y

Vessuri, H. (1994). La ciencia académica en América Latina en el siglo XX. *Rev. Estudios Soc. Ciencia*, 1(2), 41–76. https://ridaa.unq.edu.ar/handle/20.500.11807/304

Widmaier, C., Echeverría King, L., & Aquino, K. (2020). La diplomacia científica y la internacionalización solidaria: Catalizadores en tiempos de pandemia para las universidades. In G. Mutti (Coord.), *El tiempo que vivimos: COVID-19 y su impacto en nuestras sociedades* (pp. 83-89). UNR Editora. https://cutt.ly/ghO7uRV

ENDNOTES

1 History — IPCC

2 United Nations Conference on Environment and Development, Rio de Janeiro, Brazil, 3-14 June 1992

3 The Paris Agreement | United Nations

[4] Vienna Convention on Diplomatic Relations, 1961

[5] The São Paulo Innovation and Science Diplomacy School – InnSciDSP

[6] Introducción Crítica a la Diplomacia de la Ciencia

[7] Introducción a la Diplomacia Científica - Instituto de Estudios Internacionales

[8] » Curso virtual en Diplomacia científica

[9] Inter-American Institute for Global Change Research

[10] Science, Technology, Policy (STeP) Fellowship Program

[11] Secretaría Nacional de Ciencia, Tecnología e Innovación – LA SENACYT

[12] LINEAMIENTOS DE LA ESTRATEGIA DE DIPLOMACIA CIENTÍFICA, TECNOLÓGICA Y DE INNOVACIÓN

[13] Ciencia Puerto Rico

[14] DiploCientifica

[15] Sobre el PR-SPAN | Ciencia Puerto Rico

Chapter 5 Boosting Productivity Through Science Diplomacy: German and Latin American Networks

Hannia Bridg
Bridg-Networking Consultancy, Germany

ABSTRACT

German and Latin American science diplomacy fosters institutional partnerships and academic exchanges, bridging cultural and geographical divides to enhance productivity. This analysis highlights Germany's advanced research infrastructure and Berlin's dynamic innovation ecosystem, marked by strategic multi-actor frameworks, effective science communication, and robust public-private partnerships. It addresses Germany's challenges and opportunities in strengthening its economic position as a key player in scientific and technological progress. The network between Germany and Latin America is examined, showcasing initiatives that enhance international networking, drive innovation, stimulate economic activity, and generate mutual benefits. Collaboration between Germany and Latin American countries advances scientific knowledge, fostering bilateral progress and joint growth. These efforts address contemporary challenges, promote sustainability, and underscore the importance of science diplomacy.

DOI: 10.4018/979-8-3693-2746-3.ch005

1. INTRODUCTION

In today's interconnected world, countries with shared interests and values are increasingly working together to boost academic research and innovation. This collaborative approach aims to increase knowledge and economic productivity, contributing to growth and development while addressing global challenges. Countries like Germany are updating their internal strategies by fostering open and supportive environments that encourage networking and facilitate the international exchange of ideas.

Science and technology create ecosystems that advance knowledge and have a significant impact on the economic and social development of societies (UNESCO, 2016; United Nations, 2022). Technological advancements such as the internet, mobile applications and research databases have revolutionized communication and collaboration between scientists and researchers on both local and global scales (Zydney & Warner, 2016; Figueiredo & Pereira, 2017). These advances facilitate rapid connections and the exchange of ideas, enabling the formation of interdisciplinary groups and networks. They help to identify expertise and discover new resources, thereby driving rapid global progress through collaborative efforts.

Cross-border science and technology projects have become increasingly interdisciplinary and culturally diverse, involving scientists, researchers, industry, policy makers and societal stakeholders. This broad interaction highlights the critical need for improved leadership to foster communication, engagement and effectiveness. Effective project management linked to scientific endeavour supports communication and equitable participation of diverse team members, which is essential for the success of innovation networks (Skipper, 2021).

Such multi-actor networks not only advance scientific projects, but also foster new business ventures, highlight the role of science in societal decision-making processes, and support innovation processes of knowledge exchange and co-creation (Franco Ferreira et al., 2019). However, the global nature of modern collaboration poses new challenges, such as cultural differences, time zone differences, language barriers, and disparities in development and access to resources (Dusdal & Powell, 2021). These obstacles can be overcome by shifting towards dynamic networks that prioritise collaboration and complementarity over competition and isolation (Granstrand & Holgersson, 2020).

Europe and other regions of the world recognise the priority of promoting bilateral cooperation through science and technology. The European Commission supports such cooperation through initiatives such as Horizon Europe, which includes non-EU countries in its programmes. This commitment to inclusivity underlines Europe's commitment to a global scientific community (Council of the European Union, 2021). As part of the European Innovation Union, Germany is committed to strengthening knowledge-sharing networks and promoting international scientific cooperation with different regions of the world.

In line with Europe's goals to promote bilateral scientific networking, Germany seeks to strengthen geostrategic solidarity and promote strategic partnerships through science diplomacy. According to the Federal Foreign Office (2020), this approach aims to cultivate a collective identity among teams involved in national and international affairs. Despite geographical and cultural distances, Germany recognizes that strengthening cooperation with Latin America and the Caribbean uses complementary strengths to address societal challenges through science and technology, thereby promoting trade and capacity building (UNIDO, 2021). Germany's cooperation with Latin America includes academic and industrial cooperation, promoting mobility and infrastructure development (Federal Foreign Office, 2024). These initiatives not only support innovation, but also address critical global challenges, demonstrating the power of science diplomacy to bridge cultural and geographical divides.

This chapter presents a comprehensive overview of how Germany is building an innovative and dynamic ecosystem that supports international collaboration in research and innovation, setting the stage for science diplomacy aimed at boosting productivity through network collaboration between Germany and Latin America. The methodology involves a multifaceted approach, beginning with a systematic literature review to understand existing theories and frameworks related to international collaboration and innovation ecosystems. Detailed case studies of successful partnerships between Germany and Latin American countries are conducted to highlight best practices and outcomes. A comparative analysis of regional and national strategies is performed to identify effective approaches to fostering international collaboration. Additionally, data from organizations and research institutions are analyzed to support the case studies and provide context.

The work is further enriched by the unique perspective of a researcher of Latin American origin, with German citizenship and academic training in both regions, while living in Germany. This background provides first-hand insights into the contexts and dynamics of both regions. By integrating these methodologies, the chapter offers a thorough understanding of how Germany's innovative ecosystem supports international collaboration, enhances research and innovation, and drives economic growth, thereby demonstrating the transformative potential of science diplomacy in boosting productivity through collaborative networks.

The first step delves into the German National Innovation System, exploring the strategic framework that supports Germany's traditional research landscape. It highlights the role of academic and non-university research organisations and examines how funding and investment support the development of decentralised innovative regions.

Berlin, the capital, is used as an example of an innovation hub, showing its transformation after reunification. This section focuses on strategic frameworks, regional policies and key research institutions. It analyses the Innovation Region Berlin-Brandenburg (InnoBB) strategy as the cornerstone of Berlin's approach to integrating regional strengths into a unified and competitive innovation region. Then examines the role of science in Berlin, detailing initiatives to foster collaboration between universities, industry and policy makers. It describes strategies for science communication and public engagement through events such as the Berlin Brain City Campaign, the Long Night of Science and Berlin Science Week.

Germany faces internal challenges and is looking for solutions. The chapter presents the importance of substantial investments in research and development (R&D) across sectors to address global issues such as climate change, health and biomedical research. These investments not only address pressing global challenges, but also strengthen Germany's position as a leader in technological innovation and economic growth. An overview of the German higher education system, renowned for its excellence, illustrates its crucial role in supporting the country's innovation ecosystem by training future scientists and engineers and facilitating industry partnerships.

It then analyses the transformative potential of Germany-Latin America partnerships in science and technology and how science diplomacy initiatives foster bilateral cooperation. Examples in Latin America, such as CAPAZ in Colombia, the Fraunhofer Institute in Chile and the Max Planck Society in Argentina, illustrate the commitment of both sides to building strong global partnerships. Key organisations such as the German Academic Exchange Service (DAAD), the Alexander von Humboldt Foundation (AvH) and the Deutsche Gesellschaft für Internationale Zusammenarbeit (GIZ) play a central role in promoting scientific exchange and cooperation, strengthening diplomatic relations and contributing to global development. These institutions and projects demonstrate how international cooperation can address global challenges and promote sustainable development.

The prospects for science diplomacy and international cooperation lie in continuing to invest in R&D, fostering inclusive partnerships and using scientific advances to create a more sustainable and equitable future. By working together, Germany and Latin America can drive innovation, increase productivity and make a significant contribution to collective progress. Germany's relationship with Latin America is an example of how science diplomacy serves as a unifying force, bridging cultural and geographical divides and fostering innovation and productivity. By examining the strategic frameworks, regional policies and successful case studies of German-Latin American partnerships, insights are gained into the best practices and outcomes of these collaborations. The integration of interdisciplinary and international efforts underlines the importance of science diplomacy in promoting sustainable development and addressing current global challenges.

2. THE GERMAN NATIONAL INNOVATION SYSTEM

The German national innovation system adeptly structured to support the development of economy businesses. The roots of the innovation system emerge in the nineteenth century where academic and industrial companies as BASF, Daimler, Hoechst (part of Sanofi-Aventis since 2004), and Siemens start a cooperation to link research and innovation for economic performance (Allen, 2015). The following section outlines the innovation system as a whole and how it is promoted within the capital to strengthen the links of regional science and technology at an international level.

Innovation Outlook

Germany is globally recognized for its excellence in research and innovation. A heritage that is built in its well-established reputation. The innovative product portfolio is a testament of the strong foundation in science results, which includes patents, inventions, product brands, methods, and processes. Both academic and applied research efforts are a dynamic force, that driving developments that are together influencing the German economy (Göring, 2023).

Situated at the heart of Europe, Germany invests over 20 billion euros in research and development (R&D) and boasts more than 1,000 publicly funded research institutions, alongside numerous R&D centers. This investment supports a robust infrastructure that promotes the efficient transition of knowledge from academic circles to the industrial sector. With approximately 420 universities, a broad spectrum of non-university research organizations, research-oriented companies, and a comprehensive network of federal and state research institutions, Germany has established itself as a key driver in the advancement of knowledge and technology as of 2020.

Germany is home of specialized academic centers dedicated to different focus on research. The country host organizations for diverse types of research, supported by world-class universities and research centers that lead a strong innovation performance, boasting a distinguished patent and trademarks portfolio according with the Worl Intellectual Property Organization WIPO. These are the pillars of the German economy, reinforcing its status as a global leader in innovation (Federal Ministry of Education and Research, 2021).

Research Strategic Framework

Germany serves a pivotal platform that foster scientific diplomatic endeavors, recognizing that science and technology not only foster the dialogue between nations but also strengthen the academic freedom. Through its strong research and development (R&D) infrastructure, Germany promotes scientific diplomacy and international cooperation in science and technology. Its emphasis on the importance of academic dialogue between nations, underlining Germany's role as a key player in the global research innovation landscape. Programs such as Erasmus and Horizon among others, illustrate the German dedication to promote European and international values through education and research. Moreover, the country efforts to strengthen connections between the scientific community and the political and business sectors underscore its role as central hub for business and innovation.

Key organizations such as the DAAD and the Alexander von Humboldt Foundation among other exemplify how Germany is committed to support mobility of research, enhancing the mobility of students' researchers, helping to build strong and lasting international networks with Germany. This is essential for the country's research landscape, as such connections enable the exchange of knowledge and foster fruitful collaborations linked to the academia.

The German Foreign Office's call to foster the initiative that help to establish a group of state secretaries dedicated to science diplomacy, with its emphasis on fostering collaborations between science and foreign policy through innovative cooperation models, it underlines the importance of a systematic and effective integration of science and diplomacy in Germany. These measures are integral to Germany's research framework, demonstrating the efforts that the county do to invest in science and technology not only for their inherent benefits but also as strategic instrument that support the foreign policy and economic growth (Federal Foreign Office, 2020).

Research Landscape

The research landscape in Germany is strategically structured to support the country's development and industrial sectors. The academic environment is characterized by innovative spirit and scientific inquiry, this foster Germany as leading nation in patents, products, and processes. Its diverse research infrastructure supports basic research, technical formation, research-driven companies, and is complemented by extensive federal and stage research institutions, promoting excellence and networking to make knowledge more accessible to everyone (Federal Ministry of Education and Research, 2021).

The history of German universities dates to the medieval period, with a drawing inspiration of the European university models established at the Univer University of Bologan and University of Paris Today, German universities, including universities of technology (TU), universities of applied sciences (Fachhochschulen), are an integral part of the research and innovation ecosystem. They provide high-quality education across the fields of science, technology, engineering, and mathematics (STEM), contributing significantly to knowledge and development.

Annually, Germany allocates nearly 18 billion euros to research and development efforts, in collaboration with university hospitals and healthcare institutions. In Germany, roughly 420 higher education institutions cater to a wide range of academic and professional interests. This includes about 110 traditional universities, or "Universitäten," of which 20 are universities of technology, known as "Technische Universitäten" (TUs). The country also hosts 210 universities of applied sciences, or "Hochschulen für Angewandte Wissenschaften," with approximately 40 of these being technical universities of applied sciences, referred to as "Technische Hochschulen" (THs). For those passionate about the arts and music, there are over 50 specialized colleges, termed "Kunsthochschulen" and "Musikhochschulen." Furthermore, 30 institutions are dedicated to applied administrative sciences, known as "Verwaltungshochschulen." (BMBF, 2021).

These academic institutions foster inside the innovative-research landscape alliances as the TU9 Alliance (TU9-Allianz) committed to pioneering and creative research in engineering and the natural sciences. The University Alliance (Universitätsalliaz) UA <11+ promote the cooperation across science, politics, and society, aiming to explore how technology and knowledge transfer can be advanced, and new impulses for technological and social innovations can be initiated.

In Berlin, the University Alliance focuses on supporting the integration of Grand Challenges initiatives within the integrated Berlin research space. This support includes two funding lines: "Exploration Projects" for the development of novel themes, projects, and networks, and "Einstein Research Units" for large-scale interdisciplinary and inter-institutional research projects, targeting areas such as Quantum & Climate and Water Changes (CliWaC), and exploring social cohesion and global health.

Non-University Research Organizations

At the heart of Germany's research funding landscape lies the Deutsche Forschungsgemeinschaft (DFG). As a self-governing research funding organization, it plays a playing a crucial role in promoting scientific progress, enabling interdisciplinary research, and maintaining high standards of excellence in all fields of science and scholarship, from the natural sciences and engineering to the humanities. (BMBF, 2021).

With renowned institutions, for instance, the Max Planck Society which delves into basic research, exploring the fundamental questions of science. The Fraunhofer Society is focused on all about practical, applied research, finding practical solutions to bridge the gap between academia and industry, boosting economic technological advancement and economic growth. The Helmholtz Association stands for multi-disciplinary research with international recognition. The Leibniz Association that excels in cross-disciplinary teamwork, working closely with university students, industry partners, and other esteemed research groups.

Germany is addressing critical global challenges such as climate change and environment security by committing to renewable energy projects. It is aims to meet energy demands by promoting smart technologies, accelerating the energy transition by linking research institutions and industry. In the realm of health and biomedical research, the focus is on preventive care and cutting-edge development of medical technologies. Germany leads in innovation and international networking, showcasing its commitment to fostering greater involvement in science diplomacy. To enhance engagement with policymakers and society, underlining the country's dedication to interacting scientific insights into societal and policy realms (Boytchev, 2020; Fjæstad, 2010).

Universities and Technical Colleges

The history of German universities data to the medieval period, with a drawing inspiration of the European university models established at the Univer University of Bologan and University of Paris Välimaa, 2019). Today, German universities and universities of technology (TU), universities of applied sciences (Fachhochschulen) are an integral part of the research and innovation ecosystem. They provide high-quality education in the realms of science, technology, engineering, and mathematics (STEM), empowering Germany in research and innovation.

Annually, Germany invest nearly 18 billion euros in research and development, working in collaboration with university hospitals and healthcare institutions. The country's higher education system includes around 420 higher education institutions that cater to a wide range of academic and professional interests. This includes about 110 traditional universities, or "Universitäten," of which 20 are universities of technology, known as "Technische Universitäten" or (TUs). Additionally, the country hosts 210 universities of applied sciences, or "Hochschulen für Angewandte Wissenschaften" with about 40 being technical universities of applied sciences "Technische Hochschulen" (THs). For those passionate about the arts and music, over 50 specialized colleges, termed "Kunsthochschulen" and "Musikhochschulen." Furthermore, 30 institutions are dedicated to applied administrative sciences, known as "Verwaltungshochschulen." (BMBF, 2021).

These academic institutions foster inside the innovative-research landscape alliances as the TU9 Alliance (TU9-Allianz) committed to pioneering and creative research in engineering and the natural sciences. The University Alliance (Universitätsalliaz) UA <11+ promotes collaboration across science, politics, and society. It aims to explore how technology and knowledge transfer advance and initiate new impulses for technological and social innovations.

In Berlin, the University Alliance focuses on supporting the integration of Grand Challenges initiatives within the broader Berlin research space. This integration is supported through two funding lines: "Exploration Projects", for the development of novel themes, projects, and networks, and "Einstein Research Units" aimed at a large-scale interdisciplinary and inter-institutional research project, targeting areas such as Quantum & Climate and Water Changes (CliWaC), and exploring social cohesion and global health.

Public-Private Partnerships

The German Federal Ministry of Education and Research (BMBF) has initiated the "Research Campus" funding program to encourage networking among researchers from universities, research institutions and, companies by fostering partnerships "under one roof". The theoretical framework of this program considers public-private research partnerships (PPP) from the perspectives of transaction cost, innovation outcomes, regional dynamics. It facilitates the exploration of complex and multi-faceted research areas characterized by important levels of research risks and hold potential for breakthrough innovations as quantum leaps. Research topics under this program span a wide range, including new methods of infectious disease diagnosis and the transformation of the energy system to materials development and the advancement of innovative production technologies.

Public-private partnerships (PPPs) in research and innovation plays a pivotal role in leveraging expertise, resources, and creativity. By fostering collaboration, Germany and other countries can accelerate scientific breakthroughs and create a more competitive and resilient innovation ecosystem. The physical proximity of the individual partners on a research campus helps to strike a balance between their different interests and to unite them in a joint research and transfer approach. In this way future applications are from the outset of research projects, streamlining the translation of findings to innovative products, processes, and services. In addition, research campuses are also involved in university teaching or the education of young scientists, thereby ensuring the supply of skilled professionals for the economy.

Funding and Investment

Investment catalyzes innovation. The Federal Ministry of Education (BMBF) plays an essential role in supporting the funding of innovative ideas and technologies. Central of this effort are key stakeholders, including state governments, the European Union, and the Federal Government, all supporting innovation under the scope of science diplomacy, that engages, connect, and translate.

Germany actively promotes research and innovation through various EU programs, each with specific scopes and objectives. These programs include HORIZON EUROPE, COSME, EUREKA, Eurostars, ZIM, and EXIST, showcasing the country's commitment to advancing scientific discovery and technological progress (Förderberatung des Bundes). Additionally the High-Tech Strategy stands as a national initiative aimed to strengthen Germany's leadership in key domains.

The German Research Foundation (DFG) further contributes by offering grants for fundamental research endeavors. The AiF (German Federation of Industrial Research Associations), in the realm of industry-oriented innovation, supports R&D projects tailored to meet industrial needs. Moreover, collaborative efforts such as the Innovative Medicines Initiative (IMI) facilitate public-private partnerships in pharmaceutical research and development. Germany's commitment to foster sustainability lead initiatives as the Energy Research Program aimed to the exploration of new energy solutions (European Commision – Research and Innovation; BMBF).

Regions of Innovation

Over the past few years, Germany has successfully developed its Regions of Innovation by establishing a communication platform between specific geographic areas, aimed at enhancing the innovation ecosystem, promoting economic growth and competitiveness, while also networking wealthier areas with less developed ones. This collaborative approach allows for a balanced assessment of innovation performance, encouraging a keen sense of commitment within society and strengthening the strategic position of regional competitive innovation across urban and rural industries (Broekel et al., 2018).

Pfotenhauer (2021) argues that is impossible to talk about economic development or societal progress without explicitly or implicitly addressing the need for innovation. Despite innovation policy becoming scientized and institutionalized, the promises made by science and technology, initiatives to boost regions and cities nevertheless regularly fail. The geography of Germany's declared high-tech hubs remains uneven, dominated by a select few research-rich and affluent urban regions. Nonetheless, innovation is recognized as important driver of competitiveness and long-term prosperity.

Startups

Germany stands out as one of the largest, most dynamic, and appealing start-up ecosystems globally, boasting over 60,000 start-ups, specialized in different sectors. The German start-up landscape has an international character. Diversity extends across regions a showcase of the German decentralized economic structure (GTAI, 2022)

Germany's thriving startup ecosystem is supported by robust funding opportunities and the establishment of strong foundations to promote venture capital investment, recognizing it as key drivers of economic activity. According with Borlovan (2022), sectors such as Fintech, Health-tech, energy, food, and enterprise software have experienced significant deal activities. The entrepreneurial landscape benefits from a variety of support mechanisms, including incubators, accelerators, co-working spaces, mentors, and events.

Additionally, the process of transferring scientific discoveries, knowledge, or inventions from research institutions to the commercial sector for development, known as technology transfer, plays a pivotal role in driving innovation and economic growth. The Deutsche Start-Up Monitor DSM (2021) highlights the diverse range of startups in Germany and delves into strategies employed by startups to access markets and expand internationally, further highlighting their significance in driving innovation and economic prosperity within Germany (DSM, 2022).

Start-ups and companies are constantly seeking for new markets, and the federal government lays the foundation for international cooperation through science diplomacy. This enhances the pathway to market for innovative solutions. Support the importance of integrating scientific advancements with global economic and policy frameworks to cultivate environment where startups can thrive, contributing to development and well-being.

Clusters and Networks

Beyond the regional efforts to bridge the gap across regional and national boundaries, clusters and networks have emerge as key facilitators for communication and collaboration in the German innovation ecosystem on an international level with players and stakeholders. They enable the integration of diverse expertise, significantly addressing global challenges (Buhl et al., 2019). These collaborations promote a culture of open innovation, thereby creating new opportunities for Small and Medium-sized Enterprises (SMEs) by interconnecting companies, service providers, and associated institutions in specific fields that not only compete but also cooperate. The cluster is a dynamic ecosystem that benefits stakeholders by the implementation of innovation (Kieze, 2017).

According with the Federal Ministry of Education and Research (BMBF), the aim of the clusters, is to address current challenges by forming strategic partnerships with innovation regions worldwide. They should strengthen international cooperation and establish long-lasting connections, encouraging culture of open science innovation, which is great news for small and medium-sized enterprises (SMEs),(BMBF, 024) .

3. BERLIN: A SHOWCASE OF INNOVATION

Berlin, the capital city of Germany, has undergone significant transformations during the years. It stands as emerging city with a vibrant social and cultural movements, being a hub of innovation in Europe. Following the reunification of Germany on 3 October 1990, Berlin lead an interesting new transformation process aimed to establish itself as a leading center for research and innovation. The comprehensive approach to encourage critical thinking and stimulates innovation through education at all levels. The city is promoting networking initiatives bringing together entrepreneurs, researchers, and professionals from various fields with strong interest in the international community. The new business ventures are vital for a strong growing startup ecosystem. It also fosters the close relationships between academia, research institutions, and industries ensuring that research and innovation support the market needs and societal changes. These initiatives lead communication, job opportunities and economic development.

The city acknowledges that the future of science and innovation is directly related with high-quality education and active social engagement. To this end, several German institutions and foundations, such as Konrad Adenauer Foundation, Friedrich Ebert Foundation, Heinrich Böll Foundation, Einstein Foundation among others, actively support the international endeavors of Latin American diplomatic representations to enhance the international affairs policies of their respective countries. Moreover, Berlin's scientific community collaborates with local industries to mention some,like BASF, Bayer, Siemens, to facilitate the transition of scientific innovations to the marketplace (Life Science Directory, 2023).

Science communication in Berlin is significantly enhanced by marketing activities, which not only showcase the scientific community, including professors, researchers, and students. It improves the visualization of sciences and engage the society through campaigns like the Berlin Brain City. Additionally, events such as the Berlin Falling Walls Summit are pivotal in fostering global science networking. These efforts exemplify how Berlin is leading activities to transform the city, leveraging scientific knowledge as the primary source for driving innovation and production.

Supported by public and private funding the city believes in the innovation ecosystem. Next section underscore how Berlin catalyzes its socioeconomic development base on science. The community of scientists, researchers, and businesspersons drive scientific advancements by fostering international collaborations, engaging with other countries, projects, and networks.

Berlin innovation landscape build bridges between regions, supporting a mutual understanding and cooperation through science, embodying the essence of science diplomacy without explicitly stating so. The convergence of innovation and scientific advancement in Berlin, creates a platform for dialogue exchange and collaboration that transcends national borders.

Berlin's Strategic Framework

Gualini and Fricke (2019) offer insightful analysis into how economic development policies are in place to build a strategic-relational framework to stand behind the transformation of Berlin. In 2011, the governments of the Länder Berlin and Brandenburg united to collaboratively initiate the Innovation Region Berlin-Brandenburg (InnoBB) strategy. With the aim to create a conducive policy environment and boosting the global innovation competitiveness of the Capital region.

The core objectives of InnoBB emphasize the promotion of partnerships between the two regions. Berlin and Brandenburg, bridging the regional east-west political divide, and foster a strong economic strategy moving forward with the modernization of infrastructure and supporting the clustering of micro-regional enterprises for mutual collaboration.This approach seeks the integration of diverse region economic activities and foster a single socio-economic power for sustainable development. The strategy, InnoBB for cross-regional, collaboration and private sector engagement is most notably demonstrated through its creation of clusters. These clusters cultivate a culture of unity, supporting the advancement of strategic technological and innovation sectors in cooperation with the academia, moving forward projects in healthcare, media, mobility, optics, energy among others.

The InnoBB approach, grounded in the principles of neo-regionalism, aimed to elevate innovation capabilities in Berlin and Brandenburg, across various sectors. This strategy work with the realms of strategic marketing and profiling to visualize actions, projects, while prioritizing investments and resource allocation for their maximum impact. The execution phase underscores the importance of stakeholders' involvement, chambers of commerce, investment banks, and innovation agencies. The establishment of "centers of excellence", primarily led by companies, are enhancing the region's collaborative innovation landscape as we see it today (Gualini and Fricke, 2019).

The formulation of policies is overseen by the Joint Innovation Strategy Working Group, which involves ministry representatives from both states. This leadership team includes state secretaries for economy, science, and technology from Berlin and Brandenburg. Central to these efforts are key stakeholders such as chambers of commerce (IHK Berlin, IHK Brandenburg), investment banks (Investitionsbank Berlin, Investitionsbank Brandenburg), and innovation agencies like ZAB and TSB (now part of Berlin Partner), all actively participating in innovation summits since 2007.

Academia in Berlin

The Berlin Research Landscape upholds both economic and social responsibilities by promoting innovation and strategies aimed to elevate scientific standards to international levels of excellence. It actively creates bridges between the scientific community and entities, facilitating the transfer and translation of scientific findings into innovative products. Additionally, the city also leads programs aimed to enhancing the local infrastructure, investing in remodeling projects to support the lives of academics, scientist, researchers, and other individuals. These initiatives collaboratively foster the development of science and business.

Berlin's innovation scene is dynamic and intricately linked to science and research. Noteworthy is the city's interdisciplinary network, which brings together experts from different fields to drive innovation. This vibrant ecosystem is supported by a robust academic infrastructure with four major universities, including the world-renowned University Medicine Berlin "Charité", as well as specialized technical colleges, art academies and private institutions. This multifaceted educational landscape firmly establishes Berlin as a top destination for scientific research in Europe and serves as a hub for creativity and collaboration.

Berlin is home to over 70 non-university research institutions devoted to a wide range of basic and applied research pursuits. Among these prestigious entities are globally recognized organizations such as the Max Planck Society, the Helmholtz Association, the Leibniz Association, and the Fraunhofer Society. Collectively, these institutions bolster Berlin's esteemed reputation as a leading global science capital (Berlin Partner Business and Technology, 2024). The Berlin University Alliance, which brings together Freie Universität, Technische Universität, Humboldt-Universität zu Berlin, and Charité – University Medicine Berlin, is recognize as an Alliance of Excellence.

Science is the pivotal role in Berlin's identity as a global metropolis for innovation, attracting and nurturing scientists and emerging talents with a strong desire of self-improvement. The city boasts a research community, with approximately 250,000 people engaged in teaching, research, work, and study from around the world. Remarkably, 24 percent of all students in Berlin are international, with the majority coming from China, India, and Turkey, showcasing the city's international appeal. Numerous cooperative programs facilitate connections between Berlin's higher education institutions and their global partners, further enriching the city's international academic and scientific landscape (Berlin: The Capital of Science, businesslocationcenter.de).

Start-up Ecosystem in Berlin

The Berlin's innovation landscape is characterized by the seamless integration of science and business. The Startup Ecosystem is diverse, collaborative and covers a wide range of industries as Fintech startups, which call Berlin home, others thrive e-commerce, digital media and content creation, health tech, mobility solutions. (Startup Ecosystem of Berlin | StartupBlink)

The 2024 Berlin report shows an impressive number of more than 6,000 start-ups, as recorded by Explore the Berlin-Brandenburg Ecosystem | Berlin tech ecosystem map (startup-map.berlin). The breadth of investment in Berlin's start-ups mirrors the diverse industry trends that can be seen across Europe. These ventures are leveraging the city's strong public R&D base to create sustainable value. They are also attracting significant investment from established companies.

In 2022, Berlin Partner Business Location Center supported 239 projects and facilitated the establishment of 73 companies in Berlin, 49 of which were from abroad. These initiatives are expected to create a total of 8,389 new jobs, with investments amounting to approximately 1.1 billion euros. The majority of these new international companies are based in Europe, accounting for 32 percent, followed by North, Latin, and South America, which together represent about 19 percent ((businesslocationcenter.de).

The scope of investment has expanded from initial incubation efforts to creating comprehensive R&D centers and manufacturing units, exemplified by Tesla's operations. These synergies between scientific research and innovative entrepreneurial activity are central to the transformation of Berlin from its past economic and structural challenges into a thriving business center. However, despite its strengths, Berlin's innovation landscape faces obstacles, particularly the lack of large, research-focused corporations and a fledgling SME sector. These issues lead to restrained privately funded R&D activities, a largely small-scale economic framework and subdued patent performance (Dauchert & Barzik, 2022).

The academic landscape in Berlin is deeply intertwined with science diplomacy. The internationalization of the academia is supported by scholarships, partnerships, networks with academic institutions from other regions, thereby strengthening education, research and innovation at international level.

Science Communication, Engagement and Recognition

The Federal Ministry of Education and Research identifies the core functions of science diplomacy as connecting, informing, and enabling. Berlin underscores science diplomacy policy by actively working on the commitment to developing active partnerships and supporting the institutions that foster evidence-based research. It is also committed to prioritize the capability of policymakers and fostering an environment that foster independent science, research, and teaching to respond to the need of the organizations and society they serve. The scientific community in Berlin, spanning from the academy or industry is committed to build bridges at international level.

Berlin Brain City Campaign. The Berlin Brain City Campaign aims to showcase the innovation, diversity, and excellence of Berlin's science and technology sectors, emphasizing the vital connection between science, business, and society. The campaign emphasizing the importance of transferring knowledge and technology from research to real-world applications and how all the sectors are leading the efforts to address global challenges such as climate change, environmental degradation, digital transformation, demographic shifts, and public health.

As explained, the robust network of high excellence universities, applied sciences institutions, and research organizations, Berlin excels in converting scientific breakthroughs into innovative products, processes, and high-tech solutions. The city's vibrant business science start-up ecosystem, noted for its successful spin-offs, is a vivid illustration of the science transformation movement.

The campaign extends to support knowledge and technology transfer efforts by showcasing the daily exemplary efforts of Berlin's research community through interviews, reports, and open channels of communication. Furthermore, it offers a detailed look at the transfer services provided by Berlin’s academic institutions and underscores the diverse initiatives and funding mechanisms that are related. By doing so, Brain City Berlin, reinforces the international reputation of Berlin’s science-innovation hub (Berlin's science hub | Brain City Berlin).

Long Night of Science. The “Lange Nacht der Wissenschaften” (Long Night of Science) stands as a cornerstone of science communication and public engagement in Germany. This event, deeply rooted in the tradition of making science accessible to everyone, allows a wide range of scientific institutions to open their doors to the public for one special night each year in summertime. From late afternoon until early morning, universities, research laboratories and scientific organizations offer a diverse program of lectures, demonstrations, and interactive sessions to showcase their work and the wide range of research topics they explore.

During the Long Night of the Sciences, the city people can engage in experiments, attend lectures and presentations, participate in interactive workshops, and enjoy science shows. The event covers a broad spectrum of scientific fields, from natural sciences and engineering to social sciences and humanities, making complex scientific concepts accessible and entertaining to the public. It's a celebration of scientific curiosity, aiming to inspire future generations of scientists and researchers.

As a vital public relations and science communication initiative, the Long Night of the Sciences not only simplifies complex scientific concepts but also strengthens the bond between the scientific community and society at large. This event is an occasion that celebrates curiosity, fosters a culture of knowledge-sharing, and highlights the integral role of science in shaping our understanding of the world. This event encapsulates Berlin's commitment to promoting an open and inclusive dialogue on science, affirming the city's reputation as a leading hub for scientific research and education (langenachtderwissenschaften.de)

Science Week. At the heart of Europe's innovation landscape, Berlin is emerging as a key arena where the scientific quest and the art of diplomacy converge. Annually, from November 1st. to 10th, Berlin transform the city into a vibrant hub, promoting the Berlin Science Week, an event that captures the spirit of science diplomacy by bridging the gap between diverse scientific communities with the broader global public. This science week celebrates the quest of knowledge and scientific achievement and opens innovative avenues for international collaboration and exchange.

With a series of physical and virtual gatherings across the city, Berlin Science Week offers a unique blend of interactive and interdisciplinary activities. These include panel discussions on cutting-edge scientific topics, workshops designed for hands-on learning, and immersive experiences that bring science to life. The participation of internationally renowned scientists and institutions in these events underscores the event global reach impact and its role in addressing contemporary societal challenges.

Behind the scenes, the success of Berlin Science Week is the result of a collaborative effort involving numerous organizations, scientists, and artists. All bolstered by a supportive network of funders and partners. This collaboration underscores the critical importance of making science and innovation accessible to all, emphasizing that significant progress hinges on open and inclusive dialogue among all actors, including scientist, government, society, and industry.

For example, in the year 2023 of edition, the Berlin Science Week marked a milestone, with the course of ten days, more than 200 scientific divulgation events took place, attracting an audience of over 35,000 people from more than 140 countries. This unprecedented engagement is a testament to the event's ability to invigorate the scientific community, stimulate public interest in science, and showcase the dynamic nature of scientific research (https://berlinscienceweek.com/de/).

Through its commitment to nurture conversations, reflection, and collaboration, Berlin Science Week exemplifies the principles of science diplomacy. It showcases how scientific endeavors can transcend borders in the same country, encourage a sense of mutual understanding, and contribute to a more sustainable and interconnected world. In doing so, Berlin emerges not just a hub of scientific innovation, but also as a testament to the unifying strength of science diplomacy.

The Falling Walls Summit. Set against the stunning backdrop of Berlin's rich history, the Falling Walls Science Summit is a beacon of innovation, collaboration, and inspiration for the global scientific community. Taking place from 7-10 November, this wonderful summit brings together a mix of leaders from science, business, politics, arts, and society, making it a unique international communication platform that transcends the boundaries of academic disciplines. It guided by the pivotal question "What are the next walls to fall in science and society?" This question not only shapes the content of their presentations and discussions but also propels the conference's mission to explore future breakthroughs and transformative solutions, creates a multi-dimensional approach that enriches the scientific dialogue and serves as a powerful tool for conversations, making complex ideas accessible and engaging a diverse and international audience.

At the heart of the summit are various thematic areas designed to address different interests and foster cross-disciplinary exchange and connections. The Falling Walls Breakthrough Day celebrates remarkable scientific breakthroughs: The Falling Walls Circle provides a forum for high-level discussions on research policy and strategy: The Falling Walls Engage focuses on the science of communication, with an emphasis on public engagement. The Falling Walls Lab showcases emerging talent and early-career researchers: The Falling Walls Venture connects scientific innovators with investors, and The Falling Walls Female Science Talents highlights the achievements and challenges of women in science.

The Falling Walls summit acts as an accelerator, fostering the link between curiosity and pioneering thought. It champions the exchange of ideas capable of dismantling the figurative barriers restricting knowledge and advancement. It is a declaration of how collective wisdom is addressing the pressing challenges of our times. Scientists from across the world, representing all disciplines bring their groundbreaking research to the society, enhancing the discourse within both the scientific sphere and the wider community can work together (https://falling-walls.com/)

4. GERMANY'S INTERNAL CHALLENGES AND OPPORTUNITIES

Germany faces a myriad of challenges and opportunities not only in the global stage with its endeavor to maintain a leadership position in innovation but also in its effort to strengthen various economic sectors that contributes to the country prosperity and societal well-being. As emphasized by Fischer-Bollin (2019), the early identification of emerging trends and challenges is imperative for shaping Germany's future. This approach is essential for positively influencing the conditions that underpin freedom, democracy, innovation, economic competitiveness, as well as diplomatic security and risk assessment.

Germany's competitive production is foundational to the present economic strength but also for its prospective development. Recognizing the importance of enhancing the production framework, there's concerted effort across various sectors to elevate these frameworks to new heights. Moreover, the government is taking significant strides in reviewing and placing an increased emphasis on other sectors to reinforce the strategy. This includes spotlighting science diplomacy as a key strategy for nurturing international collaboration and networking. Through the reinforcement of strategic partnerships and alliances within other countries and regions, Germany can leverage its scientific expertise in innovation to foster sustainable growth. This strategy emphasizes the importance of cross-border cooperation in harnessing the benefits of scientific progress for the benefit of society worldwide.

The Challenge: The German Dilemma

Germany stands for innovation with solid investment in science, research and development. According with the Global Innovation Index, it secured the 8th among 132 economies in 2023, as reported by World Intellectual Property Organization (WIPO, 2023). This ranking is indicative of Germany's efficiency in generating significant innovation outputs from its investments, underscoring its robust innovation ecosystem. However, despite its great capacity for innovation, Germany exhibits a paradoxical performance in global competitiveness rankings, as noted by Brandt (2023) . This discrepancy is often referred to as the "German paradox", a term that describe the Germany's strength in bringing new products to market does not correspondingly elevate its competitiveness on the global stage.

Kudnani (2014) highlights this complex situation, pointing out the challenge Germany faces in translating its innovation prowess into overarching competitive advantage. This complexity, which transcends the explanatory scope of both realism and institutionalism, needs the introduction of a critical state theory. Proposed by Germann (2022), this theory examines the uneven and combined evolution of global capitalism, offering insights into the underlying dynamics of Germany's position in the global market. the theory posits that to remain competitive internationally, Germany's strategic policy framework should not solely amplify innovation but also elevate productivity levels. According to the Organization for Economic Co-operation and Development (OECD, 2022), the vitality of Germany's socio-economic fabric is anchored in its ability to uphold employment, catalyze investment, and invigorate job creation.

These insights emphasize the critical need for Germany to adopt a multifaceted strategy that promotes innovation, boosts productivity, and ensures socio-economic stability to navigate the challenges of global competition successfully. The need to collaborate with the EU and other regions to create conditions that strengthen exports and reduce high export surpluses are pivotal steps in this direction (Jungbluth, 2019). The aim is keeping pace with emerging markets, ramping up industry investments, enhancing digitalization, and encouraging entrepreneurial initiatives. To achieve these objectives, Germany aims to establish favorable framework conditions in collaboration with the EU and other regions. Additionally, Germany is committed to supporting the establishment of international economic norms that promote the expansion of free trade. Advocating for open markets and actively participating in the formulation of relevant agreements are central to Germany's strategy to enhance its economic standing and foster global trade relations (Neuhäusler, et a., 2021).

Innovation boosts investment and creates new job opportunities, driving employment, thereby playing a crucial role in Germany's socio-economic health. Embracing science diplomacy provides a strategic method for advancing Germany's innovation agenda, making a substantial contribution to addressing global challenges and strengthening its socioeconomic integrity. Through the reinforcement of international cooperation and the exchange of knowledge and resources essential for sustainable development, science diplomacy strategically strengthens Germany's economic resilience. This approach aligns Germany's innovative efforts with global goals and skillful crisis management. It ensures that the focus remains on long-term strategies during current challenges.

The Opportunity: Science Diplomacy

Historically, science has been recognized for its contribution to providing evidence-based solutions and influencing political decisions (Carrier, 2022; Aguirre-Bastos, 2023). Science diplomacy has emerged as new strategic approach that bridges the gap between international political affairs and the scientific field (Kunkel et al., 2021). This innovative framework emphasizes the intersection of academic knowledge whit international cooperation. Within this context, engaging in collaborative scientific activities is key to driving positive outcomes and underscoring the importance and respect of science and technology. When countries collaborate with a spirit of investigation and research problem-solving, they effectively tackle global challenges, leading to long-lasting partnerships (Ruffini, 2020; Turchetti and Lalli, 2020).

Science diplomacy is not just a bridge connecting industrialized countries with a group of emerging diverse economies, but also enables the flow of solutions and innovation to improve the responses to today's global challenges. The drive to embrace a global framework through academic collaboration is a key objective. This approach not only fosters development but also empower the emerge of new actors with ideas that encourage knowledge exchange. Science diplomacy is a facilitator that encourage scientific knowledge and boosts innovation applications (Margison, 2021).

Science diplomacy (SD), underscores the potential of science and collaboration to catalyze diplomatic engagements and partnerships, showcasing the value of join knowledge and problem-solving for positive outcomes. (Ruffini, 2020; Turchetti and Lalli, 2020). The Royal Society and the American Association for the Advancement of Science (AAAS) elucidate the fluid nature of science diplomacy, emphasizing its role across three dimensions of policy: science in diplomacy, diplomacy for science, and science for diplomacy. This framework supports the notion that science operates beyond national boundaries and can play a crucial role in advancing diplomatic objectives through the concept of "soft power" (Nye, 2004).

The scientific community is dedicated to tackling global challenges while also seeking to expand knowledge and push scientific boundaries, which have been central to Science Technology and Innovation (STI) collaboration since the 20th century. (OECD, 2021). With the rapid progression of digitalization and the shift towards a sustainable economy, the Federal Ministry of Education and Research highlights the increasing relevance of education, research, and innovation. Germany's response to these global shifts is a strategy that focuses on fostering international collaborations, protecting academic freedom, and ensuring scientific research remains a fundamental pillar of democratic societies. This approach is instrumental in navigating the complexities of the 21st century, demonstrating Germany's proactive stance in leveraging education and innovation for societal progress.

Science diplomacy is the link that network science and international policy to provide a strategic framework for addressing worldwide challenges. The scientific cooperation plays a pivotal role in Germany to solve some challenges, serving as a pathway to nurture international scientific partnerships, enabling academic face to face work (Federal Foreign Office, 2024). Science is widely recognized for offering a robust, evidence-based foundation that is essential for guiding political decisions (Carrier, 2022). This enduring practice, now commonly referred to as "science diplomacy," has been a fundamental aspect of global diplomacy for years, as highlighted by Aguirre- Bastos (2023)· Moreover, the growing recognition of science diplomacy today, while appearing modern, builds upon its historical role as a cornerstone of international relations worldwide (Kunkel et al., 2021).

Today, science plays a vital role that goes beyond laboratories and academic journals. It serves as a bridge, that connects countries, fostering cooperation and mutual understanding on a global level. This intersection of knowledge-seeking an international relation is aptly termed science diplomacy (SD). It leverages the practices in which science and technology intersect with foreign affairs and international relations to address common problems facing 21st century humanity and to build constructive global partnerships Within this realm, scientific collaboration becomes a powerful catalyst from positive diplomatic engagements, proving that when nations unite in the spirit of discovery and problem-solving outcomes (Ruffini, 2020; Turchetti and Lalli, 2020).

Science operates beyond national borders, facilitating new diplomatic approaches through collaboration on scientific matters to address sensitive global issues. In Germany, this diplomacy thrives on an extensive network of contributors, from universities, research institutes, to funding organizations. German Embassies around the world frequently host German Science Day events, and annual highlights include Berlin's German Science Week and the Falling Walls Conference. Germany's dedication to international scientific collaboration and its role as a center for scientific dialogue and innovation aim to enhance its contributions to the global scientific community. scientific community.

5. SCIENCE DIPLOMACY: GERMAN-LATIN AMERICAN COLLABORATION

Germany values Latin America's vast territorial expanse, rich seas, mountains and ecosystems, and abundant natural resources, recognizing its exceptional biodiversity and innovation, as well as the diverse nature and forward-thinking societies. Latin America recognizes Germany’s economy robustness, resilient spirit, and research and innovation excellence. This mutual admiration underpins a commitment to maintain strong collaborative ties in scientific endeavors to favor, science, society and economic domains. Science diplomacy, grounded in connectivity, information exchange, and communication, plays a crucial role in shaping international relations. Germany’s focus on research-innovation enhances cooperation, offering insights that not only enrich internal analyses but also foster a constructive dialogue.

The bilateral collaborations between Germany and Latin America are particularly noteworthy, supporting economic and technological progress and facilitating regional transformation through quality-driven processes. These partnerships advances are market oriented, services to cooperating countries while supporting knowledge exchange, skill development, and strategic decision-making. These bilateral efforts in cooperation, are important to fostering sustainable growth, technological improvement, structural transformations, and sustained growth through innovation access. The global consensus is that scientific cooperation is vital for forging a prosperous future for all parties involved, highlighting the importance of joint efforts in the scientific domain.

The Legacy of Humboldt and Caldas

Throughout history, the countries of Latin America and the Caribbean have been actively engaged in bilateral, regional, and global scientific cooperation with other parts of the world. This collaboration has been instrumental in strengthening and complementing their national capacities in research, technological advancement, and innovation (Soler, 2020). The scientific engagement dates to the pioneering expeditions of Alexander von Humboldt from 1799 to 1804, which took him through what are now the countries of Cuba, Venezuela, Ecuador, Colombia, Peru, and Mexico, followed by a brief visit to the United States before his return to Europe. During his travels, Humboldt engages in conversations with local scientist like Francisco José de Caldas, a Colombian scientist. Their exchange on plant geography and the classification of plants, including the sharing of manuscripts that laid the groundwork of biogeographical studies (Gutiérrez, 2020). The academic interaction between Humboldt a German scientist and his Latin American counterpart set a precedent for foundational academic exchanges, fostering a deeper understanding of our world's complexities.

Institutional Partnerships and Academic Exchanges

Science and innovation are acknowledged in Latin America and Germany, where a collaborative framework between institutions blend diverse academic expertise. This cooperation base on scientific research bridges gaps across the Atlantic, underscoring the international nature of science. The aim is to cultivate a network linking high-quality research that is not always available in the English language. This narrative underscores the role of scientific diplomacy in strengthening the economic, political, social bilateral ties between Germany and Latin America societies.

The institutional dialogue between Germany and Latin America became more engaged in the 19th century, resulting in the establishment of institutions such as the *Ibero-American Institute* (IAI) and the Ibero-American Library founded in Berlin in January 1930 (IAI) being an interdisciplinary that foster academic and cultural exchange between Germany and Latin America, Spain, Portugal, and the Caribbean. It functions as an information center, research hub and cultural diplomatic venue.

The institute aims to serve as platform for cooperation and a catalyst for intercultural and transcultural dialogue.

During the 1970s, both Latin America and West Germany cultivated a keen interest in development and dependency, united by shared goals of advancing academic freedom, fostering international solidarity, and facilitating conversations and debates when networking among Latin Americanists.

In Berlin, the *Institute for Latin American Studies at the Free University Berlin* boosts a shift towards a future research agenda that transcends traditional institutional boundaries. This approach emphasizing collaboration, brings together diverse generations and sources, and acknowledges the bilateral exchange of knowledge (Kriegesmann, 2023). These initial interactions established the foundation for friendly and respected partnership, characterized by the interest to learn from each other. The Collaboration between Germany and Latin America strengthens when German research and funding organizations open offices in Latin America aimed to network and work directly with the local scientific community.

There are good number of initiatives that boosts the academic ties between Germany and Latin America as highlighted by the Federal Foreign Office's Regional Focus on Latin America (2024). For instance, the establishment of the first *Fraunhofer Institute* in Santiago, Chile, in 2010 under Chilean law. It was the first research institution that was funded as part of a program from CORFO, a Chilean economic development agency, aimed at attracting international centers of excellence. The foundation currently operates the Fraunhofer Chile Center for Solar Energy Technologies CSET. Fraunhofer Chile Research is the only legally independent Fraunhofer affiliate in Latin America.

In Argentina, the partnership with CONICET and the Max Planck Society in 2011, with the subsequent opening Max Planck Society branch in Buenos Aires and the Max Planck along Laboratory in Rosario in 2014, exemplifies a commitment to advancing in structural biology, chemistry, and molecular biophysics. The research carried out at this laboratory focuses on identifying drug candidates for neurodegenerative diseases such as Alzheimer's and Parkinson's, aiming to elucidate their structures. Furthermore, another activities is the training of junior scientists. In support of this goal, a binational doctoral program has been established between the universities of Rosario and Göttingen with the support of the Deutsch-Argentinischen Hochschulzentrums (German-Argentinian University Centre (CUAA/DAHZ). Göttingen Nobel laureate Stefan Hell has committed to delivering lectures as part of this educational initiative.

As part of the laboratory opening, a contract was signed between the company Nikon, USA, and the university, which contains plans for the establishment of an international Excellence Centre for microscopy at the Max Planck laboratory. This center, the first of its kind in Latin America, mirrors like existing centers in the

USA and Europe, being a milestone in Latin America scientific infrastructure and international collaboration efforts.

In Brazil, São Paulo, efforts to strengthen academic relations have led the establishment of the German House of Science and Innovation. This platform serves as a hub for events and services provided by German scientific organizations, significantly contributing to the enhancement of academic collaboration. Additionally, the presence of the German Research Foundation (DFG) office with the German Centre for Research and Innovation (DWIH) *São Paulo* since 2011 has played a crucial role in bolstering partnerships with research institutions and funding organizations throughout Latin American countries.

The DFG Office Latin America is currently supported by liaison scientist in Chile and Mexico, facilitating closer ties between researchers, thereby networking local scientific communities with their counterparts in Germany. This initiative is part of the *German Research Foundation* (DFG)'s broader strategy to foster international collaboration across all its diverse funding programs, emphasizing the facilitation of cross-border mobilization when research partnerships. Operating as a politically independent organization, the DFG holds a strategic place in the sphere of international research policy. Holding memberships in several scientific and science policy organizations and committees at both the European and global levels. This involvement underscores the DFG's dedication to furthering global scientific cooperation and dialogue on a worldwide scale (DFG, 2024, March 3), advancing with the frontiers of knowledge and cooperation.

To strengthen the scientific cooperation with Germany and promote advanced studies in social sciences and humanities, the *Maria Sibylla Merian Center for Advanced Latin American Studies* (CALAS)was inaugurated in 2017. Based in Guadalajara, Mexico, CALAS also operates offices in Ecuador, Costa Rica, and Argentina, signifying a step in promoting interdisciplinary research across the region. In São Paulo, Brazil, the Maria Sibylla Merian Centre, known as (Mecila) focuses on studies in conviviality-inequality in Latin America. Mecila's work delves into the social, political, and ecological crises affecting the region, organizing it research around four pivotal clusters: "Social-ecological Transformation," "Social Inequalities," "Violence and Conflict Resolution" and "Identity" (BMBF, 2022).

Since 2017, ***German and Latin American universities***, under the coordination of the University of Bielefeld, have been developing the framework for the establishment an international research college. This collaboration involves notable institutions like the Leibniz University Hannover and the Centre for Atlantic and Global Studies, with additional participation from the universities of Kassel and Jena of the German side.

The *Heidelberg Center Latin America* (HCLA), exemplifies the collaborative efforts between Germany and Latin America. Being a Center of Excellence in Research and Teaching. Supported by the German Academic Exchange Service (DAAD) withing its program the HCLA plays a pivotal role in fostering the development of innovative projects areas of focus include Astronomy, Communication, Geosciences, International Law, Medical Informatics, Medical Physics, and Political Science. It is a showcasing in strengthening research networks across a variety of disciplines with Latin American partners promoting academic and research endeavors.

In Germany, a consortium comprising universities, technical institutions and non-academic research entities has forged successful collaborations successfully established a cooperation with homologous partners in Latin American countries. These partnerships are fully or partially supported by the German government. Among these, the *German Institute of Global Area Studies* (GIGA) stands out as key player, focusing on the political and economic developments in the countries of South and Central America and the Caribbean. The projects addressed vital issues such as the quality of political institutions and participation, challenges by conflicts and crime, and the regional integration dynamics. There is a steady bilateral influx of doctoral and visiting researchers from Latin American and Caribbean countries as from Germany, enriching academic environment and fostering diverse perspectives.

The Institute for *Latin American Studies* (LAI) at Freie Universität Berlin stands as Germany's premier institution dedicated to the study of Latin America, encompassing a wide range of disciplines from the social sciences and cultural studies, including Anthropology, Economics, Political Science, and Sociology. The institute extensive research profile "Latin America: Interdependencies, Asymmetries, Mobilities", offers a holistic examination of Latin America contemporary landscape.

The LAI delves into the historical roots and explores the dynamics, perspectives, and current trends in Latin America. It is engaged in several bilateral projects, highlighting its active role in international academic collaboration. Notable projects include the DFG-supported "Temporalities of Future" International Research Training Group and, the "Conviviality-Inequality in Latin America – Mecila" (BMBF-funded), Additionally, it contributes to "History Transfer through Entertainment Media in Latin America: GUMELAB" (BMBF), and the "trAndeS – Trans-Andean Network of Sustainability" (DAAD), further illustrating its commitment to understanding and disseminating knowledge on the complex and diverse nature of Latin American societies.

Furthermore, *cooperation programs* like *UNIBRAL* and UNIBRAL II, initiated by the Brazilian and German Education Ministries in 2000 and 2008 respectively, have played an important role in promoting collaboration between universities in both countries. Managed by German Academic Exchange Service (DAAD) and its Brazilian counterpart CAPES, these programs enabled students, doctoral, postdoctoral candidates, and professor's exchanges, while also fostering cooperation between faculties. In 2017, these initiatives were integrated into the regular DAAD programs, including the International Study and Training Partnerships (ISAP) and International Degree Programs with a Double Degree.

On January 20, 2021, the *Bavarian Academic Center for Latin America* (BAYLAT) and operating under the Bavarian State Ministry of Science and the Arts (StMWK), announced the launch of the LATinBAY Cooperation Information System. As a key facilitator of partnerships between Bavarian and Latin American universities, research institutions, and governmental entities, BAYLAT plays a crucial role in fostering academic collaboration. With Brazil as a leading partner country, BAYLAT actively supports science, cooperation, and higher education diplomacy across Bavaria and Latin America.

The introduction of this innovative digital platform represents a concerted effort by BAYLAT to enable Bavarian academics engaged in Latin American studies or collaborations to enhance their international networks across various disciplines. BAYLAT's commitment extends beyond facilitating research collaborations; it also includes providing mobility grants for Bavarian students and scholarship opportunities for Latin American students. These initiatives are part of BAYLAT's broader strategy to promote professional exchanges and internationalize academic institutions across 20 Latin American countries, thereby creating new opportunities for cooperation and knowledge exchange (DWIH-SãoPaulo, 2024)

The *German Centre for Research and Innovation* (DWIH) established on 2010, by the Federal Minister of Education and Research a global network aimed at enhancing international cooperation in research, scientific innovation, and talent exchange. As joint project of the Federal Ministry of Education and Research and the Federal Foreign Office, initially led jointly by the German Academic Exchange Service (DAAD) and the German Research Foundation (DFG). DWIH plays a crucial role in fostering global academic and scientific partnerships.

On 2017, the German Academic Exchange Service (DAAD), assume sole administrative responsibilities for the German Centre for Research and Innovation (DWIH). The DWIH is funded by the Federal Foreign Office with centers located in New Delhi, Moscow, San Francisco, São Paulo, and Tokyo. This expansive presence facilitates a broad spectrum of international partnerships. In alignment with the principles of science diplomacy, DWIH serves as a platform for transatlantic cooperation, connecting various stakeholders from North America and Germany.

A core mission of DWIH is to promote Germany as a premier destination for science, technology, and innovation. Through fostering dialogue and collaboration across a diverse range of disciplines and sectors, DWIH aims to enhance the visibility and appeal of the German research landscape on the international stage. This initiative underscores Germany's dedication to advancing scientific inquiry and innovation through global partnerships and exchanges (Wikipedia, 2024).

Strategic Framework to Foster Internationalization

In the German education landscape, international collaborations are very important in shaping a sustainable future: The Federal Foreign Office acknowledges science diplomacy as an important mechanism that bridging the gap between science, civil society, and policymakers, highlighting its role in promoting freedom and international networking. Scholarship programs play a key role in this context, designed to enhance the learning and mobility experiences of international students and researchers, facilitating the exchange of ideas, opinions, and different knowledge experiences across borders.

The Federal Ministry of Education and Research (BMBF)· has laid down the principles "connect, inform, and enable" as the cornerstone withing its third policy framework, effectively merging education with the objectives of science diplomacy. This strategic approach is aimed at bolstering international connections, fostering global stability, and catalyzing innovation for the betterment of society. The framework emphasizes the creation of sustainable models for education and cooperation, grounded on learning and research. Its goal is to bridge individuals and societies from diverse cultural backgrounds to share information and understand better how the world can improve in communication and collaboration.

The Federal Ministry of Education and Research (BMBF) outlines Germany's new policy framework with the objective of allocating 3.5 percent of its gross domestic product to research and development by the year 2025. This ambitious goa is intended to address the evolving requirements at state, political and governmental level driven by the current crises and the need for transformative processes.

Central to this initiative is the advancement of a key framework program like the "Future Strategy for Research and Innovation" a High-Tech Strategy, which aim to revolutionize current processes and spearhead what is known as the in the fields of science, research, and transfer, by fostering innovation through enhanced research and innovation efforts, supporting emerging talent and favor an agile research and innovation policy. The mission is facilitating resource-efficient and circular economy, promoting competitive industries, ensuring sustainable mobility, achieving climate protection and adaptation, food security, and biodiversity conservation and, advancing health innovation.

Germany recognizes the importance of international cooperation in advancing education, science, and research. Therefore, is dedicated to strengthening global excellence. This includes integrating global knowledge and value chains by fostering partner countries supporting mobility and professional recognition of international specialist. Withing this international framework, the Federal Government continues to be bolstering networking efforts to support the growth of small and medium-sized enterprises (SMEs), encouraging innovation and partnerships. By expanding collaborations with emerging and developing countries, Germany seeks to contribute significantly to the global knowledge society, aiming to create synergies, enhance coherence by expanding collaborations with emerging and developing countries. The mail goal is foster synergies, coherence and build bridges based on scientific knowledge (BMBF, 2016).

Bridging the Gap

The notion of "science diplomacy" within the German context is characterized by several key principles: scientific autonomy, minimal governmental interference, the federal structure of the country, and a historical tendency where the scientific communities often stay to refrain from political demands. This landscape has cultivated a model for scientific exchange that is partnership-based with goals and objectives developed through collaborative dialogue (DAAD,2023)

Science diplomacy serves as a bridge between the realms of science and international policy. It offers a strategic framework for addressing global challenges through scientific collaboration. Germany plays a pivotal role in nurturing international scientific partnerships, promoting academic mobility, and facilitating scientific cooperation. These efforts are recognized as a vital effective action of the German foreign policy (Federal Foreign Office, 2024).

The aim of Germany's international strategy is to enhance scientific cooperation, exchange, and networking, thus broadening mutual understanding and collaboration beyond borders. Such concerted efforts not only facilitate joint scientific projects but also fortify foreign policy through the establishment of significant interpersonal and intercountry relationships. Germany extends its academic outreach through global higher education partnerships, with the Higher Education Compass detailing about 37,000 collaborations with over 5,400 partner universities across more than 150 countries (Federal Foreign Office, 2024).

From Germany, the German Academic Exchange Service (DAAD), the Alexander von Humboldt Foundation (AvH), the Deutsche Gesellschaft für Internationale Zusammenarbeit (GIZ), actively support international academic collaboration by awarding fellowships and sponsoring academic programs. They work in tandem with Latin Americans in missions and abroad, launching initiatives and offering different levels of educational options, master degree, PhD graduation, postdoctoral opportunities in German academic institutions.

Within Germany, these international scientists and scholars are recognized as crucial partners and influencers. They serve as vital links within the scientific and academic communities, significantly contributing to the country's research and development endeavors. Their involvement enriches the German academic landscape, bringing diverse perspectives and expertise that drive forward scientific inquiry and innovation.

Showcases

German science diplomacy in Colombia, exemplified by the German-Colombian Peace Institute, (CAPAZ) is a Center of Excellence in Research and Teaching that facilitates dialogue, knowledge exchange, and collaborative projects to support peacebuilding efforts in Colombia. Showcases a distinct approach where science-driven objectives take precedence over geopolitical considerations. This focus on fostering dialogue, understanding, and the cultivation of intercultural competencies that transcends the immediate goals of cooperation, leading significant outcomes. Importantly, this strategy is characterized by its deliberate avoidance of direct political influence, emphasizing the role of science in facilitating constructive discourse and mutual growth.

In the realm of science diplomacy, Colombia presents a compelling case study with the establishment of the German-Colombian Peace Institute, CAPAZ. Founded in 2016, CAPAZ stands as a beacon of how scientific collaboration can extend beyond traditional boundaries to influence peacebuilding and governance. This initiative underscores the potential of science to guide political decisions and mitigate conflicts through evidence-based discussions (https://www.instituto-capaz.org/en/)

The German Colombian Peace Institute – CAPAZ, a Centre of Excellence in Research and Teaching, is a cooperation platform operating in Germany and Colombia and financed by the German Academic Exchange Service (DAAD), with funds provided by the German Federal Foreign Office. The Institute stimulates the exchange of experiences and knowledge on peacebuilding through the creation of networks between universities, research centers, civil society organizations, and governmental entities that act in the territorial arena. CAPAZ supports research, teaching, and advisory activities that create new approaches to understanding peace and conflict, transmit knowledge to society, and provide responses to the multiple challenges of a society in transition (https://www.instituto-capaz.org/en/about-us-2/)

CAPAZ embodies a multifaceted approach to peace research and education, seamlessly integrating these efforts with political consulting to foster a more peaceful society. The institute specializes in creating platforms for dialogue and knowledge exchange, leveraging scientific research to inform policy and public opinion. Through workshops and personal engagements, CAPAZ facilitates the transmission of scientific insights into the political arena, ensuring that policy formulation is grounded in rigorous research.

The role of CAPAZ in Colombia is a testament to the power of interdisciplinary scientific support coupled with political consulting in addressing complex social challenges. By bridging the gap between science and politics, CAPAZ not only contributes to the peace process in Colombia but also sets a precedent for science diplomacy in Latin America. This initiative illustrates the profound impact that scientific collaboration can have on societal well-being, offering a model for other regions grappling with similar issues.

The German-Colombian Peace Institute, CAPAZ, exemplifies a thriving model of science diplomacy, where research, education, and consultancy seamlessly interconnect. Nestled within the diverse landscapes of Colombia, CAPAZ leverages the rich experiences of both Colombia and Germany to forge a path of peace through scholarship and dialogue. This nexus of networking enables a vital exchange of knowledge and strategies, drawing on the strengths of interdisciplinary academic work.

At CAPAZ, the dialogue extends beyond theoretical discussions, focusing on tangible impacts within Colombia's diverse regions. It is here that the power of collaboration takes center stage, as experts from various fields come together to contribute to Colombia's journey toward enduring peace. The institute stands as a testament to the role that science and education play in bridging cultural and political divides, making it a beacon of hope and a prototype of international cooperation for peace and development.

By fostering an inclusive environment, CAPAZ encourages a vibrant exchange of ideas and solutions, making significant strides in how science can inform policy and shape futures. It is a unique space where academic rigor meets societal needs, where the pursuit of knowledge is intimately linked with the quest for peace, and where every collaborative effort is a step towards a more harmonious world. This dedication to interdisciplinary and cross-cultural partnership is what places CAPAZ at the forefront of science diplomacy in Latin America and beyond.

German science diplomacy in Colombia is a showcase with CAPAZ focusing on science-immanent rather than geopolitical goals, emphasizes discourse, understanding, and the development of intercultural competencies beyond the immediate aims of cooperation. This approach has yielded numerous tangible outcomes, notably because it steers clear of seeking direct political influence (DAAD, 2023).

Commissioned by the German Federal Ministry of Economic Cooperation and Development (BMZ), the Deutsche Gesellschaft für Internationale Zusammenarbeit (GIZ) operates as a leading provider for of international cooperation services for sustainable development. GIZ is deeply involved in development and cooperation projects throughout Latin America and the Caribbean. Employing a triangular cooperation model, it aims to bolster partner countries and foster more extensive and systematic collaboration among stakeholders and between the participating countries.

GIZ lead programs focused on international competence building, which are designed to promote systematic learning processes. These programs facilitate the exchange of experiences through dialogue, networking, and training stakeholders in specific projects. The ultimate goal is to support the development of intentional cooperation, leadership, management, best practices across a wide range of sectors. Countries involved in these initiatives include Bolivia, Brazil, Chile, Costa Rica, Dominican Republic, Ecuador, El Salvador, Guatemala, Honduras, Caribbean, Colombia, Mexico, Nicaragua, Paraguay and, Peru.

These programs are implemented in cooperation with country partners and other international development organizations. They aim to drive action in sustainable infrastructure, security, reconstruction and peace, social development, governance and democracy, climate, environment, management of natural resources, economic development, and employment. The target group for these initiatives varies according to the objectives of each project, ensuring that efforts are tailored to meet specific needs and challenges in each area of focus. GIZ is actively engage in the development and implementing green business models, such as circular economy approaches, that prioritize theuse natural resources sparingly while increasing business productivity.

The Deutsche Welt Hungerhilfe (WHH), a German nonprofit organization, that significantly contributes to global development and humanitarian aid. WHH has been actively involved in implementing projects across worldwide, including Latin America, where it's aimed to promote agricultural development, humanitarian assistance and supporting the increment of nutrition levels by innovative approaches. The organization's mission is closely aligned with the United Nations' 17 Sustainable Development Goals (SDGs), aiming to create a positive impact on the communities it serves.

In Latin America, WHH is currently active in Bolivia and Peru, where its projects focus on securing the right to food for smallholder farming households. This is achieved through initiatives that support these communities in asserting their land rights, improving their living conditions, and protecting the environment. The overarching value of these cross-country programs in Bolivia and Peru illustrates WHH's commitment to sustainable development and environmental conservation. Additionally, WHH extends its humanitarian and development efforts to the Caribbean, with significant projects in Haiti. In this region, the organization continues to address urgent needs and support sustainable growth, demonstrating its dedication to alleviating hunger and poverty on a global scale.

Founded with support from various entities, including the German Federal Foreign Office, the European Union, numerous enterprises, organizations, and foreign governments, the Deutscher Akademischer Austauschdienst (DAAD) is the world's largest funding organisation for the international exchange of students and researchers. Under the motto "Change by Exchange", DAAD support the education of academics and students, facilitating international exchange and collaboration. In Latin America and the Caribbean have benefitted from DAAD's commitment to fostering international education, with numerous scholarships awarded to facilitate their academic and professional growth, ensuring that universities and institutions remain competitive and integrated withing the global academic community (DAAD, Colombia, 2024)

Since 2021, the Deutscher Akademischer Austauschdienst (DAAD) has been actively supporting programs in Latin America and the Caribbean, notably including the German-Latin American Centre of Infection & Epidemiology Research & Training (GLACIER) This initiative involves key partners such as Cuba and Mexico, as well as prestigious institutions like Charité - Universitätsmedizin Berlin and Martin-Luther-University Halle-Wittenberg, alongside other significant contributors from universities and research institutes. This multidisciplinary consortium, led by pandemic research and virology experts from Charité and UNAM, in collaboration with leading vaccination and therapeutics developers in Cuba and their German and Mexican partners, is focused on strengthening surveillance of emerging infectious diseases, diagnostics, treatment resistances, as well as the development of new vaccines and therapeutics. GLACIER aims to fortify capacities in the Latin American region by serving as a think-tank for a multidisciplinary network of institutions across eight Central American countries, building local research capacity, increasing the number of well-trained experts/scientists and trainers, and involving regional and international policymakers to aid information dissemination and enable faster regional crisis responses.

In Colombia and international project supported by the Deutscher Akademischer Austauschdienst (DAAD) and by five Colombian Universities Universidad Nacional de Colombia, Universidad de los Andes, Universidad de Bogotá Jorge Tadeo Lozano, Universidad de Antioquia, and Universidad del Valle, alongside the German university Justus-Liebig-University in Gießen. The Consortium Excellence Center in Marine Sciences (CEMarin)· established in 2009, represents a pivotal academic and scientific global center, dedicated to the study and application of solutions to problems in the field of marine sciences by promoting research and teaching, enhancing the collective capacity to address and solve pressing marine and coastal challenges.

The Goethe Institute, a non-profit German cultural association, operates in numerous countries worldwide, with a particular focus on Latin America. In this region, its mission centers on strengthening cultural ties between Germany and Latin American countries. Since its establishment in 1957 in Colombia, the institute has expanded its operations to include Argentina, Bolivia, Brazil, Chile, Costa Rica, Ecuador, Guatemala, Mexico, Paraguay, Peru, Uruguay, and Venezuela. Leveraging cultural, artistic, and intellectual networks and partnerships, the Goethe Institute initiates projects in various formats to foster dialogue, artistic expression, and cross-cultural appreciation. These projects often address pressing issues such as migration, decolonization, history, racism, the crisis of the left, and urbanism, with a particular emphasis on South-South migration.

6. FUTURE PROSPECTS OF SCIENCE DIPLOMACY

We recognize that by joining forces, we can achieve more collectively than we ever could individually. Additionally, we are conscious that our existence is linked to the balance of our ecosystem with mother nature. Therefore, we need to achieve global harmony and learn how to manage the Anthropocene situation with responsibility, our coexistence is linked. Transitioning to a sustainable economy is a goal for all of us. Science serves as the common language, and through teamwork and collaboration, we strive to foster knowledge and principles that help societies to find solutions to common challenges and build a future base on education with benefits for everyone. Moving towards cooperation and science diplomacy, we embrace these ideals as pathways to a more sustainable and inclusive world.

A Strategy Linked to Progress

Germany bridges distances and fosters connections with far-reaching regions through the positive engagements of science diplomacy. It fosters strong international connections with the huge region of Latin America, including the Caribbean, underscoring its commitment to knowledge and progress by collaboration through promotion of research, mobility, dialogue, and mutual learning. Publications are becoming more international, embracing not only English but local languages as Spanish, Portuguese and German, ensuring wider access and engagement across different linguistic communities worldwide.

Seen from Latin America, the international relationship with Germany, facilitated by science diplomacy, enables the region to access to advance research and technology as Latin American students and research groups to engage in specialized training and knowledge exchange, enhancing the development of cutting-edge technologies by promoting best practices in scientific development, this collaboration foster innovative solutions and highlights the valuable contributions that Latin America offer to the global scientific arena.

Improve the Collaborative Network

The field of international cooperation is experiencing a transformative moment, market by the emergency of science diplomacy as driving force. This novel perspective is uniting not governments and policymakers, but also stakeholders, business leaders and scientists from countries that are cultural, historical, and economic different.

Germany and Latin America find through science diplomacy a common understanding, they are uniting efforts to progress together, fostering an environment of shared knowledge and, cooperation. The aim is supporting the power of complementation and joint efforts to achieve not only common goals but to enhance knowledge solutions and best productivity.

Countries in Latin America are promoting a robust networking foundation with Germany, driven by collaborative projects that blend innovation with diversity, addressing societal challenges through bilateral cooperation, offering the possibility for international collaboration, mobilization, and research on the topics of the Anthropocene, linking societal need with technological advancements. Such networking in scientific research enhances the quality of outcomes, providing solutions to global challenges.

Promote Technological and Environmental Expansion

Scientific collaboration not only promotes the integration of academia but also introduces innovation to remote areas, providing access to well-equipped laboratories and exotic natural regions, thereby improving outcomes, and positioning scientific achievements as solution for the needs of our time. It ensures access to critical expertise, which may not always originate from industrialized countries, ensuring wider recognition. This approach globalizes science, making it accessible to all and broadening the scope of knowledge and its applications in real-world scenarios.

Create and Attractive Knowledge-Base Ecosystem

Germany is strengthening its research and innovation framework through international networking. By boosting innovation and productivity, is shaping a compelling science and innovation ecosystem that draws both national and international attention, especially from Latin American partners interested in enhancing productivity by finding solutions to the region main challenges. Through science diplomacy, Germany not only strengthens its own research and innovation capabilities but also fosters synergizes with Latin American countries. This cooperative model encourages scientific progress and the introduction of leading-edge innovations, building a trust space where stakeholders from Germany and Latin America unite for the desire of progress activate the trade channels to favor production and economic growth.

A Problem-Solving Strategy

In 2023, the German Academic Exchange Service emphasized the significance of unity and cooperation in addressing complex issues. The necessity for a shared understanding and willingness to collaborate have become essential for preventing conflicts and uniting efforts to meet contemporary challenges. Through acknowledging the role of science diplomacy, Germany showcases the impact of collaborative endeavors.

Research and Conflict Prevention

Initiatives such as the German Colombian Peace Institute CAPAZ provide valuable platforms for conducting peace research and supporting conflict prevention. This institute serves as a bridge between Colombia and its international partners, creating an open door that significantly impacts both the scientific community and civil society. By producing insights that gain traction withing the scientific ecosystem, CAPAZ efficiently transfers these insights and methodologies into the wider societal context, a contribution highlighted by the German Academic Exchange Service in 2022.

German universities make significant contributions to the global discourse in politics, science, and business with their research findings, including those relevant to Latin America. Acting as bridges, they connect not only with Germany but also with European policies through open channels, utilizing their extensive networks and experience to foster scientific exchange. These exchanges, driven by geostrategic goals, aim to promote peace, freedom, and democracy, grounded in idealism. Particularly in Latin America, such efforts underscore the commitment to creating a collaborative educational and research environment that transcends geographical boundaries, as highlighted by the German Academic Exchange Service (DAAD) in 2023.

Empower the Scientific Landscape

The funding programs led by the German Federal Ministry of Education and Research (BMBF), the German Academic Exchange Service (DAAD), and various foundations play a pivotal role in promoting collaboration among scientists and individuals. These initiatives support the science diplomacy effort by encouraging international research collaborations between research institutions, universities, and teams from the private sector. Moreover, they facilitate the development of projects where one party offers expertise while the other provides access to resources that would otherwise be difficult to obtain. This arrangement fosters a conducive environment for development due to its unique performance, resulting in a mutually beneficial outcome.

7. AFTERWORD

Germany, situated at the heart of Europe and playing a pivotal role in the European Union, stands out for its robust academic tradition that nurtures its reputation for excellence. The country foster knowledge, boasting an outstanding academic infrastructure that values scientific achievements and transform them into innovative systems, products, methods through interdisciplinary collaboration. This approach fuels its economy with a strong industrial base, earning recognition in international markets.

Germany, characterized by its resilient spirit and significant contributions to scientific innovation, has cultivate a culture of transformation throughout its history. With its formidable academic infrastructure, this approach leverages the country extensive research networks, world-class universities, and a strong commitment to science and technology. As a leader in global innovation and scientific research, Germany is on a mission to merge innovation with productivity and foster scientific collaboration on the international stage. This endeavor underscores the country's dedication to advancing both its scientific prowess and its role in global development through diplomacy and collaboration.

Following the reunification, Berlin, the capital, has become a beacon for scientific innovation and collaboration, embodying creativity, freedom, and commitment. The city's transformation into an innovation hub is largely attributed to federal initiatives aimed at modernizing infrastructure and enhancing science marketing and science transformation activities. These efforts include promoting an ecosystem blend on innovation, creativity, and international connections, promoting academic performance, science translation, start-ups, clusters, networking opportunities, and science communication. The Berlin showcase is a source of inspiration for how linking diverse actors can make science visible and integrated into the industry. The academic culture of Berlin collaborates with the industry and policymakers to advance in a vision focused on was sustainable green economy.

Berlin, a showcase for science visualization and engagement, demonstrates the significant impact of strategic initiatives supported by federal funding. The city has developed notable science marketing efforts such as the Berlin Brain City campaign, the Falling Walls Summit, and the Long Night of Science. These initiatives have not only raised Berlin's international profile but also aligned with Germany's broader ambitions to foster an inclusive, innovative, and forward-looking scientific community that embraces all sectors of society. This approach underlines the vital importance of creating an environment that attracts, nurtures, and collaborates with international talents and partnerships. It embodies a collective commitment to enhancing scientific collaboration, within the realms of science diplomacy, to reinforce Germany's innovation and scientific research while integrating productivity.

Guided by Agenda 2030 and its Sustainable Development Goals, both Germany and Latin America face the critical challenge of bridging the gap between innovation and production. Science diplomacy emerges as a pivotal force in facilitating these efforts, playing an essential role in promoting a collaborative agenda and underscoring the importance of interdisciplinary and international cooperation when addressing the multifaceted challenges of the Anthropocene. The strategic engagement, where Germany contributes with its research infrastructure, technological prowess, and dedication to scientific excellence, paired with Latin America's unique cultural diversity and rich biodiversity, provides a robust environment for scientific inquiry and technological innovation.

The scientific collaboration between Germany and Latin America within the realm of science diplomacy reflects the mutual need to strive toward a better, fairer, and sustainable future for all. By working together, they embrace diversity and realize scientific excellence in new arenas. Together, Latin America and Germany have the capacity to shape a brighter future for coming generations by addressing contemporary challenges with untapped potential for development. The narratives and case studies presented herein showcase the transformative potential of the partnership between Germany, Latin America, and the Caribbean in science and technology. This alliance promises to play a pivotal role in the global research and innovation landscape, particularly addressing critical issues such as climate change, biodiversity loss, and socio-economic disparities. The collaboration between these regions highlights the crucial role of leveraging scientific expertise and innovation to create a more resilient and equitable world.

Science diplomacy is the bridge that promotes socio-economic development, environmental sustainability, and cultural understanding, thriving on nurtured education. Knowledge transcends barriers such as language, skin colour, and geographical divides between the global north and south. Science diplomacy offers opportunities for mobility, learning, and discovery. These concerted efforts lead to societal progress on a bilateral scale. The achievements of collaborative research initiatives, academic exchanges, and the establishment of cooperative institutions like CAPAZ in Colombia exemplify the significant impact of science diplomacy in facilitating peacebuilding and governance. Projects like CEMarin open doors for Germany to explore tropical sea waters, among many instances that highlight the importance of scientific collaboration in fostering improved relations and mutual comprehension across multidisciplinary teams. This commitment to shared objectives and efforts to foster solutions sets the stage for a future where science and diplomacy converge to catalyses positive global change.

Germany's commitment to science diplomacy, supported by its robust academic and research infrastructure, sets a precedent for nations seeking to strengthen their global collaborative networks. Together, Germany and Latin America are a formidable force, combining their strengths to advance science and culture. The strong scientific frameworks in both regions provide a solid foundation for diplomatic initiatives that facilitate the exchange of knowledge, foster academic and research partnerships, and promote mutual understanding and respect. This blend of science and diplomacy illustrates how investment in scientific projects can transcend national boundaries and drive global cooperation and development. This collaborative model expands scientific research and strengthens international relations, demonstrating the transformative nature of science diplomacy in addressing global challenges and advancing collective progress.

REFERENCES

Aguirre-Bastos, C. (2023). Internationalization of Science and Diplomacy, Concepts and Practices: Lessons for Developing Countries. In Ittekkot, V., & Baweja, J. K. (Eds.), *Science, Technology and Innovation Diplomacy in Developing Countries. Research for Development*. Springer. 10.1007/978-981-19-6802-0_2

Alexander von Humboldt Foundation. (2024, March 1.). Retrieved from https://www.humboldt-foundation.de/

Allen, M. M. C. (2015). The national innovation system in Germany. In Narayanan, V. K., & O'Connor, G. (Eds.), *Wiley encyclopedia of management: Technology and innovation management* (Vol. 130047). Wiley. 10.1002/9781118785317.weom130047

Bayerisches Hochschulzentrum für Lateinamerika (BAYLAT) an der Friedrich-Alexander-Universität Erlangen-Nürnberg (FAU). (2024. March 3). LATinBAY Kooperationsinformationssystem. Retrieved from https://www.baylat.org/latinbay-kooperationsinformationssystem.html?tx_tweprivacy_eprivacydialog%5Baction%5D=dialog&tx_tweprivacy_eprivacydialog%5Bcontroller%5D=Subject&cHash=93bc279ae1fdc7ca7480bc0a1d9c7943

Berlin Partner for Business and Technology. (2014). Informationen rund um den Wirtschaftsstandort Berlin. Retrieved February 23, 2024, from www.businesslocationcenter.de

Berlin Partner for Business and Technology. (2024, February 23). Explore the Berlin-Brandenburg Ecosystem | Berlin tech ecosystem map. Retrievedfrom https://startup-map.berlin/intro

Berlin Partner for Business and Technology. (2024, February 24). Science. Business Location Center. https://www.businesslocationcenter.de/en/science

Berlin Science Week. (2024, March 13) Das Internationale Wissenschaftsfestival. Retrieved from https://berlinscienceweek.com/de/

Berlin University Alliance (BUA). (2024, March 2). Retrieved from https://www.berlin-university-alliance.de/

Borlovan, P. (2023) Dive into the German Startup Ecosystem Key Players, Trends and Opportunities. SeedBlink Blog. Retrieved from https://seedblink.com/2023-10-19-dive-into-the-german-startup-ecosystem-key-players-trends-and-opportunities

Boytchev, H. (2020). How Germany retains one of the world's strongest research reputations. Nature Index. *Nature*, 587(7834), S104–S105. 10.1038/d41586-020-03318-w

Brain City Berlin. (2024, February 24). Berlin's science hub. Retrieved from https://braincity.berlin/en/science-location-berlin-1.

Brandt, M. (2023, June 20). So hat sich Deutschlands Wettbewerbsfähigkeit entwickelt. World Competitiveness Ranking 2023. Statista. Retrieved from https://de.statista.com/infografik/30229/abschneiden-von-deutschland-im-world-competitiveness-ranking/

Broekel, T., Rogge, N., & Brenner, T. (2018). The innovation efficiency of German regions – A shared-input DEA approach. *Jahrbuch für Regionalwissenschaftt = Review of Regional Research*, 38(1), 77–109. 10.1007/s10037-017-0112-0

Buhl, C. M., Meier zu Köcker, G., & Sedlmayr, B. (2019). Development of Cluster Initiatives in Germany from 2008 - 2018. VDI/VDE Innovation + Technik GmbH. Retrieved from https://www.researchgate.net/publication/337671892_Development_of_Cluster_Initiatives_in_Germany_from_2008_-_2018

Bundesministerium für Bildung und Forschung. (2022) CALAS: Wissenschaftskooperation in und mit Lateinamerika. Retrieved from https://www.geistes-und-sozialwissenschaften-bmbf.de/de/CALAS-Wissenschaftskooperation-in-und-mit-Lateinamerika-2666.html

Bundesministerium für Bildung und Forschung (BMBF). (2021) The German research landscape. Deutscher Akademischer Austauschdienst (DAAD). https://www.research-in-germany.org/website/public/epapers/german-research-landscape/#0

Bundesministerium für Bildung und Forschung (BMBF). (2024, March 2) Research. High-tech and Innovation. Research Campus – Public Partnerships for Innovation. Retrieved from https://www.bmbf.de/bmbf/en/research/hightech-and-innovation/research-campus/research-campus_node.html

Bundesministerium für Bildung und Forschung, BMBF/Federal Ministry of Education and Research. Division General for European and International Cooperation in Education and Research. (2016). Internationalisation of Education, Science and Research: Strategy of the Federal Government. Retrieved from https://www.bmbf.de/SharedDocs/Publikationen/de/bmbf/FS/31286_Internationalisierungsstrategie_en.pdf?__blob=publicationFile&v=3

Bundesministerium für Bildung und Forschung (BMBF)/Federal Ministry of Education and Research. Division General for European and International Cooperation in Education and Research. (2016). Science Diplomacy: Wissenschaftsdiplomatie. Retrieved from https://www.bmbf.de/bmbf/de/europa-und-die-welt/vernetzung-weltweit/bildungs-und-wissenschaftsdiplomatie/bildungs-und-wissenschaftsdiplomatie.html

Carrier, M. (2022). What Does Good Science-Based Advice to Politics Look Like? *Journal for General Philosophy of Science*, 53(1), 5–21. 10.1007/s10838-021-09574-2

Carrier, M. (2022). What Does Good Science-Based Advice to Politics Look Like? *Journal for General Philosophy of Science*, 53(1), 5–21. 10.1007/s10838-021-09574-2

CEMARIN. (2024, March 1). Retrieved from https://cemarin.org/

Colombia, D. A. A. D. (2024, March 1). Retrieved from https://www.daad.co.). Retrieved from https://www.daad.co

Council of the European Union. (2021). Council recommendation on a Pact for Research and Innovation in Europe (Interinstitutional File: 2021/0230 (NLE)). Retrieved from https://www.bmbf.de/bmbf/en/international-affairs/the-european-research-area/the-european-research-area_node.html

Das Ibero-Amerikanische Institut (IAI). (2024. March 3). Geschichte des IAI. Retrieved from https://www.spk-berlin.de/iai/institut/geschichte-des-iaiDas Ibero-Amerikanische Institut (IAI). (2024, March 3). Geschichte des IAI. Retrieved from https://www.spk-berlin.de/iai/institut/geschichte-des-iai

Dauchert, H., & Garzik, L. (2022). Berlin Innovation System. In Garzik, L. (Ed.), *Successful Innovation Systems. Future of Business and Finance*. Springer., 10.1007/978-3-030-80639-2_7

Deutsche Forschungsgemeinschaft (DFG). (2014, March 3). Retrieved from https://www.dfg.de/en/research-funding/funding-opportunities/international-cooperation

Deutsche Forschungsgemeinschaft (DFG). (2024, March 2). Retrieved from https://www.dfg.de/resource/blob/172938/5b6fd952251824b1… (PDF file)

Deutsche Gesellschaft für Internationale Zusammenarbeit (GIZ). (2023, March 1). Retrieved from https://www.giz.de/

Deutsche Gesellschaft für Internationale Zusammenarbeit (GIZ). (2024, March 14). Triangular Cooperation with Latin America and the Caribbean: International Competence Building. Retrieved from https://www.giz.de/en/worldwide/13032.html

Deutsche Startup Monitor (DSM). (2022, March 2). Deutscher Startup Monitor. Retrieved from https://deutscherstartupmonitor.de/

Deutsche Welthungerhilfe (WHH). (2024. March 14). Lateinamerika. Retrieved from https://www.welthungerhilfe.de/informieren/laender?tx_igxprojektkarte_projektkarte%5Baction%5D=list&tx_igxprojektkarte_projektkarte%5Bcontroller%5D=Projektkarte&cHash=3f8e08b14fd8714962bad01c8125db48

Deutscher Akademischer Austauschdienst (DAAD). (2024. March 14). Oficina Regional del DAAD en Bogota.Movilidad entre Alemania y Países de la Región Andina (Colombia, Ecuador y Venezuela). Estadistics Actuales Septiembre 2023. Retrieved from https://www.daad.co/files/2023/12/DAAD_Estadisticas_Estudiantes_en_Alemania_ColEcuPerVen_2023_Stand_06_12_23_NM_RB-1.pdf

Deutscher Akademischer Austauschdienst (DAAD). (2024. March 14).The DAAD. Retrieved from https://www.daad.de/en/

Dusdal, J., & Powell, J. J. W. (2021). Benefits, motivations, and challenges of international collaborative research: A sociology of science case study. *Science & Public Policy*, 48(2), 235–245. 10.1093/scipol/scab010

European Commission. (2024, March 2). Research and innovation, international cooperation: Why and how? Strategy on Research and Innovation, Strategy 2020-2024—Europe in the World. Retrieved February 28, 2024, from https://research-and-innovation.ec.europa.eu/strategy/strategy-2020-2024/europe-world/international-cooperation_en

European Commission. (2024, March 2). Science diplomacy in the global approach to research and innovation. Retrieved from https://research-and-innovation.ec.europa.eu/strategy/strategy-2020-2024/europe-world/international-cooperation/science-diplomacy_en

Falling Walls. (2024, February 26). Retrieved from https://falling-walls.com/

Federal Foreign Office. (2020). Latin America and the Caribbean: A strategically important region. Regional Focuses. Retrieved from https://www.auswaertiges-amt.de/en/aussenpolitik/regionaleschwerpunkte/lateinamerika/-/227548

Federal Foreign Office. (2024, March 2). Education and Research. Retrieved from https://www.auswaertiges-amt.de/en/aussenpolitik/regionaleschwerpunkte/lateinamerika/-/227554

Federal Foreign Office. (2024. March 23). Science Diplomacy (09.11.2020-Article). Retrieved from https://auswaertiges-amt.de

Federal Ministry of Education and Research. (2021) German Research Landscape. Publisher Deutsche Akademische Austauschdienst (DAAD), Section Internationl Research Marketing 86p. https://www.research-in-germany.org/dam/jcr:3e5ad44b-8e6a-4832-b64d-ff037d106018/The-German-research-landscape_2021_barrierefrei.pdf

Federal Ministry of Education and Research. (2021) German Research Landscape. Retrieved from. https://www.research-in-germany.org/en/research-landscape/university-landscape/universities.html

Federal Ministry of Education and Research. (2022, Mach 2). CALAS: Science cooperation with Latin America. Retrieved from CALAS: Science cooperation in and with Latin America | Humanities & Social Sciences (geistes-und-sozialwissenschaften-bmbf.de)

Federal Ministry of Education and Research. (2024, March 2). Funding: Research funding. Retrieved from https://www.bmbf.de/bmbf/en/ministry/research-funding/research-funding_node.html

Federal Ministry of Education and Research. (2024, March 2). International affairs. *Horizon*, 2020, •••. https://www.bmbf.de/bmbf/en/ministry/research-funding/research-funding_node.html

Federal Ministry of Education and Research. (2024, March 2). Research. Hightech and innovation. Cluster-networks-international. Retrieved from https://www.bmbf.de/bmbf/en/research/hightech-and-innovation/clusters-networks-international/clusters-networks-international_node.html

Federal Ministry of Education and Research. (2024, March 2). Worldwide networking. Retrieved February 27, 2024, from https://www.bmbf.de/bmbf/en/international-affairs/worldwide-networking/worldwide-networking.html

Federal Ministry of Education and Research (BMBF). (2024. March 2). Science diplomacy. Retrieved from https://www.bmbf.de/bmbf/en/international-affairs/science-diplomacy/science-diplomacy.html

Figueiredo, M. S. N., & Pereira, A. M. (2017). Managing Knowledge – The Importance of Databases in the Scientific Production. *Procedia Manufacturing*, 12, 166–173. 10.1016/j.promfg.2017.08.021

Fischer-Bollin, P. (2019). Die Welt und wir: Globale Trends und Zukunftsthemen für die deutsche Politik. Konrad-Adenauer-Stiftung. https://www.kas.de/documents/259121/4624147/Die+Welt+und+Wir+Globale+Trends+und+Zukunftsthemen+f%C3%BCr+die+deutsche+Politik.pdf/f3e4b449-ef03-da73-2cf1-8c915d24a740?version=1.2&t=1575999718243

Fjæstad, M. (2010, April). Research institutes in Germany: Basic and applied science institutionalized? CESIS Electronic Working Paper Series, (Paper No. 232). Centre of Excellence for Science and Innovation Studies (CESIS), The Royal Institute of Technology. Retrieved from https://static.sys.kth.se/itm/wp/cesis/cesiswp232.pdf

Förderberatung des Bundes. (2024, March 2). Förderung. Retrieved from https://www.foerderinfo.bund.de/foerderinfo/de/foerderung/foerderung_node.html

Franco Ferreira, L., Justo, A., Cotelo, C., & Arias, I. (2019). Multi-actor engagement: An open innovation process of knowledge exchange and co-creation. In *Proceedings of the 12th annual International Conference of Education, Research and Innovation*, Seville, Spain. DOI:10.21125/iceri.2019.0969

Frauenhofer-Gesellschaft. (2014, March 13). Frauenchofer in Chile. Retrieved from https://www.fraunhofer.de/en/institutes/international/north-southamerica/fraunhofer-chile.html#:~:text=The%20Fundaci%C3%B3n%20Fraunhofer%20Chile%20Research%20FCR%20was%20established,%E2%80%94%20aimed%20at%20attracting%20international%20centers%20of%20excellence

German Center for Research and Innovation (DWIH) (2024. March 2). Science Diplomacy. Retrieved from Science Diplomacy (dwih-newyork.org)

German universities of technology (TU9). (2024, March 2). Retrieved from https://www.tu9.de/

German University Alliance (UA) 11+. (2024, March 2). Retrieved from https://www.ua11plus.de/

Germann, J. (2022). Global rivalries, corporate interests and Germany's 'National Industrial Strategy 2030'. *Review of International Political Economy*. Advance online publication. 10.1080/09692290.2022.2130958

Germany Trade & Invest. (2024, March 12). Start-up location Germany. Retrieved from https://www.gtai.de/en/invest/hot-topics/start-up-location-germany

GLACIER. (2024, March 1). Homepage. Retrieved from https://glacieronehealth.org/

Göring, J. Intellectual Property Protection - German Patents and International Competition. ATZ Worldw 125, 50–53 (2023). https://doi.org/, J. (2023). Intellectual property protection - German patents and international competition. ATZ Worldwide, 125(1), 50–53. https://doi.org/10.1007/s38311-023-1537-4Göring

Granstrand, O., & Holgersson, M. (2020). Innovation ecosystems: A conceptual review and a new definition. *Technovation*, 90–91, 102098. 10.1016/j.technovation.2019.102098

Gualini, E., & Fricke, C. (2019). Who governs Berlin's metropolitan region? The strategic-relational construction of metropolitan scale in Berlin–Brandenburg's economic development policies. Environment and Planning C. *Politics and Space*, 37(1), 59–80. 10.1177/2399654418776549

Gutiérrez, A. G. (2020). Imbabura and Chimborazo at the cradle of Biogeography in the South American Andes. *Biodiversity International Journal*, 4(5), 198–200. 10.15406/bij.2020.04.00185

Heidelberg Center Latin America (HCLA). (2024, March 2). Retrieved from https://www.hcla.uni-heidelberg.de/en/the-center and https://www.hcla.uni-heidelberg.de/en/research

Institute for Latin American Studies (LAI). (2014, March 3). Freie Universität Berlin, Institute for Latin American Studies. Retrieved from https://www.fu-berlin.de/en/einrichtungen/fachbereiche/zentralinstitute/lai/index.html

Instituto, C. A. P. A. Z. (2024. March 5). About Us. Retrieved from https://www.instituto-capaz.org/en/about-us-2/

Instituto, C. A. P. A. Z. (2024. March 5). Home. Retrieved from https://www.instituto-capaz.org/en/

International Consortium of Critical Theory Programs. (2021, March 1). Homepage. Retrieved from https://directory.critic

Jungbluth, C. (2019) Roadmap 2030: 3 Ways to Ensure the German Economy's Competitiveness in the Coming Decade. Bertelsmann Stiftung. GED New Perspectives on Global and European Dynamics. Retrieved from 3 ways to ensure the German Economy 's competitiveness in the coming decade (globaleurope.eu)

Kiese, M. (2019). Regional cluster policies in Germany: Challenges, impacts and evaluation practices. *The Journal of Technology Transfer*, 44(6), 1698–1719. 10.1007/s10961-017-9589-5

Kriegesmann, K. (2023, March 25). Doing Area Studies: Present and Future Perspectives on the History of Latin American Studies at Freie Universität Berlin. History of Knowledge: Research, Resources, and Perspectives. Retrieved from https://historyofknowledge.net/2023/03/25/history-of-latin-american-studies-at-fu-berlin/

Kundnani, H. (2014). *The paradox of German power*. Hurst & Company.

Kunkel, S. (2021). Science Diplomacy in the Twentieth Century: Introduction. *Journal of Contemporary History*, 56(3), 473–484. 10.1177/00220094211006762

Lange Nacht der Wissenschaften. (2024, February 24). Retrieved from https://langenachtderwissenschaften.de

Leibniz Universität Hannover. (2019, March 13). Verbundprojekt CALAS erforscht Krisen in Lateinamerika. Presseinformation. Retrieved from https://www.uni-hannover.de/de/universitaet/aktuelles/presseinformationen/detail/news/verbundprojekt-calas-erforscht-krisen-in-lateinamerika/

Life Sciences Directory. (2023). Excellence in life sciences & healthcare: The German capital region. Life Sciences Report – Contacts 2. Retrieved from https://www.healthcapital.de/files/user_upload/LS_Report_Adressen_Nov23.pdf

Marginson, S. (2022). What drives global science? The four competing narratives. *Studies in Higher Education*, 47(8), 1566–1584. 10.1080/03075079.2021.1942822

Max-Planck-Gesellschaft. (2014, March 13). New laboratory opened in Argentina (newsroom 2014, November 25). Retrieved from https://www.mpg.de/8778738/max-planck-laboratory-argentina

Neuhäusler, P., Rammer, C., Frietsch, R., Feidenheimer, A., Stenke, G., & Kladroba, A. (2022). Neue Liste FuE-intensiver Güter und Wirtschaftszweige sowie wissensintensiver Wirtschaftszweige 2021. Studien zum deutschen Innovationssystem Nr. 13-2022. Expertenkommission Forschung und Innovation (EFI). ISSN 1613-4338. Retrieved from https://www.e-fi.de/fileadmin/Assets/Studien/2022/StuDIS_13_2022.pdf

OECD. (2021). *OECD Science, Technology and Innovation Outlook 2021: Times of Crisis and Opportunity*. OECD Publishing., 10.1787/75f79015-

OECD. (2022). *Building agility for successful transitions*., 10.1787/50b32331-

Pfotenhauer, S., & Jasanoff, S. (2017). Panacea or diagnosis? Imaginaries of innovation and the 'MIT model' in three political cultures. *Social Studies of Science*, 47(6), 783–810. 10.1177/0306312717706110285715l0

Ruffini, P. B. (2020). Conceptualizing science diplomacy in the practitioner-driven literature: A critical review. *Humanities & Social Sciences Communications*, 7(1), 124. 10.1057/s41599-020-00609-5

Ruffini, P. B. (2020). Conceptualizing science diplomacy in the practitioner-driven literature: A critical review. *Humanities & Social Sciences Communications*, 7(1), 124. 10.1057/s41599-020-00609-5

Ruffini, P. B. (2020). Conceptualizing science diplomacy in the practitioner-driven literature: A critical review. *Humanities & Social Sciences Communications*, 7(1), 124. 10.1057/s41599-020-00609-5

São Paulo, D. W. I. H. (2024. March 3). Mission. Bavarian Academic Center for Latin America (BAYLAT). Retrieved March 3, 2024, from https://www.dwih-saopaulo.org/en/dwih-sao-paulo/mission/

Skipper, M. (2021). Science lies at the heart of solutions to important problems. In Hazelton, A. (Ed.), Why is scientific collaboration key? 4 experts explain. World Economic Forum. Retrieved from https://www.weforum.org/agenda/2021/06/4-views-on-why-scientific-collaboration-is-key-for-the-future/

Soler, M. G. (2020). *Diplomacia científica en América Latina y el Caribe*. United Nations Educational, Scientific and Cultural Organization.

Startup Blink. (2024, February 24). Startup Ecosystem of Berlin. Retrieved from https://www.startupblink.com/startup-ecosystem/berlin-de

The Federal Ministry of Education and Research (BMBF). (2023. March 2.) R&D policy framework. Retrieved from https://www.research-in-germany.org/en/research-landscape/why-germany/r-d-policy-framework.html

Turchetti, S., & Lalli, R. (2020). Envisioning a "science diplomacy 2.0": On data, global challenges, and multi-layered networks. *Humanities & Social Sciences Communications*, 7(1), 144. Advance online publication. 10.1057/s41599-020-00636-2

UNESCO. (2016). How science can help to create a sustainable world. News. Retrieved from https://en.unesco.org/news/how-science-can-help-create-sustainable-world

United Nations. (2022). Science, technologies can transform global challenges, but need to be accessible to all, senior officials stress, as Economic and Social Council forum concludes. Economic and Social Council Plenary Multi-Stakeholder Forum, Am & Pm Meetings ECOSOC/7083. Meetings Coverage and Press Releases. Retrieved from https://www.un.org/press/en/2022/ecosoc7083.doc.htm

United Nations Industrial Development Organization (UNIDO). (2021, October 8). Germany and UNIDO strengthen cooperation in Latin America and the Caribbean and boost trade capacity-building worldwide. Retrieved from https://www.unido.org/news/germany-and-unido-strengthen-cooperation-latin-america-and-caribbean-and-boost-trade-capacity-building-worldwide

Välimaa, J. (2019). The emergence of universities in the Middle Ages. In *A history of Finnish higher education from the Middle Ages to the 21st century* (pp. 11–38). Springer., https://link.springer.com/chapter/10.1007/978-3-030-20808-0_210.1007/978-3-030-20808-0_2

Wikipedia contributors. (2024, March 3). Deutsches Wissenschafts- und Innovationshaus New York. Wikipedia. Retrieved from https://en.wikipedia.org/wiki/Deutsches_Wissenschafts-_und_Innovationshaus_New_York

Wissenschaftsstandort Berlin. (2024, February 23). Retrieved from https://digital.zlb.de. https://www.startupblink.com/startup-ecosystem/berlin-de

World Intellectual Property Organization. (2023). Global Innovation Index 2023. Retrieved from https://www.wipo.int/edocs/pubdocs/en/wipo-pub-2000-2023/de.pdf

World Intellectual Property Organization (WIPO). (2021). World Intellectual Property Indicators report: Worldwide trademark filing soars in 2020 despite global pandemic (PR/2021/883). Retrieved from https://www.wipo.int/pressroom/en/articles/2021/article_0011.html

Zydney, J. M., & Warner, Z. (2016). Mobile apps for science learning: Review of research. *Computers & Education*, 94, 1–17. 10.1016/j.compedu.2015.11.001

Chapter 6
Geopolitics and Science and Technology Cooperation in Argentina, Chile, and Colombia:
An Analysis of Diplomacy for Science Dynamics

Julian Prieto
https://orcid.org/0000-0002-4983-9291
Penn State University, USA

Nevia Vera
University of the Center of the Province of Buenos Aires, Argentina

Cristian Guglielminotti
https://orcid.org/0000-0001-7308-2682
University of the Center of the Province of Buenos Aires, Argentina

ABSTRACT

Science can shape beliefs, behaviors, and societal structures globally. Recently, the concept of science diplomacy (SD) has gained prominence. However, SD has faced criticism for overlooking power dynamics inherent in diplomatic agendas. Countries from the Global South have begun integrating SD strategies. Yet, there is a gap in understanding the diplomacy for science dimension captured by the formal agreements these nations are pursuing. This chapter addresses this gap by examining SD initiatives from Argentina, Chile, and Colombia with major economic powers like the United States, the European Union, and China. Using a comparative case

DOI: 10.4018/979-8-3693-2746-3.ch006

study research design this chapter identifies and categorizes formal agreements from leading scientific agencies between 2018 and 2023. It also builds a taxonomy of science diplomacy instruments. It concludes that despite a historical subordination to the science diplomacy agendas of the economic powers, Argentina, Chile, and Colombia are emerging as proactive partners, transitioning from recipients to viable collaborators in training and research.

INTRODUCTION

In April 2024, Argentina and China were on the verge of a major diplomatic dispute after the U.S. ambassador to the South American country insinuated that the Chinese space station in Patagonia was for military rather than scientific purposes. Immediately, the Chinese embassy issued a statement affirming that the Far Space Station in Patagonia is a space technology cooperation facility between China and Argentina where researchers from both countries carry out their scientific research scientific research (Noticias Argentinas, 2024). This spatial station resulted from an international agreement signed in 2014 by both governments including the construction and operation of the research facilities for a common use for 50 years.

This incident illustrates the 21st-century transformation of the geopolitical order with the re-emergence of China as a great power, presenting a clear threat to the international influence of the US and the European Union. This changing geopolitical context is the result of the economic development highly related to the advancements in Science and Technology (S&T) and industrial strategies, strongly based on the emerging technologies of the sixth techno-economic paradigm (Perez, 2010). Consequently, S&T cooperation agreements have been instrumental in promoting strategic international relations while advancing economic interests (Dolan, 2012).

In this scenario, the concept of Science Diplomacy (hereafter SD) has gained relevance. SD refers to practices at the intersection between S&T and foreign policy serving to facilitate international scientific cooperation (Diplomacy for Science), inform foreign policy with scientific advice (Science in Diplomacy), and rely on scientific cooperation to improve international relations (Science for Diplomacy). The guiding principle of SD is that science provides a non-ideological environment for the participation and free exchange of ideas between people, regardless of cultural, national, or religious backgrounds (The Royal Society & AAAS, 2010). However, far from this idealized vision, SD initiatives have become a political tool to expand hard and soft power (Fägersten, 2022; Olšáková, 2023; Snidjers, 2023). Particularly, these dynamics are evident in initiatives under the Diplomacy for Science dimension with the promotion of S&T agreements. For instance, through the implementation of SD, more advanced countries can find greater access to natural and human resources, and the possibility of data collection, in addition to exerting a positive influence on public opinion and the elites of the recipient country (Flink & Schreiterer, 2010). Similarly, less developed countries can secure material and financial resources and access to scientific infrastructures that are not available in their territories and strengthen the reputation and prestige of their national scientific ecosystem (Fedoroff, 2009).

The emerging geopolitical disputes around S&T cooperation agreements are useful to illustrate several issues related to the Diplomacy for Science dimension of SD: first, S&T agreements have become an instrumental tool in the diplomatic portfolio used by emerging economies as a strategy to foster relations with economic powers. Second, these agreements serve powerful economies to advance and maintain their influence in the international arena. Finally, S&T agreements are at the center of the geopolitical dispute and the global redistribution of power that has intensified in the new millennium.

Countries from the Global South have been developing and integrating SD strategies and Latin America is not the exception (Gual Soler, 2021). However, there is a lack of understanding of the type of S&T agreements that particularly South American countries are promoting under SD strategies and the impact of international shifts in power on their agendas. This chapter analyzes to what extent the redistribution of power at the international level has been reflected in the S&T cooperation in three Latin American countries, including Argentina, Chile, and Colombia. To do so, we identify, describe, and analyze S&T cooperation agreements from these three South American Countries between 2018 and 2023 with the United States, China, and the European Union as the three main economic powers. Our mapping of S&T agreements allows us to present a taxonomy of formal Diplomacy for Science mechanisms including framework agreements for S&T cooperation, scholarship programs for academic mobility, joint research programs, binational research centers, and agreements for data exchange and access.

Following this introduction, we start by presenting a theoretical background on SD and the emerging critical views on this field, we then explain our methods and continue by introducing the scientific ecosystems in Argentina, Chile, and Colombia and the main characteristics of the SD strategies from the United States, China, and the European Union with the region. The final section includes the findings and discussion from each case. We conclude that Diplomacy for Science, and specifically S&T official agreements, are a useful tool to explore the evolution of geopolitical transformation. Although initially positioned as resource managers and beneficiaries of knowledge transfer and capacity-building initiatives, Argentina, Chile, and Colombia are increasingly taking proactive steps to transition from being mere recipients of development aid from more advanced nations. Instead, they aim to establish themselves as credible partners capable of offering alternative destinations for training and research collaborations.

CRITICAL PERSPECTIVES ON SCIENCE DIPLOMACY

The important relationship between international relations and science exhibits contrasting and complementing political phenomena. On one side, there is strong competition between nations and research centers to be at the forefront of innovation, military, and economic competitiveness, in the context of the sixth techno-economic paradigm (Pérez, 2010). On the other side, given that interinstitutional and international collaboration between research groups and individual researchers are required to achieve significant and more generalizable knowledge (Powell et al., 2017), this competition also correlates with aspects of soft power involving prestige, attraction, and influence (Adler & Haas, 1992; Ruffini, 2017). It is precisely under this second phenomenon that SD has emerged and consolidated as an important tool of Public Diplomacy (Ruffini, 2020; Turekian et al., 2014).

In recent years, much has been discussed and written about SD, especially on the dimensions of Science in Diplomacy and Science for Diplomacy. In Science in Diplomacy, historical studies have documented the long-standing scientific efforts and influence on international ocean policy (Robinson, 2020). Other studies have researched the evolution of global environmental programs led by scientists worldwide (Rispoli & Olšáková, 2020). Another example is the case of the Intergovernmental Panel on Climate Change which has been a positive international effort that provides scientific evidence for policies on climate change (Ruffini, 2018). In Science for Diplomacy, literature has documented the pivotal role that science collaborations played in the aftermath of World War II as a peace-building tool (Miller, 2006). Similarly, the important role of scientific collaboration in avoiding a latent nuclear conflict during the Cold War (Barth, 2006; Turchetti, 2020), or more recently the creation of scientific collaboration programs between the United States and Arab countries (Campbell, 2014).

Diplomacy for Science, in contrast, is a field of less research and attention (Linkov et al., 2016). The results of Diplomacy for Science programs take the form of international S&T agreements between countries with a clear intention to support diplomatic interests (Dolan, 2012). Particularly, these mechanisms may respond to top-down strategic priorities, such as the interest between governments to promote development and innovation in certain fields of knowledge. Examples of this approach are the International Thermonuclear Experimental Reactor (ITER), the Large Hadron Collider (LHC) (The Royal Society & AAAS, 2010) or the SESAME International Research Centre. In these cases, efforts from a single nation-state would not have been sufficient to fund and advance the queries of this overarching research endeavor requiring formal diplomatic means to achieve scientific goals (López de San Román & Schunz, 2018). Additionally, some mechanisms support bottom-up collaborations between individual researchers in different fields of knowledge. These programs generally include the development of institutional tools for scientific exchange, articulating requirements for cooperation such as procedures for sharing data and clarifying authorship and patent processes, and identifying existing funding mechanisms between governments that could be paired to promote collaboration (Harden-Davies, 2018). Some examples include partnerships between national science agencies that partner with counterparts in other countries to set up funding models where each agency creates a funding program to allocate resources to institutions in their own country. For example, the Newton Fund program from the UK to promote international collaboration (Department for Business, Energy, and Industrial Strategy, 2022) or the InFEWS program from the U.S. National Science Foundation (Prieto & Scott, 2022). Another mechanism is when governments grant universities to develop and set up Science and Innovation Centers abroad like in Germany and Switzerland (Epping, 2020).

However, as it was previously stated, this vision around SD has been described as naïf and idealistic by ignoring the asymmetrical power dynamics present in these diplomatic actions and strategies. Particularly, some voices claim that the AAAS dimensions fall short of identifying and explaining the interests, motivations, and goals of the actors involved in SD, and represent more of an idealistic framework proposed by diplomats and practitioners (Fägersten, 2022). In this same vein, other authors recognize the existence of a power incentive inherent to these practices and its close relationship with geopolitics and geo-economics that also permeate the implementation of SD strategies (Flink & Schreiterer, 2010; Gluckman et al., 2017). These critical voices broaden the scope of analysis from a diplomatic cooperation strategy to examining science as a social process entwined in competitive dynamics stemming from geopolitical and geo-economic rivalries fueled by national interest motivations proposing a post-naïf approach (Olšáková, 2023; Snidjers, 2023).

Moreover, critical views from the Global South brought to attention the existence of neocolonial practices in the implementation of SD initiatives. For instance, Polejack & Coelho (2021) contend that the insufficient implementation of technology transfer guidelines outlined in the United Nations Law of the Sea Convention hampers the capacity of ocean science communities in developing nations. This oversight, they argue, has been disregarded by both scientists and practitioners in the developed Global North. Similarly, other scholars point out that SD in the Global South is extremely heterogeneous but also shares certain characteristics (Gual Soler, 2021; Vera & Echeverria King, 2022; Büyüktanir & Ruffini, 2023). SD in the Global South involves practices carried out by countries that still suffer from having been colonies, experience great inequalities, are constantly involved in the power games of the great powers and suffer from very low investment rates in the S&T sector. In addition, it has also been identified how SD is not carried out only by officials at the governmental level, but also includes actions from a diversity of non-traditional actors such as universities, subnational entities, and aboriginal communities (Echeverria King, 2020; Echeverria King et.al., 2021).

In this context, efforts to strengthen Diplomacy for Science strategies in countries in the Global South such as in South America, represent an effort by national diplomacy to create opportunities for expansion and access to resources for a country's scientific community to increase national S&T capacities. The next section describes the methods used to reconstruct the Diplomacy for Science trends of the three South American countries analyzed in relation to three powerful actors.

METHODS

This chapter is structured under a qualitative comparative case study research design that allowed us to cope with technical distinctive situations where we found more variables of interest than data points, and therefore we triangulated our data from multiple sources of evidence (Yin, 2018). The selection of cases responded to the following criteria: 1) total population, 2) researchers (in full-time equivalent) per million inhabitants, 3) the percentage of GDP investment in research and development, and 4) research output published in scientific and technical journals. Table 1 shows data on these criteria for all countries in South America excluding Brazil[1]. These indicators show Argentina and Chile as leading countries in the region according to all S&T indicators. Although Uruguay shows better aggregate results in terms of researcher investment in comparison with Colombia, Colombia contributes greatly to the world's scientific production.

Table 1. Science and Technology Indicators in South America

	Total population*	Researchers (FTE) per million inhabitants	GDP % investment on R&D***	Publications in scientific and technical journals***
Argentina	46,234,000	1,283	0.52	9,730
Bolivia	12,224,000	62		152
Chile	19,603,000	528	0.33	8,980
Colombia	51,874,000	89	0.29	9,297
Ecuador	18,000,000			2,848
Paraguay	6,780,000	142	0.16	154
Peru	34,049,000		0.17	2,702
Uruguay	3,422,000	838	0.45	960
Venezuela	28,301,000	351		657

*Data source: World Bank 2022 (https://data.worldbank.org/indicator/SP.POP.TOTL)
**Data source: UNESCO, UNESCO Institute for Statistics 2016-2021 (http://data.uis.unesco.org/index.aspx?queryid=3685)
***Data source: World Bank 2020 (https://datos.bancomundial.org/indicador/IP.JRN.ARTC.SC?view=chart)

We complement the case selection criteria with trade indicators with the US, the EU, and China under the reasoning of the influence of economic relations in geopolitics. Table 2 shows that Argentina, Chile, Colombia, and Peru have strong trade relations with these economic powers. For this study, we selected Argentina, Chile, and Colombia taking into account their leading position in terms of scientific output in the region and their commercial relations with the US, the EU, and China.

Table 2. Trade Relations from South American Countries with the US, the EU, and China

	United States*		European Union**		China*	
	Imports	Exports	Imports	Exports	Imports	Exports
Argentina	$5,922,117	$4,995,203	$9,885,083,791	$6,931,072,110	$13,525,039	$6,155,976
Bolivia*	$580,201	$545,674	$582,297,025	$619,415,199	$1,975,784	$658,999
Chile	$16,027,010	$14,932,947	$10,671,717,858	$7,684,682,550	$27,515,174	$36,523,645
Colombia	$14,191,682	$11,624,290	$7,998,760,484	$7,098,965,203	$14,795,946	$3,660,969
Ecuador	$5,668,124	$6,402,858	$2,450,164,699	$4,216,371,447	$6,036,903	$4,074,885
Paraguay	$961,766	$218,034	$903,406,185	$463,537,871	$4,067,308	$30,564
Peru	$9,562,259	$7,192,891	$4,005,310,369	$6,899,520,285	$14,611,641	$17,994,761
Uruguay	$929,693	$542,580	$2,017,441,304	$1,784,314,321	$1,997,322	$2,598,261
Venezuela			$571,852,795	$1,204,717,954		

*Data source: World Bank, World integrated Trade Solutions 2021 (https://wits.worldbank.org/CountryProfile/en/Country/)
**Data source: European Commission, Acces2Markets 2023 (https://trade.ec.europa.eu/access-to-markets/en/statistics)

Our unit of analysis is determined by the main public organizations in charge of leading and regulating the scientific sector including the promotion of an SD strategy. Although we found significant variations among all three cases concerning the institutional structure, we were able to identify a main institution with these responsibilities in all countries. In the case of Argentina, we focus our analysis on the Ministry of Science, Technology and Productive Innovation (hereafter MINCYT) and the National Scientific and Technical Research Council (CONICET), in Chile we prioritized the National Agency for Research and Development (hereafter ANID) and the Ministry of Science, Technology, Knowledge and Innovation (MinCiencia), and in Colombia on the Ministry of Science, Technology, and Innovation (hereafter MinCiencias).

We collected data between 2018 and 2023 of official international agreements signed by the aforementioned institutions with partners from the US, the EU, and China. Official international agreements are important instruments and a strong manifestation of will that provide evidence of the formal international relations that law subjects pursue with their counterparts (Acharya et al., 2023). Despite that in many cases these agreements may represent just an intention of will and not achieve real actions, we consider that the bureaucratic effort to achieve them is a sufficient sign that allows us to capture the development of Diplomacy for Science strategies. Following this reasoning, our sources included official reports from MINCYT,

ANID, MinCiencias, and other institutions belonging to each country's scientific ecosystem. In addition, we included reports from the Ministries of Foreign Affairs, publications on Embassies webpages, official reports from US, EU, and Chinese institutions, and media publications.

Brief Historical Context of Scientific Ecosystems in Argentina, Chile, and Colombia

The scientific ecosystems from the three case studies have important parallels both historically and institutionally. S&T public institutions were created in the 1950s and 1960s. Argentina has a scientific ecosystem with a multi-institutional structure including government agencies and specialized research institutes. Chile and Colombia, in contrast, have a similar S&T ecosystem that was initiated with specialized agencies (CONICYT and ColCiencias) and recently adopted a central Ministry structure. Although the three countries have had their regulatory and strategic development at different times, they all acknowledge the importance of having an international cooperation strategy. Chile and Colombia have explicitly stated the need to develop an SD strategy that responds to national interests, while in Argentina there is not a direct mention of a central SD strategy, the different S&T institutes and organizations implement solid international cooperation strategies. In what follows we present a brief description of each case.

Science Ecosystem in Argentina

The first organized scientific community in Argentina was the Argentine Association for the Progress of Science (AAPC) in 1933, but the institutionalization of S&T as a coherent, systematized, programmatic public policy linked to economic planning took place towards the end of the 1940s (Hurtado, 2010). During that time, the government created several S&T institutions considered strategic, which were mostly linked to the incipient productive apparatus and the Armed Forces, including the National Atomic Energy Commission (CNEA, 1950), the Argentine Antarctic Institute (1951), the Center for Scientific and Technical Research of the Armed Forces (1954), among others. This process continued in the 1950s and 1960s with new agencies such as the National Institute of Agricultural Technology (INTA, 1956), the National Institute of Industrial Technology (1957), the National Council for Scientific and Technical Research (CONICET, 1958), and the National Commission for Space Research (1960). These specialized research institutions had government support with the aim of responding to the social and productive needs of the country (Hurtado, 2010).

After some decades of obscurantism and stagnation marked by the 1976-1983 civil-military dictatorship, the foreign debt crisis of the 1980s, and the technological *laissez-faire* of the neoliberal stage of the 1990s, in the 21st century the Argentine S&T sector regained relevance (Chudnovsky & López, 1996). One of the most important milestones of that period was the creation of MINCYT in 2007, intending to become the leading institution in charge of the articulation between the different agencies of the S&T ecosystem. Amid the economic and social crisis that the country went through in 2018, MINCYT was de-hierarchized and merged with the Ministry of Culture and Education becoming the Ministry of Education, Culture and Science and Technology (MECCYT). MINCYT was re-hierarchized in 2019 and finally turned into a sub-secretariat and defunded at the end of 2023. Despite not having a centralized SD strategy, S&T cooperation has often been subject to the agendas of each of the institutions of the national S&T sector and the various universities. However, in recent years the issue has entered the country's scientific and diplomatic discussion agenda and several efforts have been made to develop a systematized program.

Science Ecosystem in Chile

The S&T ecosystem in Chile entered the public arena in 1960 with the creation of the National Commission for Scientific and Technological Research (CONICYT). CONICYT was an autonomous and decentralized public agency with the goals of (a) strengthening the country's S&T base; (b) promoting the formation of advanced human capital; and (c) promoting an S&T culture among the population, including an international cooperation strategy (Zárate et al., 2022). In 2018, the S&T ecosystem went through significant transformations when the Ministry of Science, Technology, Knowledge and Innovation (MinCiencia) was created and CONICYT was transformed into the National Agency for Research and Development (ANID) entering into operation in 2020 (MinCiencia, 2020a). ANID has been the institution leading the design and execution of scientific, technological, and innovation policies guided by the National Advisory Committee for Science, Technology, and Innovation (CTCI). The CTCI published a strategy for the newly created S&T ecosystem and included the importance of international cooperation referring to the SD AAAS definition to promote partnerships that enhance the impact of research, enable the development of large-scale infrastructure, and favor collaboration in training and development of people's careers (CTCI, 2022). Some examples are the efforts carried out by the Directorate of Energy, Science and Technology and Innovation of the Ministry of Foreign Affairs, for the generation of strategic alliances to develop large international research projects and to strengthen national capacities. Moreover, Chile has recently included a formal SD training strategy within the diplomatic academy (Vera & López, 2023).

Science Ecosystem in Colombia

S&T in Colombia entered the national policy agenda in 1968 when the Department of Science, Technology, and Innovation (ColCiencias) was created as an agency attached to the Ministry of Education to lead, guide, and coordinate the national scientific policy (Ministerio de Ciencia, Tecnología e Innovación, n.d.-a). In 2019, ColCiencias was transformed into the Ministry of Science, Technology, and Innovation (MinCiencias). MinCiencias created a scientific advisory group to support developing its mission, vision, and strategic planning. This advisory group made recommendations on the administrative structure, the financial requirements, and the emphasis on developing the scientific efforts around three main challenges including biodiversity, productivity and sustainability, and equity. The expert panel proposed to address these challenges through five missions including recommendations on specific objectives, programs, and goals including a reference to the importance of including an SD strategy (Ministerio de Ciencia, Tecnologia e Innovacion, 2020).

In 2021, the National Planning Department published the National Policy for Science, Technology and Innovation 2021-2030 where an internationalization policy and an SD strategy were included prioritizing scientific mobility programs, articulation with international scientific networks, strengthening national capacities through technology transfer, increasing collaborations with the scientific diaspora, and developing joint S&T funding programs to support the missions recommended by the advisory group. *(Misión de Sabios)* (Departamento Nacional de Planeación, 2021). To promote the implementation of the SD strategy, MINCIENCIAS coordinates the Multisectoral Table of Science Diplomacy.

Science Diplomacy Context From the Economic Superpowers Toward South America

The history of SD of the US, the EU, and China towards Latin America in general and with the country case studies in particular, presents varied trajectories: while the region's relations with Europe date back to colonial times, and with the US more than 200 years, the establishment of diplomatic ties and cooperation with China is much more recent.

The US has historically recognized the relevance of S&T and SD for its foreign policy (De Greiff, 2014). Particularly, under the Cold War the US included within its international cooperation agenda the advancement of knowledge as a tool to reinforce its image as a benevolent scientific power, but at the same time gaining strategic geopolitical positioning over the scientific and technical capabilities of its partners (Whitesides, 2012). After the reunification of Europe, the US has used SD to strengthen relations with several countries even as a protection strategy in a global context marked by the terrorist threat and as a tool to build bridges with countries such as Cuba, North Korea, and regions such as the Middle East (Vera & Colombo, 2020; Ruffini, 2017). Since 2000, this strategy has been supported by an S&T adviser to the Secretary of State, which added to the existing Science Advisor to the President, a position created in the 1970s. Despite the halt that SD suffered after 9/11, it took impulse after Obama's 2009 speech (Campbell, 2014). In a recent speech in May 2024, Secretary of State, Anthony Blinken expressed the importance of Science and Technology as an instrument for America's Foreign Policy, stating that "Today's revolutions in technology are at the heart of our competition with geopolitical rivals. They pose a real test to our security. And they also represent an engine of historic possibility – for our economies, for our democracies, for our people, for our planet" (U.S. Department of State, 2024).

The US implements its strategy of cooperation in S&T fields through various areas and agencies, stemming from the White House Office for S&T Policy and State Department leadership, and with a prominent -but not exclusive- role by the Bureau of Oceans and International Environmental and Scientific Affairs, which also supervises US S&T links abroad (Ruffini, 2017). Implementation of SD initiatives is carried out through i) the signing of S&T cooperation agreements; ii) the establishment of joint funds, and iii) scholarship programs oriented to specific topics and sponsored by various agencies and organizations, as well as by foundations.

Diplomatic relations with the EU could be explained through two main phases, the first one from 1960 to 1990 initiated through the regional diplomatic actions between the EU and the integration efforts in the regions such as the Andean Nations Community and MERCOSUR, and the second one as standalone bilateral relation between the EU and individual countries (Tremolada & Quiroga Chacón, 2022). The first phase was characterized by a diplomatic agenda that was affected by the difficulties in the consolidation process of the integration efforts in Latin America explained by excessive presidentialism, populist patterns, and economic challenges (Domínguez, 2013). Despite these challenges, the EU managed to establish regional agreements with all integration initiatives in the region. This diplomatic agenda was greatly influenced by the EU-CELAC conferences that started in 1999 in Brazil and consolidated in 2012 with the agreement with the Central American countries (Roy, 2013).

The EU-CELAC launched the Joint Initiative for Research and Innovation (JIRI) in Latin America in 2010 prioritizing four main pillars including 1) mobility of researchers, 2) cooperation among the research infrastructures, 3) global challenges, and 4) innovation (Belli & Morín Nenoff, 2022). The diplomatic agenda on S&T has been aligned with the EU Research and Innovation Framework Programs, such as Horizon Europe and its seven predecessors (hereafter Framework Programs). These frameworks have given direct funding access to most Latin American countries as they belong to the low-medium income country list and therefore can join research initiatives through Official Development Aid (ODA) contributions (Horizon Europe, 2023). Within the Horizon Europe Program (the most recent framework) Latin America is included with an expected investment of 45 billion euros (European Commission, 2023). The main strategy to promote the active participation of Latin American countries in these Framework Programs has been through the establishment of a liaison office in each country and a solid network of National Points of Contact (NCPs) who are in charge of carrying out information sessions and capacity building workshops (Belli & Morín Nenoff, 2022).

Finally, concerning China, its first S&T agreements date from the Cold War, specifically from the 1950s to 1970s, when the Asian country sought to deepen its relationship with the communist bloc. In the following decades, China's accelerated S&T development was supported by the expansion of its scientific cooperative ties with several countries (Vera & Colombo, 2020). International partnerships are led primarily by the Ministry of Science and Technology (MOST), but other agencies also contribute to these efforts, such as the Chinese Academy of Sciences, the National Natural Science Foundation of China, and the Chinese Association for Science and Technology (Wagner & Simon, 2023). Areas such as agriculture, health, environment, natural resources, oceans, meteorology, and Earth sciences, as well as space and nuclear technology, represent the thematic priority of these international partnerships (Vera et al., 2023).

In the new millennium, China launched the Belt and Road Initiative (BRI) initiative presenting itself to the world as a great power that seeks the development of all states, establishing horizontal links and relationships that result in mutual benefits. The BRI includes a SD component that is captured by the active actions of the MOST including the signature of 114 intergovernmental S&T agreements, and the development of cooperative ties with 161 countries worldwide (Wagner & Simon, 2023). As of 2023, 22 countries in Latin America have officially joined the BRI, including Chile in 2018 and Argentina in 2022. The diplomatic strategy with Latin America has been facilitated by the creation of the China-CELAC Forum, including a working group on Scientific-Technological Innovation. Such convergence spaces have highlighted the need for cooperation on issues linked to energy, materials, biotechnology, agriculture, the space industry, biomedical research, 5G technology, and artificial intelligence (Vera et al., 2023).

DIPLOMACY FOR SCIENCE IN ARGENTINA, CHILE AND COLOMBIA

Argentina

Argentina-US Scientific Cooperation

S&T cooperation has always been included in the diplomatic agenda of Argentina and the US, even during periods of tense bilateral relations (Vera & López, 2023). As stated on the website of the US Embassy in Argentina, "the bilateral relationship is based on shared interests that include [S&T], education, trade, regional peace and stability, non-proliferation, cultural exchange, human rights, and social inclusion" (US Embassy in Argentina, n.d). More specifically in the S&T sector, priority agendas between the two countries include space and nuclear technology, biotechnology and the environment, and a joint technology working group established in 2010.

We found almost no evidence of a broad cooperation agenda between the former MINCYT and the US until 2023 with regard to formal agreements. A single meeting was recorded in 2020 where the Argentine Minister of S&T and the US Ambassador renewed their commitment to S&T cooperation. In 2022, as a mechanism to support the Pampa Azul Project, we found information on the signature of the Declaration of the Atlantic Alliance for Ocean Research and Innovation at a ministerial meeting at the All-Atlantic Ocean Research Forum in Washington DC, to boost Atlantic cooperation in the fields of research, climate change, and natural resources.

However, in 2023 there was a revitalization of S&T ties between the two countries. On the one hand, MINCYT participated through the Undersecretary of Science, Technology, and Innovation Policies, in a seminar co-organized by the Argentine Foreign Ministry and the US Department of State and Department of Commerce called "Advancing Bilateral Dialogue on Digital Policy: Data Protection, privacy and artificial intelligence (AI)" (Gobierno de Argentina, 2023a). In addition, the administrator of the National Aeronautics and Space Administration (NASA) visited Argentina and met with the country's President, the Minister of S&T, and the Chancellor to sign agreements that would allow "deepening the joint peaceful space activity between both countries and formalizing Argentina's adhesion to the Artemis Agreements". The Minister of S&T especially signed an agreement "for cooperation in the civil exploration and peaceful use of the Moon, Mars and other astronomical objects" between NASA and the Argentine Commission for Space Activities (CONAE).

Regarding mobility programs, the Argentine Fulbright Commission has played a pivotal role[2]. Historically they have been able to partner with various public institutions including MINCYT, CONICET, and INTA, among others. Between 2018 and

2023 we found that the Argentine Fulbright Commission signed two agreements with CONICET. The first in 2018, to support the CONICET international postdoctoral grants program, awarding 16 Argentine researchers for a three-month stay in US universities. The 2022 call offered grants for assistant or adjunct researchers, with funding of US$3,000 for each selected beneficiary, of which the Fulbright Commission covered health insurance, 50% of the stipend and airfare, and CONICET covered the other half of the remaining expenses.

On a final note, despite the chapter's focus on the links between MINCYT and similar agencies with the US, much of the cooperation is framed within the agendas of other specialized agencies such as the National Atomic Energy Commission (CNEA) or CONAE, which predate the creation of MINCYT, and which were not considered in this chapter.

Argentina-EU Scientific Cooperation

By the end of the 20th century Argentina became the first Latin American country to sign an S&T cooperation treaty with the EU. This was reinforced in 2005 with the creation of the Argentina-EU Liaison Office. Therefore, the EU currently deploys a wide variety of programs and instruments that provide a framework for Argentina's SD. According to the information found in the former MINCYT, the cooperation with the EU is carried out through bilateral agreements with EU member countries, or via programs promoted by the EU Liaison Office. In this respect, "Argentina's participation in the [Framework Program] has always been extremely successful, with a high effectiveness rate that positions the country as a leader in the region." The country took part in the Horizon 2020 Program where "it achieved an outstanding participation leading the region, with 23.92% effectiveness and is among the top 10 non-European countries in participation [in this program]" (Oficina de Enlace Argentina-Unión Europea, n.d). Currently, the country is focused on opening possibilities for the participation of its scientists in the Horizon Europe Program.

In 2018 the former MINCYT and the EU -through the European Commissioner for Research, Science, and Technology- signed the "Administrative Agreement on Cooperation for Innovation and Marine Research" to "promote human capital formation, scientific exchange and the formation of working teams in strategic areas", and to focus research on issues related to climate change, oceans, biodiversity, polar research, and food security (European Commission, n.d). In turn, within this framework, the Minister of S&T participated in the tenth Meeting of the Joint Consultative Committee between Argentina and the EU, where bio-economy, marine sciences, and space were identified as priority topics for the cooperation agenda between both actors.

At the same time, technology transfer activities were planned between the Joint Research Center (JRC) of the European Commission and CONICET, mobility mechanisms for scientists were discussed through the Marie Skłodowska-Curie Actions and the European Research Council, and reciprocity policies were designed around the installed research infrastructures (European Commission, n.d). It should be noted that although the JRC cooperates with Argentina in several sectors such as agriculture, climate, water, and communication technologies, the area where collaboration is most advanced is the nuclear area (European Commission, n.d).

For the celebration of the twentieth anniversary of Argentine-EU S&T cooperation in 2019, multiple joint activities were carried out to promote European S&T in Argentina, as well as the opportunities it offers (Gobierno de Argentina, 2019). As the Ministry of Foreign Affairs stated, "Bilateral cooperation with the [EU] is one of our country's most solid, effective and strategic alliances in [S&T] and innovation. As a result of this relationship, Argentine scientists participate with a high success rate in international calls for proposals involving collaborative work and receive funding from the European Commission" (Embajada Argentina ante la Unión Europea, 2019).

Among the activities carried out was the Visibility Workshop "Technology Transfer in Argentina" whose goal was to "bring together the main actors of the Argentine technology transfer ecosystem to exchange good practices and discuss strategies to better internationalize such transfer and linkage". In this context, it was sought that the European representatives could transmit their experiences regarding the rapprochement between the scientific-technological and productive sectors, one of the greatest obstacles in Latin America (Álvarez, Natera & Castillo, 2019). It is interesting to note that it was also within the framework of this multi-day event that the Argentine S&T Secretary presented the workshop "Science Diplomacy: towards the construction of a national concept", the first event of its kind in the country, which brought together future diplomats and national S&T leaders.

Finally, it should be noted that the Argentine SD agenda with the EU is complemented by multiple bilateral agreements carried out, especially but not exclusively with Spain, France, Italy, Portugal, and Germany. These were not included because they are not the subject of this chapter, but they certainly illustrate the dynamics of Argentine Diplomacy for Science, and the fact that Europe is still Argentina's most important S&T partner, even without taking into account bilateral agreements.

Argentina-China Scientific Cooperation

Diplomatic relations between Argentina and China date back to 1972, including the first S&T cooperation agreement in 1980 (Vera, 2018). However, a robust implementation agenda was only found in the 21st century when Argentina developed a comprehensive Diplomacy for Science agenda towards China, with diverse formal agreements between various S&T institutions from both countries.

In 2018, both countries signed an Agreement on Educational, Cultural, and Scientific Cooperation, which was fundamental for the subsequent creation of the Argentine-Chinese Center in Social Sciences - CACCS (which joined the already existing Argentine-Chinese Center in Food Science and Technology created in 2008). This binational center aims to "promote scientific-technological exchange and cooperation in the field of social sciences, facilitating researchers from the public and private sectors to carry out joint research programs on issues of mutual interest, through a digital platform" (Gobierno de Argentina, 2018). CACCS launched a call for binational projects and joint research on topics identified as relevant by both parties in 2022, related to i) poverty alleviation and development; ii) participation of emerging economies in global governance, and iii) social and cultural transformations in urban and rural areas. The call originally proposed the funding of two joint projects, which would be funded by MINCYT (contributing with US$10,000) and the Chinese Academy of Social Sciences (with a contribution of 150,000 yuan). In both cases, the funds would be used to cover the costs of visits between the two countries, activities of the research teams of each country, and organization of face-to-face and virtual exchanges, among others.

In 2021, MINCYT signed an MoU with MOST for the creation of the Chinese-Argentina Center for the Study of Innovation Policies, which finally launched in 2022. In addition, both institutions agreed to share information on the COVID-19 pandemic, on issues related to biotechnology and the development of S&T parks. Joint projects already underway were evaluated, from space cooperation (taking into account the contribution made by the Chinese space station in Neuquén mentioned in the introduction to the landing of one of the Asian probes on the moon), to the exchange of scientists and the CART telescope project. Furthermore, MINCYT signed an MoU with the Chinese Ministry of Natural Resources to create a research center for natural resources and energy transition. This initiative included joint funding mechanisms for binational projects on issues related to the subject, exchange of scientists, joint work on oceanic and Antarctic matters, etc. (Gobierno de Argentina, 2023b).

CONICET has also been a strategic institution to support the SD agenda with China. In 2018, CONICET and the National Natural Science Foundation of China held a joint workshop called "Desertification, Desert Ecology and Restoration" as a follow-up activity of the agreement signed by both institutions in 2015, where they exchanged knowledge to identify lines of research of common interest in topics related to desert ecology, ecosystem restoration and water and soil conservation (CONICET, 2018). In 2019, CONICET and the Chinese University of Southwest Jiatong signed a framework agreement to deepen academic exchange actions and S&T

cooperation between both Institutions, in subjects related to railway transportation. This agreement was a result of the "Latin American Chinese Laboratory for Railway Transport" Project, financed by MOST (CONICET, 2019). In addition, that same year CONICET and the Chinese Academy of Sciences agreed on the creation of a Joint S&T Center oriented to priority areas such as astronomy, biomedicine, earth and atmospheric sciences, and polar studies, to develop joint activities involving researchers from both countries (research projects, workshops, events, binational laboratories) (CONICET, 2019). This center would finally materialize in 2023 (CONICET, 2023).

2022 and 2023 represented a dramatic increase in joint actions between the two countries. In 2022, the China-Argentina Binational Center for the Study of Policies for Innovation and Technology between MINCYT, CONICET, and MOST was presented, with the objective of "addressing technological changes and strategies for innovation and the promotion of economic, industrial, technological, social and environmental development of both countries". In addition, a "'Smart Platform' was created to provide support for academies, scientific institutions, and companies of both countries, as well as to encourage and study the recent [S&T] and innovation policies of China and Argentina and their impact on economic and social development" (CONICET, 2022).

Finally, in these last two years, in addition to the signing of several MoUs, various seminars, joint meetings, and joint activities were held between a variety of institutions of both countries, such as the First Meeting of the Joint International Research Center on Globalization and Society between the Center for Labor Studies and Research (CEIL, CONICET) and Shanghai University (SHU); the IV Meeting of the S&T Subcommission of the Permanent Commission between China and Argentina with the participation of MINCYT and MOST; the First Workshop on Earth and Atmospheric Sciences between the Joint S&T Center between CONICET and the Chinese Academy of Sciences; the II Seminar of the China-Argentina Binational Center for the study of Innovation Policies between CONICET, MINCYT, MOST and the Chinese Academy of S&T for Development (CASTED), and a binational Workshop on Oceanic, Antarctic, and Conservation Issues held by the Pampa Azul Project (in which several Argentine ministries participate, including the Ministry of Foreign Affairs and MINCYT).

Chile

Chile-US Scientific Cooperation

S&T cooperation has been a main component of the diplomatic agenda between the US and Chile. The US embassy emphasizes that in the bilateral relationship, it works "together with Chile in the development of [S&T] to address global challenges, to protect the environment and for our mutual security" (Embajada de Estados Unidos en Chile, 2024). Likewise, it is presented as one of the longest and most productive articulations within the Western Hemisphere, thanks to the strengthening through the Basic Framework Agreement on [S&T] signed in 1992, consolidated in the present century with the creation of the Chile-US Council for Science, Technology, and Innovation (STIC) in 2018. This body is presented as the first public-private forum to achieve the binational strengthening of the S&T, and innovation ecosystem.

S&T cooperation with the US dates back to 1955 when the Fulbright Commission in Chile was created as the first commission in Latin America (Fulbright Chile, 2024). In addition, the National Commission for Scientific and Technological Research (CONICYT) reports that between 1974 and 2017, the US was its main international partner including 31 cooperation agreements (CONICYT, 2024). This includes agreements for mobility programs between the Fulbright Commission and CONICYT. Since 2015, their partnership has operated through the framework of the Chile Scholarship System and the Advanced Human Capital Formation Program. Thus during the period under study, joint calls for proposals were made in 2018 and 2019 for the Equal Opportunity Doctoral Scholarship (BIO). Under the ANID operation, this program has had continuity. A total of 166 students have been beneficiaries of this program (ANID, 2023a).

In 2018 Chile created the Ministry of Science, Technology, Knowledge and Innovation (MinCiencia), which throughout the first year presented the Data Observatory, a non-profit organization to generate a new pole of economic development based on data science, in collaboration with the public, private, and academic sectors of the country, and with the participation of Amazon Web Services (MinCiencia, 2019a). Also that year, MinCiencia was part of the first experience of the "Science Academy for Teachers and Managers" through which 25 teachers were trained by scientists from Chile and the US (MinCiencia, 2019b).

In 2021, MinCiencia authorities and NASA's Stratospheric Observatory for Infrared Astronomy (SOFIA) mission held the first meeting to operate a telescope aboard an airplane, in flights that took place the following year (MinCiencia, 2021a). Finally, in 2023 an agreement was reached to bring the training program of the multinational company International Business Machines (IBM) called "IBM

SkillsBuild" free of charge to women over 18 years of age to develop their skills in emerging technologies (MinCiencia, 2023a).

Chile-EU Scientific Cooperation

Chile and the EU signed an Association Agreement in 2002 where science was included as one of the main priorities (Hernández and Lebret, 2019). Consequently, CONICYT and the EU signed an S&T Cooperation agreement in 2003 as the main framework to advance an SD agenda. At the 9th EU-Chile Joint Steering Committee Meeting, research areas related to health, energy (especially aspects related to green hydrogen), bioeconomy and sustainable cities, open science, and oceanic and Antarctic research were identified as priorities. Recently, in December 2023 Chile became the first country in the region to conclude a new agreement with the EU, called the Advanced Framework Agreement, which contains a novel vision of potential areas for cooperation in S&T and innovation (European Union, 2023). Despite being a more political and commercial agreement that seeks to "boost the EU's long-standing partnership with an important likeminded partner, strengthen both countries' economic resilience and support them in tackling current global challenges" with high geopolitical content, it promises S&T cooperation especially in the digital transition sector.

The buoyant relationship with the EU is captured by what ANID reports on its official webpage through joint calls for funding with the EU and its country member states. Particularly, with the EU the European Research Area (ERA) program stands out. It has opened funding programs for joint research in 2019 and co-funding in mineral raw materials (ERA-MIN) of Joint Research, with a triannual budget of 75 million Chilean pesos for each award. Moreover, in 2021 additional calls were launched including one with the ERA-NET co-funded by the ERA-PerMed, with a total of 75 million Chilean pesos per project for three years of implementation; and the first exclusively national call for applications for the Promotion of International Linkages for Regional Research Institutions (ANID, 2024). In 2021, an S&T cooperation agreement was signed between ANID and the European Southern Observatory (ESO) to build and operate the Extremely Large Telescope (ELT) (MinCiencia, 2021b). This also includes the process that the Chilean government has initiated to become an associate state of the European Center for Nuclear Research (CERN) (MinCiencia, 2023b).

The SD agenda with the EU is complemented by bilateral agreements between ANID and institutions from France, Germany, Spain, Italy, among other countries (CONICYT, 2024). Even though bilateral cooperation agreements are not the object of this chapter, all these initiatives explicitly state the aim to strengthen the articulation with specialists and scientific institutions of EU member states in their respective calls. Considering the EU agreements and all the bilateral agreements with individual EU member countries, they reach a higher number of agreements showing more seniority than that evidenced by the US.

Chile-China Scientific Cooperation

Chilean Diplomacy for Science with China has been deployed especially since the 21st century. It should be noted that in parallel to this process, the Asian country has become Chile's main trading partner since 2014 (Ministerio de Relaciones Exteriores de Chile, 2015), a situation that has been sustained over time (Serrano Moreno, 2022), especially as a destination for national exports (Servicio Nacional de Aduanas de Chile, 2023).

As for CONICYT, the organization reported that a total of twelve agreements were signed between 2001 and 2018 with various Chinese agencies (CONICYT, 2024), which accounts for a remarkable difference concerning what was evidenced with the US and the countries that make up the EU, whose first formal links can be traced back to the first half of the 1970s. However, this significant delay has not prevented the rapid development of activities within the framework of Diplomacy for Science. A series of calls for proposals have been promoted during the period under study based on some of these agreements with Chinese academic, scientific, and technological institutions.

One of the most relevant results was the cooperation between CONICYT and the National Natural Science Foundation of China (NNSFC) whose first MoU was signed in 2013. In 2018, a call for Joint Research Projects in Natural Disaster Management in Chile and China was made, through which Chilean and Chinese researchers were funded through these two organizations, with four proposals being selected with counterpart funding of 140 million Chilean pesos (CONICYT, 2018). The following year, a new call for CONICYT - NNSFC Joint Research Projects was carried out and eight beneficiaries were selected with funding of 200 million Chilean pesos per proposal (CONICYT, 2019).

This relationship in turn has been strengthened since the creation of ANID in 2020. One of its highlights is the MoU for Scientific Collaboration and the Development of Astronomical Research, signed with the Chinese Academy of Sciences (CAS) in December 2021. Its first precedent can be traced to the MoU on Cooperation in S&T between CAS and CONICYT in 2012. Within this framework, a collaboration program has been established through the constitution of the CAS-ANID Fund, which led to the call in 2022 for postdoctoral applications for joint activities between researchers and scientific institutions, aimed at promoting collaborative scientific research in the field of astronomy (ANID, 2022, 2023b). This initiative has also counted on the collaboration of the South American Center for Astronomy of the

Chinese Academy of Sciences (CASSACA). As a result of this first joint call, five proposals were selected and awarded around 56,200 and 59,200 million pesos by the Chilean Government (ANID, 2023b).

In the mapping of MinCiencia we found some specific actions carried out with China's health sector, especially in 2020 and 2021, the years when the COVID-19 pandemic experienced its peak. In that period diplomatic actions were carried out to favor and guarantee access to vaccines and other health supplies. Consequently, in 2020 only one agreement concerning the collaboration between the Pontificia Universidad Católica de Chile (PUC), through the Millennium Institute of Immunology and Immunotherapy and the biopharmaceutical company Sinovac Biotech Limited of China (Sinovac) is mentioned. This agreement made it possible for Chile to participate in the analysis of the effectiveness of the vaccine against COVID-19 and to improve access to this vaccine. The agreement was framed within the national strategy that at the time sought the timely and equitable supply of a vaccine against the virus (MinCiencia, 2020b). This articulation has continued with a meeting held in 2023 between the MinCiencia, Chilean universities, and authorities of the Sinovac company to deepen the transfer of this Chinese company in the country, mainly in the development of research and innovation in vaccines, considered a strategic issue (MinCiencia, 2023c). Previous collaborations between this university and Chinese medical companies were important in facilitating this agreement, in addition to highlighting the role of Cultural Diplomacy of the three existing Confucius Institutes: at PUC, Universidad Santo Tomás and Universidad de la Frontera.

Colombia

Colombia-U.S. Scientific Cooperation

Colombia has 200 years of diplomatic relations with the US showing diverse and dynamic relations according to the evolving political agendas in both countries (Dangond Gibsone, 2022). A constant priority in the diplomatic agenda has been economic development, strengthening of democratic institutions, military cooperation, and security related to the war on drugs. Since 2010 both countries have coordinated their diplomatic agenda through the High-Level Dialogue (HLD) organized around seven thematic groups including one in Public Diplomacy where education and science cooperation are discussed, and which has counted with the participation of delegations from MinCiencias from Colombia and their counterparts in the US (Cancilleria Colombiana, 2023a).

In 1957 the Fulbright Commission in Colombia was created through a US Congress mandate as a formal organization to promote academic mobility as an effective mechanism for cooperation and peace among countries (Fulbright Commission

Colombia, n.d.). In 2022, by the 65 anniversary, the Fulbright Commission reported an alumni network of more than 5.000 grantees from both countries. These results show the commitment of both governments through their financial commitment to all the different scholarship programs. The academic exchange at the doctoral level has been majorly a commitment of MinCiencias. Through annual agreements between 2018 and 2023, financial commitment has supported around 245 Colombian doctoral students in the US with an average of 40 grantees per year.

Another program that has entered the stage on the diplomatic agenda is the 100.000 Strong in the Americas Innovation Fund (State Department, n.d.). Colombia leads in public contributions to the innovation fund in Latin America having supported 88 institutional grants engaging more than 140 institutions in 33 US states and 20 Colombian Departments since 2015 (Partners of the Americas, n.d.). The agreements with MinCiencias have committed resources from the program Nexo Global that aims to promote and strengthen scientific and research skills among undergraduate students (Ministerio de Ciencia, Tecnologia e Innovacion, n.d.-b). There have been two rounds in 2019 and 2022 under the Nexo Global program.

Complementing the mobility opportunities for Colombian researchers, in 2022 MinCiencias also signed an MoU with the private company VaxThera (Ministerio de Ciencia, Tecnologia e Innovacion, n.d.-c), to create research mobility programs in the field of biotechnology with a special emphasis in vaccine development and other sanitary technologies. Nevertheless, this MoU has not been supported with financial resources to date to support exchange programs.

Another important agreement has been with NASA and its ARTEMIS program. Colombia has been the third country in Latin America to join this program through the adoption of principles, guidelines, and best practices to strengthen the governance of exploration and use of outer space (Ministerio de Ciencia, Tecnologia e Innovacion, n.d.-d). By joining the ARTEMIS program, MinCiencias aims to promote national research on space exploration and open possibilities for researchers to access information and create and use technologies in partnership with other members of the program.

MinCiencias has also signed bilateral MoUs with US universities. This is the case of the Massachusetts Institute of Technology (MIT) with an agreement with its Environmental Solutions Initiative, with the Health Institute from Wisconsin-Madison University, and the Department of Psychiatry from the General Hospital of Massachusetts. No further information was found about implementation agreements from these MOUs.

Finally, in 2020 in the geosciences research royalties funding program for the hydrocarbon sector, MinCiencias included a selection criterion to partner with international research organizations in the field. Particularly, the call included explicitly the American Association of Petroleum Geologists (AAPG), the Society of Petroleum Evaluation Engineers (SPEE), and the Society of Exploration Geophysicists (SEG). Even though there was not a formal agreement with these organizations, MinCiencias explicitly created incentives for international collaborations with these US-based organizations.

Colombia-EU Scientific Cooperation

Bilateral diplomatic relations with the EU show an important development since 1990 with the opening of the European Commission Embassy in Bogotá in 1993 and through the Roma declaration in 1996, a bilateral MoU in 2009, and the Free Trade Agreement that came into force in 2013 (Union Europea, n.d.). The priorities of the diplomatic agenda have been around economic development and poverty reduction, strengthening of the democratic institutions, peacebuilding, human rights, and sustainable development (Lillo González & Santamaría García, 2009; Tremolada & Quiroga Chacón, 2022).

In the field of scientific cooperation, Colombia has been taking part in the EU-CELAC Joint Initiative for Research and Innovation (JIRI). However, there has not been a formal S&T cooperation agreement to expand the scientific cooperation agenda (European Commission, 2023). Scientific cooperation in contrast has been promoted through diplomatic efforts that enhance collaborations in higher education through institutional grants and mobility programs (Tassara, 2019).

Since 2007, ColCiencias (current MinCiencias) has been officially recognized by the Research General Directorate of the European Commission as the liaison office in Colombia of the nine National Contact Points (NCPs). In addition, these efforts have been complemented by the role of the EURAXESS Regional Coordinator who has been based in Bogota since 2020. The participation of Colombian institutions in the EU Framework programs captures 83 grants within the Horizon 2020 program and 35 grants in the Horizon Europe program including the mobility of 179 Colombian researchers to the EU and 541 researchers from the EU to Colombia. Particularly, MinCiencias is an official partner in the Horizon 2020 RESINFRA EU-LAC project for which they have signed an agreement for its implementation in 2021. MinCiencias has also promoted an agreement with the European Commission in 2021 for the implementation of the DeSIRA project which aims to carry out capacity-building activities for black, indigenous, and marginalized communities through research in agroecology and bioeconomy.

Two other diplomatic efforts have been activated by MinCiencias that look to gain access to strategic information. The first one is the agreement with the Copernicus initiative in 2018 that allows member countries to access data on terrestrial, oceanic, and atmospheric monitoring, as well as in the prediction, management, and mitigation of natural disasters (European Union, n.d.). In addition, MinCiencias has also signed an agreement with the Director General of the Defense Industry and Space – DEFIS to promote the development of actions aimed at supporting innovations in Global Navigation Satellite Systems for applications in aeronautical and maritime mobility, telecommunications, location and mapping services, precision agriculture and telemedicine (Cancilleria Colombiana, 2021).

The SD strategy from MinCiencias with Europe has been complemented by two other agreements that are not directly with the EU. The first one is with CERN which allows Colombian researchers to partake in the ATLAS, ALICE, CMS, and LHCb from the Large Hadron Collider program. The second one, is with ACIS (Colombian Scientific Diaspora Association in Switzerland) signed in 2023, to promote collaborations and mentorship with other Colombian diaspora groups in Europe and to promote brain circulation activities. Furthermore, similarly to the US, in 2020 in the geosciences funding program for the hydrocarbon sector, MinCiencias included the European Association of Geoscientists & Engineers (EAGE) as a suggested partner.

Information found on bilateral agreements between MinCiencias and individual EU member and states and countries such as Germany (DAAD and BMBF), France (EcosNORD, CNRS), Turkey (TUBITAK), United Kingdom (Newton Fund and Francis Crick Institute) are not expanded in this section as they do not respond directly to diplomatic actions at the EU level. Most of these agreements include the commitment of financial resources from MinCiencias allowing them to have more power in setting the scope of the programs and the research priorities.

Colombia-China Scientific Cooperation

Formal diplomatic relations between Colombia and China were established in 1980 once Colombia recognized the "One China" Policy through which the Colombian government supported the People's Republic of China (PRC) as the sole legal government (Borda & Paz Berger, 2012). The first diplomatic agreement was based on economic cooperation in 1985 and since then the diplomatic agenda has been primarily centered in the economic field. An intensified diplomatic strategy was developed between 2010 and 2016 reaching more than 27 formal agreements in the fields of economic cooperation, technology, culture, and agriculture (Pastrana Buelvas et al., 2016). Renovated commitments were achieved in 2023 when the Colombian president made an official visit to China subscribing to twelve new diplomatic mechanisms (Cancilleria Colombiana, 2023b).

Regarding S&T cooperation, we found the first MoU signed in 2010 between ColCiencias and MOST intending to promote mobility programs for doctoral studies and researchers (Colciencias, 2010). The prioritized fields of knowledge were agriculture, mining, energy, biodiversity, renewable energy, infrastructure, and information technology; however, no evidence was found on any further implementation agreement or reports on mobility results. In 2019, a new MoU was signed with the Chinese National Academy of Science with similar objectives to promote scientific exchange and the creation of joint research programs. No additional information was found about its implementation.

A renewed MoU for scientific cooperation was signed in 2023 between MinCiencias and MOST. Specifically, it was agreed that the implementation of this agreement would be carried out through the Chinese Young Scientist Exchange Program (YSEP) for research stays between 6 and 12 months. In this new agreement, China is responsible for providing funding for this program and Colombia is in charge of selecting the candidates.

Scientific cooperation has been supported mobility programs for graduate students including the agreement between the China Scholarship Council and the reciprocity program from ICETEX. Since 2018, this alliance has committed resources for supporting over 30 Colombian students and 6 Chinese students per year (Ministerio de Educación Nacional, 2018). In addition, University bilateral agreements and city sisterhood diplomatic agreements complement the portfolio through which higher education institutions and researchers from both countries find ground for collaborations (Embajada de Colombia en China, 2023).

DISCUSSION

Through mapping the official international agreements between 2018 and 2023 we found evidence that within the bilateral diplomacy agendas with the US, scientific cooperation has played a leading role in the three case study countries. Particularly, the strategy of SD is largely promoted through the Fullbright program with a strong emphasis on academic mobility and research programs. The funding mechanisms to operate the Fulbright involve economic contributions from both the US and its partner country. In recent years, this strategy has been complemented by more ambitious programs that seek to promote joint research initiatives. A cross-cutting issue is cooperation on space sectors, climate change, and natural resources. With Argentina in particular, there is interest in advancing oceanic and Antarctic cooperation agreements, while with Chile priority is given to gender issues, and with Colombia to biotechnology and the development of vaccines.

Concerning the EU, the inclusion of most of the South American countries in the list of low-income countries has allowed their direct participation supported by Official Aid Contributions. Both Chile and Argentina have managed to develop a more solid SD strategy by signing bilateral S&T cooperation agreements with the EU, which has allowed them to make joint and exclusive calls for proposals for their countries. Colombia has managed to have wide participation in research consortia, but not through exclusive calls. The regional articulation offices and the National Contact Points have played a key role in expanding the SD agenda and in linking researchers to projects financed through the EU. It should be noted that cooperation with the EU must also incorporate bilateral cooperation agreements with member countries, as these represent a large number of initiatives that complement the SD agenda of Argentina, Chile, and Colombia. As with the US, a cross-cutting topic in the three cases studied is cooperation in planetary and ocean science, especially in terms of access to and exchange of information. Likewise, technology transfer and capacity-building are present in all three cases.

Regarding China, as it moves forward with its trade agenda and the implementation of the BRI, there is evidence of accelerated progress in S&T cooperation. With Argentina, great advances have been made with in-depth diplomatic mechanisms including the creation of joint research centers, collaborative research programs, and mobility programs. Chile has also developed a rich agenda with China with important technology transfer projects, while Colombia has incipient partnerships with China. Despite this inactivity on the part of MinCiencias in Colombia, we found that the Chinese government has advanced independently with bilateral agreements with universities through which they promote joint research programs on topics of interest to China. For Argentina and Chile, priority topics for agreements include planetary sciences, atmospheric sciences, and astronomy, as well as a focus on Antarctica, natural disasters, and cooperation in vaccine development and health issues.

The case studies presented here are instrumental in understanding how Diplomacy for Science initiatives is promoted through official international agreements. These agreements have been used to increase the opportunities for local scientists to access scientific infrastructure, funding resources, and international scientific networks. Moreover, the mapping of Diplomacy for Science initiatives in Argentina, Colombia, and Chile allows us to identify clear mechanisms of official cooperation with which we can propose a taxonomy of SD instruments. These include framework agreements for S&T cooperation, scholarship programs for academic mobility, joint research programs (of two types, one where both parties contribute resources, and another where only one party contributes resources), the creation of binational research centers, and finally collaboration agreements for data exchange and access. In Table 3 we record the evidence of the SD agendas for the case study countries.

Table 3. Science Diplomacy Mechanisms between Selected South American Countries and the United States, the European Union, and China.

		S&T Cooperation Agreement	Mobility Scholarships	Joint Research Programs	Bilateral research Centers
Argentina	United States		X		
	European Union	X	X	X	
	China	X	X		X
Chile	United States		X		X
	European Union	X	X	X	
	China	X	X		
Colombia	United States		X		
	European Union		X		
	China		X		

Finally, it is necessary to mention that in recent years the countries analyzed have used SD initiatives to support their national objectives. It is the case of the Pampa Azul[3] program in Argentina, the Scientific Mission Oriented policy promoted by the *Mision de Sabios*[4] in Colombia, and the international researchers recruitment program in Chile. In addition, initiatives to attract highly qualified scientists and human resources from other latitudes such as Study in Argentina, Study in Colombia, and Study in Chile, are gaining relevance and being met with relative success. Yet, these trends and programs should be the subject of further research.

CONCLUSION

As illustrated by the anecdote included in the introduction, developing countries are the object of disputes over the influence and material power of the great powers. Thus, the purpose of this chapter was to generate a first approach to future discussions on the impact of power shifts in the international system on the S&T relations of developing countries such as Argentina, Chile, and Colombia and seeks to raise future questions regarding the possibilities for maneuver and opportunities that these countries can envision in a scenario that is becoming increasingly complex.

This initial descriptive effort served to appreciate that, since its re-emergence as a great power, China has expanded its ties with developing countries, including the three under study, and that this has generated a broadening of scientific-technological cooperation ties that, despite having originated in the 1980s, have deepened in recent times. This is consistent with the fact that S&T have become a fundamental axe of the dispute between China and the West, especially taking into account the emergence and consolidation of the sixth techno-economic paradigm underway.

In geopolitical terms, the region appears as a subordinate to SD strategies from the US, the EU, and China, under the premise that these diplomatic agendas help countries in the region to support their low GDP investment in S&T. It can be observed that in several calls for proposals, the topics are usually associated with the exporting profiles of the respective raw materials, which in turn are demanded by these centers. Likewise, we found some relation between the development of SD strategies in parallel with trade relations.

However, from a more optimistic perspective it was also possible to observe, as mentioned in the discussion, that many of these calls promoted in countries such as Chile or Colombia have had as an objective the access of scientists to infrastructures, resources or materials not available in their territories, which lends credibility to our previous statement about the use of Diplomacy for Science strategies to increase national capacities, even when they are not explicit or embodied in formal documents.

It can also be noted that despite their beginnings as managers of resources and recipients of knowledge transfer and capacity-building, Argentina, Chile, and Colombia are becoming more active in their efforts to stop being recipients of SD from more advanced countries and to show themselves to be valid partners that can be presented as alternative destinations for training and research. Unfortunately, the progress in developing SD strategies is highly affected by changes of presidential administrations. If SD were to become State policies rather than government initiatives, Diplomacy for Science strategies may contribute to addressing common challenges in the region such as poverty, hunger, and access to clean water and energy, among others. Future research may establish the impact of Diplomacy for Science strategies and how SD could help to increase the possibilities of improving the national capacities of Latin American countries and increase their room for maneuver in setting agenda priorities.

REFERENCES

Acharya, A., Estevadeordal, A., & Goodman, L. W. (2023). Multipolar or multiplex? Interaction capacity, global cooperation and world order. *International Affairs*, 99(6), 2339–2365. 10.1093/ia/iiad242

Adler, E., & Haas, P. (1992). Conclusion: Epistemic Communities, World Order, and the Creation of a Reflective Research Program. *International Organization*, 46(1), 367–390. 10.1017/S0020818300001533

Álvarez, I. Natera, J. M. & Castillo, Y. (2019). Generación y transferencia de ciencia, tecnología e innovación como claves de desarrollo sostenible y cooperación internacional en América Latina. Documentos de Trabajo. Fundación Carolina 19/2019.

ANID. (2022). Fondo CAS-ANID 2022. Available at: https://ayuda.anid.cl/hc/es/articles/7973636144532-Fondo-CAS-ANID-2022

ANID. (2023a). Bases del concurso Becas de Doctorado Igualdad de Oportunidades Fulbright–ANID Convocatoria 2023. Available at: https://s3.amazonaws .com/documentos.anid.cl/BecasChile/2023/fullbright/Bases_Doctorado_BIO_Fulbright_2023_LA_28062023.pdf

ANID. (2023b). Se adjudica Concurso Fondo CAS-ANID, propuestas de investigación de posiciones postdoctorales, Programa Nacional para el Desarrollo de la Astronomía y Ciencias Afines. Convocatoria 2022. Resolución Nº: 8903. Available at: https://s3.amazonaws.com/documentos.anid.cl/rec /2022/CasAnid/REX8903-2023_adjudica_Concurso_Fondo_CAS-ANID_2022_sF4LzC6QbVwN1XhT3eR-tY7uIaM5oKvP.pdf

ANID. (2024). Página institucional. Disponible en: https://anid.cl/

Barth, K. (2006). Catalysts of Change: Scientists as Transnational Arms Control Advocates in the 1980s. *Osiris*, 21(1), 182–206. 10.1086/507141

Belli, S., & Morín Nenoff, J. (2022). Cooperation in Science and Innovation between Latin America and the European Union. *Journal of Open Innovation*, 8(2), 94. 10.3390/joitmc8020094

Borda, S., & Paz Berger, M. (2012). Relaciones bilaterales China y Colombia: 1990-2010. *Colombia Internacional*, 75(75), 83–129. 10.7440/colombint75.2012.04

Büyüktanir, D., & Ruffini, P. B. (2023). Science Diplomacy in the Global South – an introduction. *Science & Public Policy*, 50(4), 742–748. 10.1093/scipol/scad028

Campbell, C. A. (2014). US Science Diplomacy with Arab Countries. In *Science diplomacy: New day or false dawn*. World Scientific.

Cancilleria Colombiana. (2021). *Unión Europea y Colombia firman acuerdo de colaboración relativo a la cooperación en el ámbito de los sistemas mundiales de navegación por satélite | Cancillería*. Available at: https://www.cancilleria.gov.co/newsroom/news/union-europea-colombia-firman-acuerdo-colaboracion-relativo-cooperacion-ambito

Cancilleria Colombiana. (2023a). *Colombia y Estados Unidos sostendrán el X Diálogo de Alto Nivel en el que abordarán los temas prioritarios de la relación bilateral | Cancillería*. Avaialbale at: https://www.cancilleria.gov.co/newsroom/news/colombia-estados-unidos-sostendran-x-dialogo-alto-nivel-abordaran-temas-prioritarios

Cancilleria Colombiana. (2023b). *Colombia y China suscribieron 12 instrumentos de cooperación | Cancillería*. Available at: https://www.cancilleria.gov.co/newsroom/news/colombia-china-suscribieron-12-instrumentos-cooperacion

Chudnovsky, D., López, A., & Porta, F. (1997). Market or policy driven? The Foreign direct investment boom in Argentina. *Oxford Development Studies*, 25(2), 173–188. 10.1080/13600819708424128

Colciencias. (2010). *Colombia y China firmaron acuerdo para la cooperación en Ciencia, Tecnología e Innovación | Colciencias*. Available at: https://legadoweb.minciencias.gov.co /noticias/colombia-y-china-firmaron-acuerdo-para-la-cooperaci-n-en-ciencia-tecnolog-e-innovaci-n

Colombiana De Ciencias Exactas. Físicas Y Naturales. Retrieved March 16, 2024, from https://accefyn.org.co/libros-mision-de-sabios/Ministerio de Ciencia, Tecnologia e Innovacion. (n.d.-c). *Linea de Tiempo Minciencias*. Retrieved in March 16, 2024, from https://minciencias.gov.co/sites/default/files/ckeditor_files/Linea%20de%20Tiempo%20Minciencias.png

CONICET. (2018). Científicos chinos y argentinos intercambiaron conocimientos sobre desertificación, ecología de desiertos y restauración. CONICET (03/10/2018). Available at https://www.conicet.gov.ar/cientificos-chinos -y-argentinos-intercambiaron-conocimientos-sobre-desertificacion-ecologia-de-desiertos-y-restauracion/ Retrieved in December 2023.

CONICET. (2019). El CONICET y la Academia China de Ciencias acuerdan la creación de un Centro Conjunto de Ciencia y Tecnología. CONICET (03/12/2019). Available at https://www.conicet.gov.ar/el-conicet-y-la-academia-china -de-ciencias-acuerdan-la-creacion-de-un-centro-conjunto-de-ciencia-y-tecnologia/ Retrieved in December 2023.

CONICET. (2019). Argentina y China fortalecen lazos en materia científica. CONICET (21/06/2019) Available at https://www.conicet.gov.ar/argentina-y-china-fortalecen-lazos-en-materia-cientifica/ Retrieved in December 2023.

CONICET. (2022). Se presentó el Centro Binacional Chino-Argentino de estudio de políticas para la Innovación y la Tecnología. CONICET (09/05/2022). Available at https://www.conicet.gov.ar/se-presento-el-centro-binacional-chino -argentino-de-estudio-de-politicas-para-la-innovacion-y-la-tecnologia/.Retrieved in December 2023.

CONICET. (2023). Se establece el Comité Directivo y Científico del Centro Conjunto de Ciencia y Tecnología entre el CONICET y la Academia de Ciencias de China. CONICET (27/01/2023). Available at https://www.conicet.gov.ar /se-establece-el-comite-directivo-y-cientifico-del-centro-conjunto-de-ciencia-y-tecnologia-entre-el-conicet-y-la-academia-de-ciencias-de-china/ Retrieved in December 2023.

CONICYT. (2018). Adjudica el Concurso para Proyectos de Investigación Conjunta en el manejo de Desastres Naturales en Chile y China, CONICYT Chile – NSFC China. Convocatoria 2018. Resolución N° at: https://www.conicyt.cl/pci/files/2018/10/REX.100438.del_.25.10.2018.NSFC_.adjudicacion.pdf10.048. Available

CONICYT. (2019). Aprueba Fallo Y adjudica el Concurso para Proyectos de Investigación Conjunta en Chile y China, CONICYT Chile – NSFC China. Convocatoria 2019. Resolución N° at: https://www.conicyt.cl/pci/files/2019/11/REX.10707.del_.12.11.2019.NSFC_.Fallo_.pdf10.707. Available

CONICYT. (2024). Convenios Internacionales. Available at: https://www.conicyt.cl/sobre-conicyt/convenios/convenios-internacionales/

CTCI. (2022). *Estrategia Nacional De ciencia, Tecnología, Conocimiento e Innovación para el Desarrollo de Chile—2022*. Consejo Nacional de Ciencia, Tecnología, Conocimiento e Innovación para el Desarrollo.

Dangond Gibsone, S. (2022). *Colombia-Estados Unidos—200 años de relaciones diplomáticas*. Gobierno de Colombia.

De Greiff, A. (2014). La norteamericanización de la tecnociencia en América Latina: diplomacia científica y hegemonía cultural. In Kreimer, P. (Ed.), *Vessuri, H.; Velho, L. y Arellano, A. (coords.). Perspectivas latinoamericanas en el estudio social de la ciencia, la tecnología y la sociedad* (pp. 194–207). Siglo XXI Editores.

Departamento Nacional De Planeación. (2021). *CONPES. Política Nacional de Ciencia, Tecnología e Innovación 2022-2031*. https://colaboracion.dnp.gov.co/CDT/Conpes/Econ%C3%B3micos/4069.pdf

Department for Business, Energy and Industrial Strategy. (2022). *Partner Country Case Study: China Final Evaluation of The Newton Fund* (Partner Country Case Study, p. 77) [Final evaluation]. Department for Business, Energy and Industrial Strategy.

Dolan, B. M. (2012). Science and Technology Agreements as Tools for Science Diplomacy. *AAAS Center for Science and Diplomacy*. https://www.sciencediplomacy.org/article/2012/science-and-technology-agreements-tools-for-science-diplomacy

Domínguez, R. (2013). El pragmatismo en las relaciones UE y Colombia/Perú. In *Después de Santiago: Integración regional y relaciones Unión Europea—América Latina*. Miami-Florida European Union Center of Excellence.

Echeverría King, L. F. (2020). Diplomacia científica e internacionalización de la investigación: Nuevos horizontes de trabajo colaborativo con base en dos experiencias de Colombia. *Revista Educación Global*, 24, 11–21.

Echeverría King, L. F., González, D. A., & Andrade-Sastoque, E. (2021). Science Diplomacy in Emerging Economies: A Phenomenological Analysis of the Colombian Case. *Frontiers in Research Metrics and Analytics*, 6(636538), 636538. Advance online publication. 10.3389/frma.2021.63653833997599

Embajada Argentina ante la Unión Europea (2019). 20° Aniversario de la Cooperación en Ciencia, Tecnología e Innovación entre la Argentina y la Unión Europea. Ministerio de Relaciones Exteriores, Comercio Internacional y Culto (26/06/2019). Available at https://eceur.cancilleria.gob.ar/es/20%C2%BA-aniversario-de-la-cooperaci%C3%B3n-en-ciencia-tecnolog%C3%ADa-e-innovaci%C3%B3n-entre-la-argentina-y-la-uni%C3%B3n-0. Retrieved in December 2023.

Embajada de Colombia en China. (2023). *Invest in Colombia*.

Embajada de Estados Unidos en Chile. (2024). Nuestra relación. Available at: https://cl.usembassy.gov/es/our-relationship-es/policy-history-es/

Embajada de Estados Unidos en Chile. (2024). Educación y Cultura. Consejo Chile-Estados Unidos para la Ciencia, Tecnología e Innovación (STIC). Disponible en: https://cl.usembassy.gov/es/education-culture-es/consejo-chile-estados-unidos-para-la-ciencia-tecnologia-e-innovacion-stic/

Epping, E. (2020). Lifting the smokescreen of science diplomacy: Comparing the political instrumentation of science and innovation centres. *Humanities & Social Sciences Communications*, 7(1), 111. 10.1057/s41599-020-00599-4

European Commission. (2023). *International cooperation with Latin America and Caribbean in research and innovation.*https://research-and-innovation.ec.europa.eu/strategy/strategy-2020-2024/europe-world/international-cooperation/regional-dialogues-and-international-organisations/latin-america-and-caribbean_en

European Commission. (n.d). Research and Innovation. Argentina. Available at https://research-and-innovation.ec.europa.eu/strategy/strategy-2020-2024/europe-world/international-cooperation/bilateral-cooperation-science-and-technology-agreements-non-eu-countries/argentina_en. Retrieved in March 2024.

European Union (2023). La UE y Chile se comprometen a una asociación con visión de futuro. European External Action Service. Available at: https://www.eeas.europa.eu/eeas/la-ue-y-chile-se-comprometen-una-asociaci%C3%B3n-con-visi%C3%B3n-de-futuro-0_es

European Union. (n.d.). *Comisión Europea firma Acuerdos de Cooperación para observación espacial con Colombia, Chile y Brasil | EEAS*. Retrieved March 15, 2024, from https://www.eeas.europa.eu/node/40977_es

Fägersten, B. (2022). *Leveraging Science Diplomacy in an Era of Geo-Economic Rivalry* (1; p. 22). The Swedish Institute of International Affairs. https://www.ui.se/globalassets/ui.se-eng/publications/ui-publications/2022/ui-report-no.-1-2022.pdf

Fedoroff, N. V. (2009). Science Diplomacy in the 21st Century. *Cell*, 136(1), 9–11. 10.1016/j.cell.2008.12.03019135879

Flink, T., & Schreiterer, U. (2010). Science diplomacy at the intersection of S&T policies and foreign affairs: Toward a typology of national approaches. *Science & Public Policy*, 37(9), 665–677. 10.3152/030234210X12778118264530

Fulbright Chile. (2024). Quiénes somos. Disponible en: https://fulbrightchile.cl/quienes-somos/

Fulbright Commission Colombia. (n.d.). *Fulbright Commission Colombia—Our history*. Fulbright Colombia – Becas Para Colombianos. Retrieved in March 12, 2024, from https://fulbright.edu.co/nuestra-historia/

Gluckman, P. D., Turekian, V. C., Grimes, R. W., & Kishi, T. (2017). *Science Diplomacy: A Pragmatic Perspective from the Inside*.

Gobierno de Argentina. (2018). Argentina y China colaborarán mutuamente en Educación, Cultura y Ciencia. Gobierno de Argentina, Noticias (20/11/2018). Available at https://www.argentina.gob.ar/noticias/argentina-y-china-colaboraran-mutuamente-en-educacion-cultura-y-ciencia. Retrieved in December 2023.

Gobierno de Argentina. (2019). 20 años de cooperación entre Argentina y la Unión Europea en Ciencia y Tecnología. Gobierno de Argentina, Noticias (11/06/2019). Available at https://www.argentina.gob.ar/noticias/20-anos-de-cooperacion-entre-argentina-y-la-union-europea-en-ciencia-y-tecnologia. Retrieved in December 2023.

Gobierno de Argentina. (2023a). El Ministerio de Ciencia participó en un seminario de diálogo sobre política digital entre Argentina y Estados Unidos. Gobierno de Argentina, Noticias (30/06/2023). Available at https://www.argentina.gob.ar/noticias/el-ministerio-de-ciencia-participo-en-un- seminario-de-dialogo-sobre-politica-digital-entre. Retrieved in February de 2024.

Gobierno de Argentina. (2023b). Argentina y China acuerdan crear un centro de investigaciones en RRNN y transición energética. Gobierno de Argentina, Noticias (07/09/2023). Available at https://www.argentina.gob.ar/noticias/argentina-y-china-acuerdan-crear-un-centro-de-investigaciones-en-rrnn-y-transicion Retrieved in December 2023.

Gual-Soler, M. (2021). Science Diplomacy in Latin America and the Caribbean: Current Landscape, Challenges, and Future Perspectives. *Frontiers in Research Metrics and Analytics*, 6, 670001. 10.3389/frma.2021.67000134222772

Harden-Davies, H. (2018). The next wave of science diplomacy: Marine biodiversity beyond national jurisdiction. *ICES Journal of Marine Science*, 75(1), 426–434. 10.1093/icesjms/fsx165

Hernández, B. & Lebret, P. (2019). Dos décadas de cooperación europea en Chile: una asociación para el desarrollo. *Revista Estudios internacionales* 51(192), 35-56. Chile. 10.5354/0719-3769.2019.52812

Horizon Europe. (2023). *List 3rd country participation horizon-Europe*. Avialable at: https://ec.europa.eu/info/funding-tenders/opportunities/docs/2021-2027/common/guidance/list-3rd-country-participation_horizon-euratom_en.pdf

Hurtado, D. (2010). *La ciencia Argentina un proyecto inconcluso 1930-2000.*

Lillo González, M., & Santamaría García, A. (2009). La Unión Europea y Colombia. Historia y desafío. *Anuario de Estudios Americanos*, 66(1), 289–319. 10.3989/aeamer.2009.v66.i1.311

Linkov, I., Basu, S., Fisher, C., Jackson, N., Jones, A. C., Kuklja, M. M., & Trump, B. D. (2016). Diplomacy for science: Strategies to promote international collaboration. *Environment Systems & Decisions*, 36(4), 331–334. 10.1007/s10669-016-9614-5

López de San Román, A., & Schunz, S. (2018). Understanding European Union Science Diplomacy: Understanding EU Science Diplomacy. *Journal of Common Market Studies*, 56(2), 247–266. 10.1111/jcms.12582

Miller, C. A. (2006). "An Effective Instrument Of Peace": Scientific Cooperation As An Instrument Of U.S. Foreign Policy, 1938–1950. *Osiris*, 21(1), 133–160. 10.1086/507139

MinCiencia. (2019a). Data Observatory: AWS y UAI serán miembros fundadores clave de Chile en la administración de datos. Available at: https://www.minciencia.gob.cl/noticias/data-observatory-aws-y-uai-seran-miembros-fundadores-clave-de-chile-en-la-administracion-de-datos/

MinCiencia. (2019b). Profesores se formarán con científicos norteamericanos para liderar la enseñanza de la ciencia. Available at: https://www.minciencia.gob.cl/noticias/profesores-se-formaran-con-cientificos-norteamericanos-para-liderar-la-ensenanza-de-la-ciencia/

MinCiencia. (2020a). Política Nacional de Ciencia, Tecnología, Conocimiento e Innovación. Available at: https://minciencia.gob.cl/politicactci/documentos/Politica-Nacional-CTCi_Chile-2020.pdf

MinCiencia. (2020b). UC e Instituto Milenio de Inmunología e Inmunoterapia firman alianza con laboratorio chino que está desarrollando vacuna contra el COVID-19. Available at: https://www.minciencia.gob.cl/noticias/uc-e-instituto-milenio-de-inmunologia-e-inmunoterapia-firman-alianza-con-laboratorio-chino-que-esta-desarrollando-vacuna-contra-el-covid-19/

MinCiencia. (2021a). Ministro de Ciencia recibe a representantes de NASA para explorar operación de la misión astronómica SOFIA desde Chile. Available at: https://www.minciencia.gob.cl/noticias/ministro-de-ciencia-recibe-representantes-de-nasa-para-explorar-operacion-de-la-mision-astronomica-sofia-desde-chile/

MinCiencia. (2021b). ESO y Chile firman convenio de cooperación científica y tecnológica con el que se convertirá en el telescopio más grande del mundo. Available at: https://www.minciencia.gob.cl/noticias/eso-y-chile-firman-convenio-de-cooperacion-cientifica-y-tecnologica-con-el-que-se-convertira-en-el-telescopio-mas-grande-del-mundo/

MinCiencia. (2023a). Ministerio de Ciencia e IBM invitan a mujeres de todo el país a cursos de formación gratuita en tecnología. Available at: https://www.minciencia.gob.cl/noticias/ministerio-de-ciencia-e-ibm-invitan-mujeres-de-todo-el-pais-cursos-de-formacion-gratuita-en-tecnologia/

MinCiencia. (2023b). Presidente Boric y ministra Aisén Etcheverry anuncian inicio del proceso para que Chile sea estado asociado del Centro Europeo para la Investigación Nuclear, CERN. Available at: https://www.minciencia.gob.cl/noticias/presidente-boric-y-ministra-aisen-etcheverry-anuncian-inicio-del-proceso-para-que-chile-sea-estado-asociado-del-centro-europeo-para-la-investigacion-nuclear-cern/

MinCiencia. (2023c). Declaración pública sobre Sinovac. Available at: https://www.minciencia.gob.cl/noticias/declaracion-publica/

Ministerio de Ciencia, Tecnología e Innovación. (2020). Libros Misión de Sabios. *Academia (Caracas).*

Ministerio de Ciencia. Tecnología e Innovación. (n.d.-a). *Línea de Tiempo Minciencias*. Retrieved in March 16, 2024, from https://minciencias.gov.co/sites/default/files/ckeditor_files/Linea%20de%20Tiempo%20Minciencias.png

Ministerio de Ciencia. Tecnología e Innovación. (n.d.-b). *Nexo Global*. Minciencias. Retrieved March 12, 2024, from https://minciencias.gov.co/nexo-global

Ministerio de Ciencia. Tecnología e Innovación. (n.d.-c). *MinCiencias y VaxThera firmaron un Memorando de Entendimiento para fortalecer las capacidades en biotecnología del país*. Minciencias. Retrieved March 12, 2024, from https://www.minciencias.gov.co/sala_de_prensa/minciencias-y-vaxthera-firmaron-un-memorando-entendimiento-para-fortalecer-las

Ministerio de Ciencia. Tecnología e Innovación. (n.d.-d). *Conozca por qué Minciencias hace parte del programa Artemis de la Nasa*. Minciencias. Retrieved March 12, 2024, from https://minciencias.gov.co/sala_de_prensa/conozca-por-que-minciencias-hace-parte-del-programa-artemis-la-nasa

Ministerio de Educación Nacional. (2018). *Ministerio de Educación Nacional y el Gobierno de China dan a conocer el programa binacional de becas | Ministerio de Educación Nacional*. Available at: https://www.mineducacion.gov.co/portal/salaprensa/Noticias/377886:Ministerio-de-Educacion-Nacional-y-el-Gobierno-de-China-dan-a-conocer-el-programa-binacional-de-becas

Ministerio de Relaciones Exteriores de Chile. (2015). Informe Anual de Comercio Exterior de Chile 2014-2015. Dirección General de Relaciones Económicas Internacionales. Available at: https://www.subrei.gob.cl/docs/default-source/default-document-library/informe-anual-de-comercio-exterior-de-chile-2014-2015.pdf?sfvrsn=325fde69_0

Noticias Argentinas. (2024). La Embajada china en Argentina sostiene que la base en Neuquén es solo "científica". *Diario Noticias Argentinas*, edición del 1/04/2024. Disponible en https://noticiasargentinas.com/politica/la-embajada-china-en-argentina-sostiene-que-la-base-de-neuquen-es-solo--cientifica-_a660b463e2d551ce9d8e1b965

Oficina de Enlace Argentina-Unión Europea, n.d. Official Website. Available at https://www.argentina.gob.ar/ciencia/seppcti/enlace_arg-ue/programas-marco

Olšáková, D. (2023). A Review of Science Diplomacy: Theoretical Evolution to a Post-Naïve Approach and Its Relevance for the Czech Republic. *Czech Journal of International Relations.*

Partners of the Americas. (2023). *The 100,000 Strong in the Americas Innovation Fund Announces New Higher Education Partnerships and Training Programs between 17 colleges and universities in the United States and Colombia – 100,000 Strong in the Americas.* https://www.100kstrongamericas.org/c4p-winners-2023/

Pastrana Buelvas, E., Castro, R., & González, P. A. (2016). Política exterior colombiana: Escenarios y desafíos en el posconflicto. In E. Pastrana Buelvas & H. Gehring (Eds.), *La proyección de China en América Latina y el Caribe* (Primera edición). Editorial Pontificia Universidad Javeriana.

Pérez, C. (2010). Technological revolutions and techno-economic paradigms. *Cambridge Journal of Economics*, 34(1), 185–202. 10.1093/cje/bep051

Polejack, A., & Coelho, L. F. (2021). Ocean Science Diplomacy can Be a Game Changer to Promote the Access to Marine Technology in Latin America and the Caribbean. *Frontiers in Research Metrics and Analytics*, 6, 637127. 10.3389/frma.2021.63712733912786

Powell, J. J. W., Fernandez, F., Crist, J. T., Dusdal, J., Zhang, L., & Baker, D. P. (2017). Introduction: The Worldwide Triumph of the Research University and Globalizing Science. In Powell, J. J. W., Baker, D. P., & Fernandez, F. (Eds.), *International Perspectives on Education and Society* (Vol. 33, pp. 1–36). Emerald Publishing Limited.

Prieto, J., & Scott, C. A. (2022). Scientific diasporas and the advancement of science diplomacy: The InFEWS US-China program in the face of confrontational "America First" diplomacy. *Frontiers in Research Metrics and Analytics*, 7, 944333. 10.3389/frma.2022.94433336277735

Rispoli, G., & Olšáková, D. (2020). Science and Diplomacy around the Earth. *Historical Studies in the Natural Sciences*, 50(4), 456–481. 10.1525/hsns.2020.50.4.456

Robinson, S. (2020). Early Twentieth-Century Ocean Science Diplomacy. *Historical Studies in the Natural Sciences*, 50(4), 384–410. 10.1525/hsns.2020.50.4.384

Roy, J. (Ed.). (2013). *Después de Santiago: Integración regional y relaciones Unión Europea –América Latina*. Miami-Florida European Union Center of Excellence.

Ruffini, P. B. (2017). *Science and Diplomacy. A New Dimension of International Relations*. Springer.

Ruffini, P.-B. (2018). The Intergovernmental Panel on Climate Change and the Science-Diplomacy Nexus. *Global Policy*, 9(S3), 73–77. 10.1111/1758-5899.12588

Servicio Nacional de Aduanas de Chile. (2023). *Reporte De Comercio Exterior 2023*. Enero – Julio. Available at:https://www.aduana.cl/aduana/site/docs/20181217/20181217125337/reporte_comex_julio_final.pdf

Snidjers, D. (2023). Reassessing STI Diplomacy in the age of Techno-politics. *Science & Diplomacy*, 7(2), 16–18.

State Department. (n.d.). *100,000 Strong in the Americas*. United States Department of State. Retrieved March 12, 2024, from https://www.state.gov/100000-strong-in-the-americas/

Tassara, C. (2019). La cooperación internacional de la Unión Europea en Colombia. Buenas prácticas y lecciones aprendidas. In E. Tremolada Álvarez (Ed.), *Colombia y la Unión Europea: Una asociación cada vez más estrecha* (Primera edición). Universidad Externado de Colombia.

The Royal Society, & AAAS. (2010). *New frontiers in science diplomacy. Navigating the changing balance of power*. The Royal Society.

Tremolada, E., & Quiroga Chacón, G. A. (2022). La Diplomacia Pública de la Unión Europea en Colombia: ¿Contribuye a una mejora en su percepción? *Revista Internacional de Cooperación y Desarrollo*, 9(2), 81–89. 10.21500/23825014.6071

Turchetti, S. (2020). The (Science Diplomacy) Origins of the Cold War. *Historical Studies in the Natural Sciences, 50*(4), 411–432. Union Europea. (n.d.). *La Unión Europea y Colombia | EEAS*. Retrieved March 15, 2024, from https://www.eeas.europa.eu/colombia/la-union-europea-y-colombia_es?s=160

Turekian, V. C., Macindoe, S., Copeland, D., Davis, L., Patman, R., & Pozza, M. (2014). The Emergence of Science Diplomacy. In Davis, L. S., & Patman, R. G. (Eds.), *Science diplomacy: New day or false dawn*. World Scientific.

U.S. Department of State. (2024). *Technology and the Transformation of U.S. Foreign Policy* [Government website]. United States Department of State. https://www.state.gov/technology-and-the-transformation-of-u-s-foreign-policy/

Vera, N. (2018). *Cooperación nuclear sino-argentina en el nuevo rol de China como potencia emergente del siglo XXI. ¿Lógica de cooperación sur – sur o norte – sur? (2003 - 2015*). XIII Congreso Nacional y VI Congreso Internacional sobre Democracia. Rosario, 10 a 13 de septiembre de 2018. Available at https://es.scribd.com/document/394877936/7-Memorias -Politica-Internacional-Ok

Vera, N., & Colombo, S. (2020). Reflexiones en torno a la Cooperación Internacional en Ciencia y Tecnología y la Diplomacia Científica: los casos de Estados Unidos y la República Popular China. In *López, M. P. (comp.). Perspectivas sobre la cooperación internacional en ciencia, tecnología y universidad: políticas, prácticas y dinámicas a principios del siglo XXI* (pp. 13–38). UNICEN.

Vera, N., & Echeverría King, L. (2022) *Hacia una diplomacia científica para América Latina.* En Oregioni, S. (comp.) Derribando estructuras, construyendo puentes, tejiendo Futuro. Aportes para pensar la internacionalización desde una perspectiva situada en el Sur. Editorial-Cooperativa El Zócalo, Buenos Aires, Argentina, pp. 213-236.

Vera, N., & López, M. P. (2023). Diplomacia Científica en la Argentina Contemporánea: Un mapeo de iniciativas en los ministerios de Ciencia, Tecnología e Innovación y de Relaciones Exteriores (2019-2022). *Estudios Internacionales (Santiago)*, 55(204), 153–185. 10.5354/0719-3769.2023.67792

Vera, N., Salvati, A., De Ángelis, I., & Pantović, B. (2023). Diplomacia Científica Transregional y prioridades en la agenda de cooperación científico-tecnológica de la Comunidad de Estados Latinoamericanos y Caribeños (2013-2023). In *López, M. P. (comp.). Ciencia, tecnología y cooperación internacional en clave de desarrollo. Reflexiones sobre Argentina y América Latina* (pp. 73–124). CEIPIL – UNICEN.

Wagner, C., & Simon, D. (2023). China's use of formal science and technology agreements as a tool of diplomacy. *Science & Public Policy*, 50(4), 807–817. 10.1093/scipol/scad022

Whitesides, G. (2012). *Science and American Foreign Relations since World War II.* Cambridge University Press.

Yin, R. K. (2018). *Case study research and applications: Design and methods* (6th ed.). SAGE.

Zárate, M. S., Sierra, D., & Goldflam, M. (2022). Orígenes y desarrollo de una política científica nacional en Chile: Conicyt, 1967-1981. *Historia, Ciencias, Saude—Manguinhos*, 29(4), 953–972. 10.1590/s0104-59702022000400005365 42032

ENDNOTES

1 Brazil represents an outlier in all the criteria and therefore it is not comparable with the rest of countries in the region.

2 The US and Argentina signed the first agreement to finance educational exchange programs in 1963, and since 1997 there has been a strong and sustained collaboration with the Fulbright Commission, from which numerous mobility grants and research stays for Argentine scientists have been financed. Since 2003 the Exchange of Teaching Assistants program has been carried out, through which US graduates assist in the teaching of English in Argentina and vice versa. Such collaboration initiatives were reinforced as of 2016 when both countries signed agreements to "diversify and significantly increase educational exchanges", adding to the previous policies, those of Friends of Fulbright aimed at undergraduate students, and a second one oriented to the training of teachers in US universities (US Embassy in Argentina, n.d). As can be seen, these latter initiatives are oriented more towards undergraduate and teacher internationalization than to specific S&T sectors.

3 The Pampa Azul Project is an interministerial initiative that coordinates scientific research, technological development, and innovation efforts to support national ocean policies. This includes bolstering national sovereignty over marine territories, promoting conservation, and ensuring the sustainable utilization of marine resources, such as the establishment and management of marine protected areas.

4 The Mision de Sabios was organized under eight thematic groups. After delivering the final reports MinCiencias has prioritized five science missions, including, 1) Bioeconomy and Territory, 2) Human Right to Food, 3) Energy Transition, 4) Health Sovereignty and Social Welfare, and 5) Science for Peace.

Chapter 7
Environmental Science Diplomacy and Traditional Knowledge: Diplomacy Within National Borders in Latin America and the Caribbean

Juan David Reina-Rozo
https://orcid.org/0000-0001-8146-674X
Universidad Nacional de Colombia, Colombia

Angie Vanessa Vargas Junco
https://orcid.org/0009-0005-7547-2015
Universidad Militar Nueva Granada, Colombia

ABSTRACT

Environmental science diplomacy involves scientists, researchers, and diplomats working together to address global environmental issues, share and co-create scientific knowledge to develop national interests, and promote evidence-based policies for environmental protection and conservation. Science diplomacy has brought about change within national borders, which means building relationships with actors beyond Western science and traditional diplomacy, and recognizing the diversity of epistemologies, knowledge and approaches. In particular, there is a crucial element of diplomacy within national borders, the negotiation between environmental science and traditional ecological knowledge in Latin America and the Caribbean. Challenges include overcoming cultural barriers and promoting equitable collaboration between diplomats, scientists, and ethnic and local communities. Future

DOI: 10.4018/979-8-3693-2746-3.ch007

directions include promoting inclusiveness, intercultural dialogue, and negotiation of mutual respect to more effectively address global and environmental challenges with actors not represented in diplomacy.

INTRODUCTION

A brief review of science diplomacy (SD) in Latin America and the Caribbean (LAC) reveals an emerging opportunity for action, but a gap in the environmental sphere is prevalent (Soler, 2021b; Isaza, 2020), also from the perspective of the Global South (RIS, 2020). This chapter aims to propose the concept of Environmental Science Diplomacy (ESD) as an essential tool to address environmental challenges in relationship to Traditional Ecological Knowledge (TEK). Science diplomacy remains a dynamic concept that can be applied in terms of diplomacy, science, technology, and innovation, which can be seen from three points: 1. Science in diplomacy: which refers to the scientific orientation to achieve foreign policy objectives; 2. Diplomacy for science: which refers to the promotion of scientific cooperation at the international level; and 3. Science for diplomacy: which refers to scientific collaboration to strengthen ties between actors in the international system (The Royal Society & AAAS, 2010). In addition to the conventional view, it should be kept in mind that "for diplomacy there are other goals beyond peace. They involve a very hard, continuous, and silent work of conflict. It is the work of confronting powers to seek their balance and stability, in the search for cooperation between peoples for their greater integration" (Lascano y Vedia, 2021, p. 41). This means that diplomacy has a broader vision than the use of science in the search for peace and conflict resolution.

Other visions can be analyzed from a scientific point of view, such as the confrontation of national interests and agendas of different states and international organizations. This includes the management of resources, research in areas such as renewable or alternative energies, and the search for a transnational consensus that benefits humanity; also the search for balance and stability because science can be an element of power when stabilizing international relations, through cooperation in joint research projects. In other words, diplomacy in general is more than just the pursuit of peace, it can be seen especially through science diplomacy as it implies a continuous effort to confront and balance powers, as well as to foster integration and cooperation through science.

SD should be redefined as a tool for building peaceful and prosperous international relations from multiple knowledge regimes and data (Turchetti & Lalli, 2020). In addition to resolving conflicts, diplomacy may promote cooperation and build a more equitable, sustainable, and stable international order based on knowledge and transnational dialogue. In particular, the Diplomacy for Science category within SD is called to generate the conditions for transdisciplinary dialogue. From the perspective of environmental thinking, knowledge dialogue is a process of informal negotiation that relates to people and the planet (Leff, 2004). As such, it allows for the negotiation of scientific knowledge (SK) agreements between countries and societies (Vila-Viñas & Crespo, 2015). Science diplomacy has made a difference in international negotiations and within national borders, which means building relationships with actors beyond Western science and the conventional diplomatic sphere (Carpenter & Tsykarev, 2021; Cruz Chavarro & Arevalo Robles, 2021; Foyer & Kervran, 2020; Pugin, 2023). In particular, the negotiation between environmental science and traditional ecological knowledge (TEK) in Latin America and the Caribbean is a crucial element of diplomatic efforts within national borders and transnational relations (Science Panel for the Amazon, 2021). Thus, the central research question for our inquiry is What is the relationship of Science Diplomacy and Traditional Ecological Knowledge in Latin America?

In this context, it is important to interrogate the practice of SD in the complexity of the region, especially from epistemologies beyond the West and critical perspectives on coloniality (Polejack, 2023). The notions of paradiplomacy (Martínez, 2019) and transdiplomacy (Arevalo Robles, 2017) become relevant in this context, as they imply the active participation of local actors and cosmovisions, respectively, in diplomatic activities. For example Decision-making processes are closely linked to the goals of the 2030 Agenda for Sustainable Development, which promotes social inclusion, equity, and participation of all sectors of society. Engaging local communities and other non-traditional actors in environmental science diplomacy will contribute to improving the effectiveness of solutions due to the knowledge of the terrain that communities have and the practical solutions that can be offered from this point of view; promoting inclusion and equity by including actors that ensure that projects and policies reflect the real needs of stakeholders; and strengthening cooperation as this participation can enrich knowledge. This phenomenon is not only a necessity in environmental science diplomacy but also highlights how other actors such as ethno-peasant communities can play a crucial role in solving environmental problems from an alterity perspective (Vessuri, 2013). In this case, Resguardos Indígenas, Consejos Comunitarios Afrodescendientes, and Zonas de Reserva Campesina are vital in the Colombian context. Their deep knowledge of local issues, close relationships with communities, and flexibility make local governments and communities valuable actors in working with scientists, researchers, and diplomats from other countries (Virchez et al., 2017).

Ongoing efforts in Latin America and the Caribbean can pave the way for a more inclusive and effective approach to environmental management, where science diplomacy becomes a catalyst for positive change (Soler, 2021a). By opening up to the participation of local governments and non-state actors, environmental science diplomacy represents a more inclusive and collaborative approach to addressing environmental challenges as a boundary space (Kaltofen & Acuto, 2018). The integration of transdiplomacy in this field allows for the use of local knowledge, promoting citizen and democratic participation, and overcoming the limitations of conventional approaches (Arevalo Robles, 2017). It may also include participation in international conferences and cooperation on environmental issues (Manero, 2018; Cruz Chavarro & Arevalo Robles, 2021).

In Latin America and the Caribbean, science diplomacy has played an emerging and complex role due to the limitations of government institutions and the weak articulation between diplomatic careers and national science focal points (Echeverría-King et al., 2023). However, efforts such as the Inter American Institute (IAI) and UNESCO's Programa Iberoamericano de Ciencia y Tecnología para el Desarrollo (CYTED) have been made to strengthen scientific and technical capacity on environmental issues, to promote the formation of international and collaborative networks, and to encourage global research and the publication of studies related to the environment at the regional level (UNESCO; 2020). The text is divided into five parts. The second section offers the methodological design based on a literature review of science diplomacy and environmental concepts in the context of Latin America and the Caribbean. The next section proposes the conceptualization of environmental science diplomacy based on audiovisual materials and literature. The fourth section analyzes the role of TEK as a key component of ESD, based on the experience of an IDEAM on glaciology in the Sierra Nevada de Santa Marta (Colombia). Then, the discussion section offers insights into the barriers and opportunities of ESD in the regional context. Finally, some conclusions are shared in the last section.

METHODOLOGY

This study on scientific diplomacy with an environmental approach and its relationship with traditional ecological knowledge began with a clear definition of the topic and the formulation of objectives aimed at understanding the interaction between these elements. The methodology includes an exhaustive documentary review which, given the limited availability of sources, is complemented by the inclusion of audiovisual resources, particularly videos relevant to the topic. A comparative analysis will be carried out between the information extracted from the literature and the selected audiovisual content, to identify possible affinities or differences. In case of information gaps, the possibility of interviewing experts and holders of TEK will be considered. It is proposed to systematically document the information gathered, establishing a protocol for the inclusion of references that reflect the multidisciplinary nature of the research.

This research used a qualitative approach based on a documentary case study. According to Stake, this type of methodology "is the study of the particularity and complexity of a single case in order to understand its activity in important circumstances" (1998, p. 11). This methodology allows for the in-depth study of a specific phenomenon in its natural context. This study focuses on science diplomacy as a strategy for environmental governance in Latin America and the Caribbean. The objective is to identify how science diplomacy has contributed to strengthening environmental governance in the region, as well as the opportunities and challenges presented by this approach.

In order to carry out this article, bibliometric analyses were carried out in various academic databases such as WOS (Web of Science) and SCOPUS. In other words, statistical methods were used to analyze the available literature on the subject. For this purpose, the word "diplomacy" was first searched independently to verify that the databases used contained information on the topic of interest. Several sources of information were found, but they were too broad so that the specific topic did not seem very related, so several Boolean codes were elaborated, combining the words "environment", "science" and "diplomacy" (which can be seen in Tables 1 and 2), where it was found that there is not much information on environmental science diplomacy in the global south.

Table 1. Search Algorithms in Scopus

Search algorithms	Númber of results
Diplomacy	21.558
Diplomacy AND Science	1.406
Diplomacy AND Environment	1.063
Diplomacy AND Science AND Environment	112
Diplomacy AND Science AND Global AND South	60
Diplomacy AND Environment AND Global AND South	55
Diplomacy AND Science AND Latin	46
Diplomacy AND Science AND Environment AND global AND South	8

Note: Own Elaboration based on Scopus results.

Table 2. Search Algorithms in WOS (Web of Science)

Search algorithms	Number of results
Diplomacy	26.380
Diplomacy AND Science	5.560
Diplomacy AND Environment	1.334
Diplomacy AND Science AND Environment	588
Diplomacy AND Science AND Global AND South	313
Diplomacy AND Environment AND Global AND South	107
Diplomacy AND Science AND Latin	38
Diplomacy AND Science AND Environment AND global AND South	63

Note: Own elaboration based on the results in WOS (Web of Science)

In order for this process to be successful, several steps were taken: first, the databases were checked to see if they yielded results appropriate to the topic under study; then, a review was carried out to select the documents that specifically corresponded to scientific diplomacy; and finally, the abstract and keyword sections were read to ensure that the documents found were consistent and coherent with the topic. Despite the rigor of this process, it became clear that the information in the Global South is scarce since when the search was focused on the region, only 8 results were found in SCOPUS and 63 in WOS (Web of Science), none of which had information directly related to environmental science diplomacy, and several of those that had this information did not refer to the Global South or to Latin America and the Caribbean. Figure 1 reflects the scientific production of the country, where a large part of the documents have been produced by the Global North, that is, knowledge production is about and for developed countries, while in the Global South knowledge production is lower, a situation that is reflected in the graph where knowledge is produced in Brazil, Chile, India, and some Asian countries. This gives an overview of information in the region and the performance of scientific activity.

Figure 1. Bibliographic Production by Country

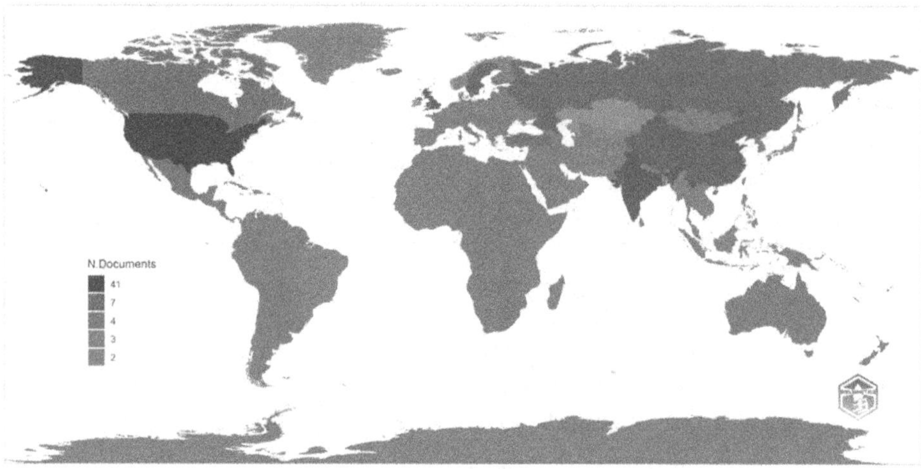

Note: Own elaboration from Bibliometrix with data from WOS using the Boolean code diplomacy AND scientific.

On the other hand, this analysis produced a graph of the most relevant sources (see Figure 2). In first place is The Hague Journal of Diplomacy, a European journal, where the knowledge generated has a Western global stance. In second place is The British Journal of Middle Eastern Studies (BJMES), which is part of Western educational centers, as it is also an interdisciplinary journal from a European educational center, the University of Glasgow. In third and fourth place are the journals Geopolitics and International Studies Quarterly. These are still part of the knowledge generated in the center, which is not always applicable or implementable in the peripheral states.

Figure 2. Bibliographic Production by Academic Journal

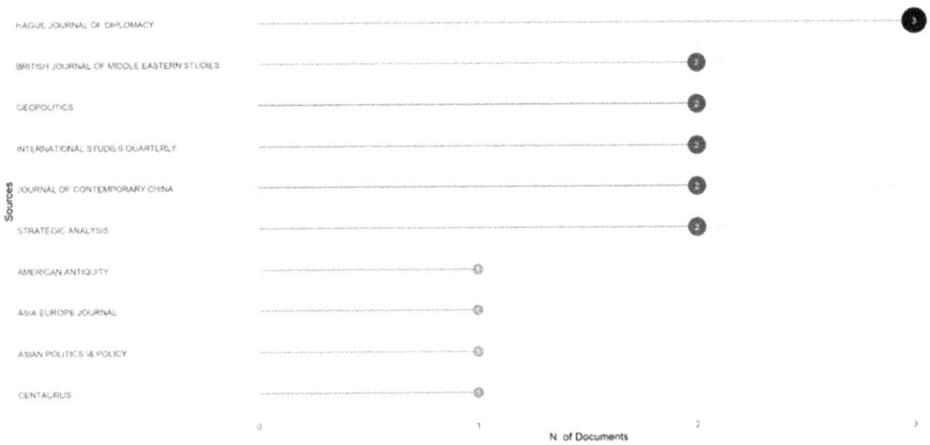

Note: Own elaboration based on Bibliometrix.

This was a difficulty for the research since the knowledge generated in Latin America and the Caribbean is in Spanish and Portuguese. Meanwhile, the databases in which the bibliometric analysis could be carried out are oriented towards knowledge generated in English. For this reason, an exhaustive search was carried out in Spanish and Portuguese in alternative databases such as Scielo, Redalyc, Miar, and Google Scholar. This allowed us to find more information, but it is limited considering that non-industrialized countries do not yet have a large amount of knowledge on the subject. Moreover, without the possibility of carrying out a bibliometric analysis, because for LAC, this topic is emerging. Through the search, a state of the art on the concept of science diplomacy in Latin America and the Caribbean has been elaborated, where functional definitions for the dynamics of the region can be found.

The papers and books found in the literature review revealed 8 resources, coming from authors located mainly in the Global North, with 3 Books, 3 Book chapters, and 2 journal papers. Their main interest is the relationships of the U.S., Germany, India, and China with other countries, based on "natural resources management". Only, one paper found interesting connections between Latin America and Africa (Nadir et al., 2020), one drew analysis on the Artic (Argüello & Rafaly, 2023); one book works on the concept of Hydrodiplomacy (Colglazier et al., 2021), and finally, the others are related to State - Environmental issues, such as India and fossil fuels (Pradhan, 2020); China and U.S. policies (Lum, 2011); Germany ocean diplomacy (Lahl et al., 2023); around more global environmental policies (Cancilla & Gargano, 2011), and North-South relationships (Rachid et al., 2023).

Towards a Path to Environmental Science Diplomacy

In recent years, SD in relation to Indigenous Peoples or the Global South has become a field of academic inquiry (See Arevalo Robles, 2017; Brigg & Graham, 2023; Carpenter, & Tsykarev, 2021; Cisneros, 2013). Journals such as Science and Public Policy and its special issue on Science Diplomacy in the Global South (Büyüktanir Karacan & Ruffini, 2023) and the Toward Indigenous Diplomacy: Relational challenges to International Relations and foreign policy survivalism issue of the Australian Journal of International Affairs (Brigg & Graham, 2023), are examples of this interest. In the context of this interest, collaboration between scholars and diplomats is fundamental and necessary, as is the application of transdisciplinary approaches to collaborative activities and strategies to address complex issues around the world. History and context have led various authors to question the role of science in foreign policy.

In this regard, it should be noted that the relationship between science and foreign policy is a topic that has generated questions over time. This is evident due to the evolution and application of the concept, as well as the different perspectives on the relationship between science and international relations, especially in environmental governance (Virchez et al., 2017; Prieto et al., 2024). This relationship is rooted in the need to adapt to an ever-changing global environment (Orsiri, 2018; Özkaragöz Doğan et al., 2021). Understanding how science influences foreign policy and vice versa can identify opportunities to promote international cooperation, address global challenges, and harness the potential of science to contribute to sustainable development and international security (Echeverría-King et al., 2020). Moreover, recognizing and considering other perspectives can foster an inclusive and constructive dialogue. Therefore, this article focuses on the relationship between environmental science diplomacy and traditional ecological knowledge. The need to resort to multimedia media stems from the emerging academic documentation in Latin America and the Caribbean, given the existence of several interviews and conferences with Latin American experts in SD as an emerging field. Thus, use is made of definitions given in video format, especially in the Latin American context, as these states are part of the periphery.

In this order of ideas, the definition of Carlos Nobre, a Brazilian scientist, plays an important role, where he proposes that science diplomacy is understood as "the use of scientific cooperation of many nations to address common problems, to build very effective and constructive international relations" (Diplocientífica, 2022, min, 1). With this, we can understand how Nobre proposes to use science diplomacy to address environmental problems commonly. On the other hand, he takes up the AAAS definition of "science for diplomacy", where he mentions that "science can provide information to support foreign policy, on issues such as climate change" (min, 3);

"diplomacy for science", which according to Nobre is "how diplomacy can promote scientific activities such as the Amazon Treaty" (min, 4); and finally "science in diplomacy", where he refers to the improvement of international relations through science, which this scientist presents as a great challenge for science diplomacy.

The definition of the climatologist Carlos Nobre is important since he has had a direct approach with the Amazon region and its communities, about this the scientist speaks of the historical destruction that has been done in this region in front of the indigenous communities that were there before the colonization, he also proposes "the construction of many indigenous universities, since there are 2.2 million indigenous communities, some of which do not want to have contact" (min, 12). He proposes that these universities maintain the TEK to interact with this region, for the conservation and preservation of the region, he also proposes "*that the Amazonian countries subsidize and create universities that bring the knowledge (...) that is transmitted by indigenous women and that is passed from generation to generation*" (Min, 12). He goes on to explain that, although it is a great challenge, this activity should be carried out to transform the relationship between this TEK with SK, for example using dialogue spaces. He mentions how this knowledge has not been taken into account in diplomacy and how, even though there are ways to protect and incorporate this knowledge, it is not included in public policy and foreign policy. For Orsiri (2018) environmental diplomacy addresses issues, such as trade (trade in protected species), intellectual property (rights of indigenous and local populations in relation to the use of natural genetic resources), energy (reduction of greenhouse gases, use of biofuels), health (impacts of consumption of genetically modified organisms (GMOs) on health and even security (consequences of global warming on transnational migration).

In this case, we define environmental science diplomacy as an emerging SD field where scientists, researchers, diplomats, and ethnic-peasant communities work together to address global environmental issues, weaving, sharing, and co-creating knowledge and methodologies to develop regional and national interests and promote policies for environmental protection, security, governance, and peace- building, such as resilience and adaptation to climate change. Having outlined the importance of weaving TEK with SK in environmental diplomacy, it is crucial to highlight the need to apply this perspective in specific contexts, such as the conservation of the Amazon. The Amazon region faces major environmental challenges that affect both national and global levels. It is crucial to implement a science diplomacy strategy to effectively address these issues. Therefore, the document Recommendations for a Science Diplomacy strategy to safeguard the Amazon is an important insight, which addresses four important points for the implementation of such a strategy: "1) Transdisciplinary scientific collaboration; 2) Strengthening of a common monitoring platform; 3) Creation of a science-policy hub; and 4) Development of capacities for science diplomacy" (Prieto et al., 2024 p. 5). In order to achieve these points in the light of Nobre's words, within the first point, the collaboration between the Amazon Cooperation Treaty Organization (ACTO) and the Association of Universities (UNAMAZ) is discussed and deepened, so that this document is a step-by-step document through suggestions given by the authors.

Diplomacy and Traditional Ecological Knowledge

ALC has stimulated a growing political interest in indigenous ecological knowledge. In particular, the proportion of scientific articles focusing on TEK has increased in all Latin American countries. In countries such as Bolivia, Colombia, Guatemala, and Nicaragua, it is much higher than in industrialized countries (UNESCO Science Report: Towards 2030, 2015). However, Büyüktanir Karacan and Ruffini (2023) suggest that the lack of recognition and/or denigration of traditional ecological knowledge by scientists from the North or the South is a challenge for SD, particularly the environmental dimension. The latter has led to calls for the "decolonization of science" and the creation of non-Western frameworks for analyzing global and transnational issues (Georghiou, 2018; Polejack, 2023).

Paradiplomacy and transdiplomacy emerge as analytical frameworks for studying the participation of ethnic and peasant communities in SD. Through the lens of paradiplomacy, Virchez et al. (2017) analyzed the case of the Inuit Circumpolar Council in the Canadian Arctic. The author examined the ICC's international actions to promote their identity and cultural practices but also to defend their economic interests as the indigenous people of the circumpolar world. On the other hand, Arevalo Robles (2017) explores the concept of transdiplomacy to inquire about the emergent relationship between diplomacy and plural ontologies (from peasant, Afro-descendant, and Indigenous peoples) in the Global South. In this context, several researchers have written about indigenous diplomacy as a field of international affairs (Brigg & Graham, 2023; Carpenter & Tsykarev, 2021; Cisneros, 2013; Pugin, 2023). An example of this in the Colombian case is the Escuela Intercultural de Diplomacia Indígena (Barraza García, 2014).

Traditional ecological knowledge is a crucial element of ESD. In particular, this regime of empirical knowledge has contributed immensely to shaping human well-being. Its influence spans a wide range of sectors, including agriculture, health, education, and governance (Subramanian & Pisupati, 2013; Vila-Viñas & Crespo, 2015). At the national level, TEK has been a platform to protect territories, dynamize tourism and the bioeconomy regionally (Pérez Cubillos, 2022), and create sustainable pathways for the future of these epistemic communities (Science Panel for the Amazon, 2021). However, it faces barriers at the national level to enter into dialogue with academic knowledge and at the level of protection by nation-states (Tamayo Ortiz

& Dilas Jiménez, 2021). There are current debates within policy design and implementation, an emerging one is diplomacy. Thus, ¿how can be effectively integrated with conventional science and policy in the international sphere? is a key question. TEK faces complex challenges such as biopiracy and equity in trade negotiations (Taubman, 2007). Scenarios such as the World Intellectual Property Organization (WIPO) are contested spaces for the interests of different actors (Fernández, 2012; Karlsson, 2023), and we suggest that environmental science diplomacy could occupy these conversations and tensions.

For the Colombian case, we share an empirical experience from the Colombian Institute of Hydrology, Meteorology, and Environmental Studies (IDEAM), with the aim of connecting the theoretical points. Specifically in the area of ecosystems - glaciology. The study of science diplomacy with an environmental focus and its relationship with traditional knowledge is incomplete without the experience of glaciologist Jorge Luis Ceballos. His work, focused on Colombia's glaciers, a crucial ecosystem for the analysis of climate change, provides invaluable information for the understanding of this issue at the national and international levels, particularly the links between the Andean glaciers and Antarctica. According to Ceballos, Colombia has "99% of the glaciers on the equator" (Presidencia de la República de Colombia, 2023, Min. 4). In this regard, an IDEAM report clarifies that to date, Colombia has 6 geographical areas with glacier coverage, distributed in 4 volcanoes: Nevado del Ruiz volcano, Nevado Santa Isabel volcano, Nevado del Tolima volcano, Nevado del Huila volcano and 2 snow-capped mountains: Sierra Nevada de Santa Marta and Sierra Nevada de Cocuy or Güicán. (IDEAM, 2023).

In addition, Ceballos commented that he is the only glaciologist in Colombia, due to the small number of glaciers in the country. In this interview, he mentions that the indigenous communities told him "that this is a sacred, spiritual place (...)". (Presidencia de la República de Colombia, 2023, Min, 16), and then he mentions how they reached an agreement that would allow him to go up there in the future to measure climatic and cryospheric variables. At this point, he mentions that the meeting he had with the indigenous leaders did not allow him to take the necessary equipment to conduct climatic research. And here lies the need to weave environmental science diplomacy with traditional ecological knowledge, especially within natural boundaries. This is because the study of some geographical areas is not taking place, in one way or another, because there is no meeting point between TEK and the way environmental issues are currently studied.

The latter is key and is an opportunity for this form of paradiplomacy by local organizations and research institutes, especially in the context of national interest in climate and ecosystem research. The ancestral knowledge of indigenous communities, such as those living in the Sierra Nevada de Santa Marta, plays a fundamental role in the protection of glaciers. Their holistic view of the environment and their connection to ecological cycles allow them to understand and manage these ecosystems sustainably (Science Panel for the Amazon, 2021). In this context of the limited inclusion of ethnic communities in diplomacy, the case of Colombia's ambassador to the United Nations in New York City, a female indigenous leader of the Arhuaco people, is noteworthy. Here, a question for further research could be the promotion of inclusiveness, intercultural dialogue, and mutual epistemological respect in order to more effectively address global and especially environmental challenges with actors who are not represented in traditional diplomatic institutions and practices.

Finally, one of the largest international negotiation spaces on environmental issues is the Conference of the Parties to the Convention on Biological Diversity (COP 16 CBD), which will take place from October 21 to November 1, 2024 in Cali, Colombia. According to the CBD, three areas need to be accelerated in order to achieve the momentum of action: 1. Translating the Biodiversity Action Plan into decisive national action; 2. Mobilizing and strengthening the means of implementation; and 3. Accelerating progress on access and benefit sharing. This is the next COP following the adoption of the Kunming-Montreal Global Biodiversity Framework.

From an ESD perspective, it could have several implications for regional and global environmental negotiations. Some of these implications are detailed below: Opportunities for cooperation: COP16 would serve as a space where scientists, diplomats, and representatives of indigenous peoples and local communities (IPLCs) can work together to find ways forward for the environment. The formation of strategic alliances and joint projects can strengthen environmental science diplomacy; knowledge exchange: COPs and similar events allow countries and actors to share knowledge and practices to enrich their understanding of environmental issues in transnational management, thereby promoting the adoption of solutions based on science and TEK. Finally, international commitments and agreements: The signing of new international commitments and agreements to address key environmental issues, mainly related to biodiversity, would strengthen the global framework for environmental action and international cooperation that could emerge from COP16.

DISCUSSION

This exploratory text offered an emerging perspective on environment and diplomacy, which focuses on reviewing non-traditional perspectives, such as indigenous communities, Latin American scientists, and non-traditional visions of diplomacy and science diplomacy in the environment, in order to find new or alternative visions in this emerging field of knowledge. In Latin America and the Caribbean, science diplomacy has played an emerging and complex role due to the limitations of government institutions and the weak articulation between diplomatic careers and national science focal points. However, efforts have been made to strengthen scientific and technical capacities in environmental issues, to promote the formation of international and collaborative networks, and to encourage global research and the publication of studies related to the environment at the regional level. It can promote the valorization and preservation of TEK by recognizing its importance in solving societal problems and facilitating the participation and the power of making binding decisions of local communities in scientific decision-making.

ESD is an activity that seeks to strengthen international relations and promote cooperation between countries through the interaction of science, technology, research, and diplomacy. The main outcome of this manuscript is the conceptualization of ESD and its relationship with TEK. Collaboration between environmental science diplomacy and TEK holders has significant potential to contribute to environmental conservation in Latin America and the Caribbean, especially for the Amazon as a transnational territory (Prieto et al., 2024). By recognizing the importance of these epistemologies, environmental science diplomacy can act as a bridge between modern science and traditional ecological knowledge. This will not only strengthen the capacity of local communities to address environmental challenges but also enrich international conservation strategies and national research initiatives (Science Panel for the Amazon., 2021). Another element is the fragmentation of the nation-state and the emergence of non-state actors that challenge its hegemony in international relations, leading to new forms of governance (Virchez et al., 2017).

The case of the Sierra Nevada de Santa Marta glaciers illustrates the need for scientific diplomacy that integrates the environmental lens and TEK. The meeting between Ceballos and the Indigenous community, although fruitful in terms of mutual understanding, highlighted the lack of mechanisms for effective negotiation, collaboration, and cooperation. On the other hand, it should be noted that strengthening environmental science diplomacy, in this case, requires several things: 1. an intercultural dialogue that makes it possible to promote spaces for knowledge exchange and trust building between scientists and scientific communities in ecosystems that have transnational implications; 2. inclusive methodologies that are developed to integrate TEK with Western science, or, as mentioned at the beginning and following Nobre's vision, to create a communication network that promulgates or disseminates TEK, for example, from the perspective of the Amazon region; and 3. the recognition and valuation of traditional ecological knowledge in national policies and international affairs.

Amid emerging efforts to decolonize science and technology, Southern researchers and practitioners are voiceless. This reveals how the colonial attributes of technoscience are still prevalent in the practices of science diplomacy. In this context, the links between decolonization and diplomacy need to be uncovered (Georghiou, 2018). Challenges could include overcoming cultural barriers and promoting equitable collaboration between diplomats, scientists, and ethnic and local communities, overcoming historical practices of exclusion in science and diplomacy, especially of ethnic communities.

CONCLUSIONS AND FUTURE RESEARCH

While conventional diplomacy has been the cornerstone in the representation of national interests by governments, ESD expands its boundaries by engaging non-state actors, especially local governments, local organizations, and research institutions, in the search for collaborative and effective solutions to local and global challenges. Recognizing the diversity of epistemologies and cosmologies in the ESD framework could reconnect local organizations, governments, and scientific organizations to create collaborative spaces at local and international levels. It is necessary to deepen the various literature reviews and expand the sources of information. It is recognized that the databases have a strong bias towards literature produced in the global north and generally in English.

Although Latin America and the Caribbean lack programs and strategies in science diplomacy, they have made efforts to maintain scientific and technical capacities in the environmental field. The region is at a new and complex moment in which collaboration between local governments, rural and ethnic communities (peasant, Indigenous, and Afro-descendant), and scientific actors can make a difference in achieving new paths toward knowledgue dialogues. The active participation of these local actors can be fundamental in addressing environmental issues and promoting sustainable development in the region. Effectively addressing these issues involves several key objectives. First, the exchange of scientific knowledge and the promotion of evidence-based public policies will be an essential pillar. Working with local governments, civil society institutions, and rural communities, environmental science diplomacy can facilitate the transfer of knowledge between scientists, policymakers, and citizens, thereby promoting the adoption of knowledge-based practices to address environmental challenges. Second, there is an urgent need to foster international environmental cooperation. By establishing direct relationships with their counterparts in other countries, local governments can foster partnerships that transcend national boundaries and promote joint efforts to address common environmental problems; some seminal examples are the Amazon, the Antarctic, and the Caribbean. By engaging these local actors, environmental science diplomacy becomes a key facilitator of international environmental cooperation, especially between communities and researchers.

Third, valuing and preserving traditional knowledge is emerging as a critical objective in environmental management. Environmental science diplomacy can be an effective tool for integrating local community knowledge into conservation, buen vivir, and sustainable development strategies. By working closely with scientists and diplomats, local communities can actively participate in the collection, preservation, and dissemination of their ancestral knowledge about the environment. The methodology for this proposal is a systematic literature review on science diplomacy with a focus on the environmental field. Different experiences in the region generate a preliminary map of practices that can be compared in future research. Finally, COP16 in Colombia could be a turning point for environmental science diplomacy in the region. The Colombian government and the CBD should carefully assess these factors and be adequately prepared to maximize the benefits and reduce the potential challenges associated with the organization of the event.

REFERENCES

Arevalo Robles, G. A. (2017). Indigenous diplomacy: A transdiplomatic approach. *SI SOMOS AMERICANOS-REVISTA DE ESTUDIOS TRANSFRONTERIZOS*, 17(1), 141–169. 10.4067/S0719-09482017000100141

Argüello, G., & Rafaly, V. (2023). Science diplomacy and Asian states: Transforming the governance landscape in the Arctic. *The Polar Record*, 59, e41. 10.1017/S0032247423000281

Barraza García, R. (2014). Escuela Intercultural de Diplomacia Indígena: Academia, posdesarrollo y diálogo de saberes. *Desafíos*, 26(1), 13–56. 10.12804/desafios26.1.2014.01

Brigg, M., & Graham, M. (2023). Approaching First Nations diplomacy from the Australian continent. *Australian Journal of International Affairs*, 77(6), 585–589. 10.1080/10357718.2023.2265849

Büyüktanir Karacan, D., & Ruffini, P.-B. (2023). Science diplomacy in the Global South—An introduction. *Science & Public Policy*, 50(4), 742–748. 10.1093/scipol/scad028

Cancilla, R., & Gargano, M. (2011). *Global environmental policies: Impact, management and effects*. Nova Science Publishers, Inc.

Carpenter, K., & Tsykarev, A. (2021). Indigenous Peoples and Diplomacy on the World Stage. *AJIL UNBOUND*, 115, 118–122. 10.1017/aju.2021.7

Cisneros, G. A. T. (2013). Diplomacia indígena: Transitando del problema a la solución. *Revista Mexicana de Política Exterior*, 98, 197–232.

Colglazier, E. W., Vafai, H. A., Lansey, K. E., & Bryson, M. D. (2021). *Sustainable Development for the Americas: Science, Health, and Engineering Policy and Diplomacy*. CRC Press. 10.1201/9781003220503

Cruz Chavarro, N., & Arevalo Robles, G. A. (2021). Indigenous Summits: Politics and Ancestral Diplomacy in Latin America. *NOVUM JUS*, 15(1), 133–160. 10.14718/NovumJus.2021.15.1.7

DiploCientifica. [@diplomaciacientifica]. (2022). *Hablemos de Diplomacia Científica con Carlos Nobre*. Tomado de Youtube. https://www.youtube.com/watch?v=QPogfuMQvgM (Abril 14 de 2022).

Echeverría-King, L., Widmaier Müller, C., & Aquino Valle, K. (2020). La importancia de la Diplomacia Científica y los Objetivos de Desarrollo Sostenible: Desafíos y Oportunidades. Revista Internacional De Cooperación Y Desarrollo, 7(1), 166-179. 10.21500/23825014.4570

Echeverría-King, L. F., Fossati, A., Raja, N. B., Bonilla, K., Urbani, B., Whiffen, R. K., & Vizinová, T. (2023). Scientific collaborations between Latin America and Europe: An approach from science diplomacy towards international engagement. *Science & Public Policy*, 50(4), 794–806. 10.1093/scipol/scad025

Fernández, E. G. (2012). Conocimientos Tradicionales, Propiedad Intelectual y Negociaciones Internacionales: ¿Gestación de Nuevos Derechos? *Revista Propiedad Intelectual*, 11(15), 204–226.

Foyer, J., & Kervran, D. D. (2020). Mettre en récit les savoirs traditionnels Une diplomatie scientifique alternative à la COP21 [bonus—Article en ligne]. Terrain, 73. *Terrain*, (73). Advance online publication. 10.4000/terrain.20607

Georghiou, C. A. (2018). Decolonization and Diplomacy. In The Encyclopedia of Diplomacy (pp. 1–9). John Wiley & Sons, Ltd. 10.1002/9781118885154.dipl0071

Instituto de Hidrología. Meteorología y Estudios Ambientales - Ideam. (2023). Informe del estado de los glaciares colombianos 2021. Bogotá: Ideam.

Isaza, J. P. (2020). Breve repaso del contexto institucional de la diplomacia científica, tecnológica y de innovación en Colombia 2004-2006. Innovación y ciencia, 27(2). https://innovacionyciencia.com/articulos_cientificos/breve-repaso-del-contexto-institucional-de-la-diplomacia-cientifica-tecnologica-y-de-innovacion-en-colombia-2004-2006

Kaltofen, C., & Acuto, M. (2018). Science Diplomacy: Introduction to a Boundary Problem. *Global Policy*, 9(S3), 8–14. 10.1111/1758-5899.12621

Karlsson, Á. G. (2023). The Lion, the Spider, and the Laid-Off Janitor: Tales from the World Intellectual Property Organization. Narrative Culture, 10(1), 88–107. *Scopus*. Advance online publication. 10.1353/ncu.2023.0003

Lahl, R., Ferse, S., & Bleischwitz, R. (2023). Reflecting on a Research Institute's Role as a Tool for Ocean Science Diplomacy—ZMT's Mission to Collaborate. In Ittekkot, V., & Baweja, J. K. (Eds.), *Science, Technology and Innovation Diplomacy in Developing Countries: Perceptions and Practice* (pp. 311–328). Springer Nature., 10.1007/978-981-19-6802-0_20

Lascano & Vedia. J. R (2021). Relaciones internacionales, política exterior y diplomacia. En: Colotta, M., Degiorgis, P., Lascano y Vedia, J., Rodriguez, Á. (compiladores) (2021). Manual de relaciones internacionales, Buenos Aires de: https://www.teseopress.com/manualderelacionesinternacionales/chapter/relaciones-internacionales-politica-exterior-y-diplomacia/#footnote-41-1

Leff, E. (2004). Racionalidad ambiental y diálogo de saberes. *Polis Revista Latinoaméricana*, 7, 2–29.

Lopera, W. (2020). Colombia en la antártida: Aproximaciones a una investigación científico/social … Un llamado a la academia/estado/sociedad. *Revista Boletín Redipe*, 9(12), 12. 10.36260/rbr.v9i12.1145

Lum, T. (2011). *U.S.-China relations: Policy issues*. Nova Science Publishers, Inc.

Manero, M. L. A. (2018). Actores sociales en la Cumbre del Clima de París: El mensaje de pueblos indígenas, grupos religiosos, mujeres y jóvenes. *Revista Mexicana de Ciencias Politicas y Sociales*, 233(233), 221–254. 10.22201/fcpys.2448492xe.2018.233.59458

Martínez, L. M. (2019). La paradiplomacia. Una aproximación a las nuevas teorías. *Oasis*, 30(30), 145–162. 10.18601/16577558.n30.09

Nadir, M., Siqueira Tybusch, J., & Bonesso de Araujo, L. E. (2020). A diplomacia Ambiental no Sul Global: Um olhar sobre África e América Latina. *JUSTIÇA DO DIREITO*, 34(1), 314–363. 10.5335/rjd.v34i1.11135

Orsini, A. J. (2018). La Diplomatie Environnementale. In *Balzacq, Dans Thierry; Charillon, Frédéric; Ramel, Frédéric. Manuel de Diplomatie* (pp. 263–283). Presses de Sciences.

Özkaragöz Doğan, E., Uygun, Z., & Akçomak, İ. S. (2021). Can science diplomacy address the global climate change challenge? *Environmental Policy and Governance*, 31(1), 31–45. 10.1002/eet.1911

Pérez Cubillos, C. M. (2022). El Conocimiento Tradicional Ecológico Indígena y su papel en el blindaje de territorios étnicos y en el ordenamiento territorial de Leticia, Amazonas – Colombia. *Espacio y Desarrollo*, 39(39), 1–35. 10.18800/espacioydesarrollo.202201.004

Polejack, A. (2023). Coloniality in science diplomacy—Evidence from the Atlantic Ocean. *Science & Public Policy*, 50(4), 759–770. 10.1093/scipol/scad027

Pradhan, S. K. (2020). *India's quest for energy through oil and natural gas: Trade and investment, geopolitics, and security*. Springer Singapore., 10.1007/978-981-15-5220-5

Presidencia de la República de Colombia. (2023, September 1). Entre-Vistas con Alma de País hoy: Jorge Luis Ceballos, Glaciólogo IDEAM. Retrieved from https://youtu.be/7KcIBSKgoF4?si=Kvk_nulbw2acN_Bw

Prieto, J., Cumba, L. M., Echeverria-King, L. F., Benavides, S., & Alarcón, C. (2024). Recomendaciones para una estrategia de Diplomacia Científica para la salvaguarda de la Amazonía (p. 15). IAI. Retrieved from 10.5281/zenodo.10695010

Pugin, M. (2023). Indigenous Australian diplomacy and the United Nations declaration on the rights of Indigenous peoples. *Australian Journal of International Affairs*, 77(6), 625–631. 10.1080/10357718.2023.2268022

Rachid, A., Moussaoui, M., & Benbba, B. (2023). Rethinking Natural Resource Development: A Sustainable and Equitable Approach to North-South Relations. In Moussaoui, M. (Ed.), *Global Science's Cooperation Opportunities, Challenges, and Good Practices. First* (pp. 50–85). IGI Global. 10.4018/978-1-6684-7874-5.ch003

RIS. (2020). Southern Perspectives on Science Diplomacy. Nueva Delhi: Research and Information System (RIS) for Developing Countries. Retrieved from https://cutt.ly/bxDkZq0

Science Panel for the Amazon. (2021). Connecting and sharing diverse knowledge systems to support sustainable pathways in the Amazon. In Amazon Assessment Report 2021 (First). Science Panel for the Amazon. Retrieved from https://www.theamazonwewant.org/wp-content/uploads/2022/05/Chapter-33-Bound-May-16.pdf

Soler, M. G. (2021a). *Diplomacia científica en américa latina y el caribe*. UNESCO. Retrieved from https://forocilac.org/wp-content/uploads/2020/11/PolicyPapers-DiplomaciaCientifica-ES.pdf

Soler, M. G. (2021b). Science Diplomacy in Latin America and the Caribbean: Current Landscape, Challenges, and Future Perspectives. *Frontiers in Research Metrics and Analytics*, 6, 670001. 10.3389/frma.2021.67000134222772

Stake, I. E. (1998). Investigación con estudios de Caso. Veracrúz. Universidad Veracruzana. Retrieved from https://www.uv.mx/rmipe/files/2017/02/Investigacion-con-estudios-de-caso.pdf

Subramanian, S., & Pisupati, B. (Eds.). (2013). *Traditional Knowledge in Policy and Practice: Approaches to Development and Human Well-Being (First).* United Nations Press., 10.18356/0e3bca6c-en

Tamayo Ortiz, C. V., & Dilas Jiménez, J. O. (2021). Conocimientos tradicionales y recursos genéticos: Una revisión conceptual, importancia y marco legal en Ecuador y Perú. *Revista de Investigación Científica y Tecnológica Alpha Centauri*, 2(3), 2–14. 10.47422/ac.v2i3.37

Taubman, A. (2007). Cereal offenders: Access and equity in trade negotiations on knowledge resources. In J. P. Kesan (Ed.), Agricultural Biotechnology and Intellectual Property: Seeds of Change (First, pp. 97–131). CABI.

The Royal Society-AAAS. (2010). New frontiers in science diplomacy. Royal Society.org. https://royalsociety.org/~/media/Royal_Society_Content/policy/publications/2010/4294969468.pdf

Turchetti, S., & Lalli, R. (2020). Envisioning a "science diplomacy 2.0": On data, global challenges, and multi-layered networks. *Humanities & Social Sciences Communications*, 7(1), 1–9. 10.1057/s41599-020-00636-2

UNESCO Science Report: Towards 2030. (2015). United Nations Scientific, Cultural and Education Organization.

UNESCO Science Report: Towards 2030 (2020). Latin America's Place in the World. Available online at: https://en.unesco.org/sites/default/files/latin_americas_place_in_the_world_usr15.pdf

Vessuri, H. (2013). El nuevo mantra de la diplomacia científica internacional: ¿Co-diseño de conocimiento? ¿Investigación integrativa? *Universitas Humanistica*, 76, 25–50. https://revistas.javeriana.edu.co/index.php/univhumanistica/article/view/5905

Vila-Viñas, D., & Crespo, J. M. (2015). Saberes y conocimientos ancestrales, tradicionales y populares. In Buen Conocer / FLOK Society: Modelos sostenibles y políticas públicas para una economía social del conocimiento común y abierto en Ecuador (Digital, pp. 551–616). Asociación aLabs. https://book.floksociety.org/ec/3/3-2-comunidades-saberes-y-conocimientos-ancestrales-tradicionales-y-populares/

Virchez, J., Zepeda, R., & Marquina-Márquez, A. (2017). Gobernanza y Paradiplomacia en el Ártico canadiense: El caso del Consejo Circumpolar Inuit. *Revista Internacionales*, 3(6 (julio-diciembre)), 14–37.

Chapter 8
Strategic Actions and Prospective Hydrodiplomacy in the Context of Transboundary Waters

Bertha Cecilia Garcia Cienfuegos
National University of Tumbes, Peru

Felix Enrique Alcoser Torres
National University of Tumbes, Peru

ABSTRACT

More than 60% of the world's flow of fresh water is shared, of the 33 countries that make up the Latin American and Caribbean region, 22 share watercourses, lakes and aquifers. The basin is considered the ideal unit to carry out the sustainable management of water resources, that contributes to territorial management and social balance at the local, national and cross-border level. The Puyango-Tumbes river basin (Peru-Ecuador) it constitutes one of the most important binational basins in South America. As the quality of water degrades or the available quantity of water has to meet increasing demands, competition among water users intensifies; nowhere is this situation more destabilizing than in river basins that cross political borders. The most critical situation began in 1980s due to mining activity in Portovelo and Zaruma area, with the presence of informal companies that were later relocated to the Calera and Amarillo rivers, the main tributaries of the Puyango-Tumbes river, dumping the mining effluents there, focusing the contamination problem and compromising the populations settled downstream of the basin. This water source

DOI: 10.4018/979-8-3693-2746-3.ch008

presents a serious environmental problem. On the one hand, it is the common recipient of mining effluents and on the other hand of wastewater, urban solid waste and contamination from the excessive use of agrochemicals in the agricultural sector. In this context, it is required to consolidate water security through diplomacy and geopolitics based on strengthening hydrodiplomacy, considering the institutional frameworks for governance in transboundary watercourses and the fundamental principles of international water law.

AN INTRODUCTION TO SCIENTIFIC HYDRODIPLOMACY IN TRANSBOUNDARY WATERS

Natural resources constitute the environmental supply and the source of use from which populations satisfy their essential needs; of all the elements that constitute the environmental offer, the water resource is of vital importance; for all forms of life, it is a scarce, vulnerable, strategic and indivisible resource (Garrido & Hernández, 2008).

In effect, water constitutes a vital resource that has always been the basis of civilizations, pointing out that the living conditions and income of populations, especially in rural areas, depend directly on the availability of this good (Baraka, 2022).

At a global level, the perspective related to this important resource proposes elevating water to the highest levels of national political agendas; therefore, water must not only be at the center of efforts to promote inclusion and human development, but also in climate change adaptation and mitigation policies. Consequently, governments, citizens and private actors in the region must consider water as a strategic, finite and manageable good, which knows no borders (World Bank, 2022).

Water has been used as a diplomatic tool to promote peace and collaboration between nations, in fact, the action of sharing rivers or transboundary waters fairly is an effective way to promote peaceful cooperation, since transboundary waters (the aquifers and lake and river basins shared by two or more countries) support the life of a large number of communities and ecosystem (Pohl, B. 2014).

Transboundary waters account for approximately 60% of global river flow; the territories of 148 nations fall within international basins that include the political boundaries of two or more countries; 21 countries lie entirely within them. In addition, there are 276 transboundary river and lake basins, and around 300 transboundary aquifers worldwide. Transboundary basins cover 46% of the earth's land Surface (UN-Water 2014). Of the 33 countries that make up the Latin American and Caribbean region, 22 share waterways, lakes and aquifers (UN - Water, 2021).

The basin is considered the ideal unit to carry out the sustainable management of water resources and a powerful tool that contributes to territorial management and social balance at the local, national and cross - border levels. Indeed, water resources know no borders or geopolitics; Water, considered an everlasting good, is far from lending itself to that concept and its sustainable use increasingly entails more challenges and threats for states (Cap - Net & LA - WETnet, 2013); Its scarcity is part of the loss of numerous natural resources as a consequence of misuse by humans and, above all, inaction in the face of climate change.

As water quality degrades or the available quantity of water has to meet increasing demands, competition among water users intensifies; nowhere is this situation more destabilizing than in river basins that cross political borders. It is considered that by the year 2050, between 4.8 and 5.7 billion people will live in areas with potential water scarcity for one month a year (UN - Water, 2021).

Transboundary Waters: Exploring the Right to Water

Transboundary waters represent 60% of the world's freshwater flows, and there are 153 countries that have territory within at least one of the 310 transboundary hydrographic, river and lake basins; This shared water resource has been and still can be a trigger for conflicts (Lynn & Aaron, 2023).

Therefore, the most significant indicator of cross-border tensions arises when one of the parties, normally located upstream, wants to build some type of infrastructure, without negotiating the impacts with downstream neighbors.

More than 3 billion people around the world depend on water that crosses national borders, however, only twenty-four countries have cooperation agreements for all the water resources they share. When water is scarce or contaminated, or when people have unequal or no access, tensions between communities and countries can increase.

The most common situation of conflicts over transboundary water occurs when this is the trigger due to a clash of interests between different users, or when the actions of one party cause the quantity and quality of the water of the other to decrease. The best - known case in Africa is the dispute over the waters of the Nile River. Egypt, Sudan and Ethiopia have had prolonged conflicts over the years over the use of Nile water that still persist today. The construction of the Grand Ethiopian Renaissance Dam is the sticking point currently, as downstream countries Egypt and Sudan are concerned about the impact on their water supply (UN - Water, 2021). In fact, and despite the fact that in December 2023, the negotiations to find a technical and legal solution that could benefit the three parties were terminated, Ethiopia continues with the project and the mega infrastructure entered its final filling phase, in January 2024.

Towards Efficient Integrated Management of Water Resources

Integrated water resources management - IWRM, has dominated public water policies worldwide since the 1990s. With the arrival of the millennium, criticism also came from social scientists because despite its apparent neutrality, citizen participation and goals of equity, efficiency and sustainability, it is considered that this management model does not always achieve its goals. This is mainly due to the fact that it is implemented on a water landscape conditioned by infrastructure and policies that do not necessarily prioritize equity and sustainability, but rather an idea of order and national development. Likewise, the Integrated Management of Water Resources requires that the rights to water be recognized and formalized by the State; Although formalization can provide water users with water security, on the other hand, it runs the risk of making invisible valid traditional rights that users have exercised long before the State intervenes in water management.

Indeed, integrated water resources management promotes the coordinated management of water, land and other related resources, maximizing the economic and social well-being of populations, in an equitable manner, and without compromising the sustainability of vital ecosystems (GWP), 2000). This coordination requires powerful water governance mechanisms, both at the local and regional levels, to ensure that sufficient water is available for all users in a river basin.

Therefore, cross - border cooperation constitutes an essential element of water security and regional integration and stability. Likewise, sustainable water management is a pillar for achieving the SDGs, because clean, safe and reliable water is inherent to health, food security and economic opportunities.

In recent years, the science - politics - society relationship in the management of water resources is becoming more marked; and it has been possible to conceive how researchers can do science that is truly relevant to society, how society's contribution to science and the sustainable management of water resources can be valued, as well as the actions necessary to achieve a plural and constructive dialogue between the different actors involved in water management. Furthermore, how is it possible to value intangibles in decision-making related to water management.

The conjunction and interaction between water science and water policy can be considered, especially with regard to conflict prevention and its resolution. An immense and growing threat worldwide is the demand for water of our populations and economies, it is increasing; while availability remains the same as always. Pollution and overexploitation of aquifers threaten both quality and quantity, and climate change endangers social and ecological systems around the world. The processes necessary to address these threats are even more difficult when water is shared across political borders, sectoral or ethnic divisions, or competing users within a basin, all of which are of concern (Lynn & Wolf, A. 2023).

Relationship Between Hydrodiplomacy and Science

Water diplomacy or hydrodiplomacy consists of dialogue and negotiation processes to reconcile interests between countries around water resources; it constitutes a process that operates under the sovereign authority of the States, but that also triggers cooperation between multiple actors, including municipalities and civil society organizations.

Water diplomacy promotes cooperation and conflict prevention in relation to shared water resources, and contributes to regional security and stability beyond the water sector, so that these resources are shared safely. Therefore, according to the Postulates of Schmeier (2018), it constitutes a diplomatic instrument to resolve existing or emerging disagreements on shared water resources, between two or more countries, with the objective of resolving or mitigating these problems for the good of cooperation, regional stability and peace. Likewise, scientific hydrodiplomacy makes use of science to prevent conflicts and crises, bringing this concept closer to the challenges related to transnational waters (IUCN, 2017).

Definitely, water diplomacy can contribute constructively to the collaborative management of available water, improving the water resources and governance capacities of conflicting parties, promoting cooperative decision-making processes and inclusive policy institutions, and facilitating peaceful dispute resolution. In this sense, there is a growing recognition of the importance of water diplomacy, especially in the case of transboundary waters, and peacebuilding strategies to mitigate the risks of conflicts related to water resources (Wilder, 2019).

A specific and important case is the Indus Waters Treaty, signed by India and Pakistan in 1960, after numerous disputes between both countries and the final intervention of the World Bank to reach an agreement. The Indus River basin is vital to the economy and agriculture of both countries, supporting millions of people and generating a considerable economy. Thus, said Treaty regulates the distribution of water from the Indus River and its tributaries, assigning the western rivers (Indus, Jhelum, Chenab) to Pakistan and the eastern rivers (Ravi, Beas, Sutlej) to India, and grants India the right to certain non-consumptive uses of water in western rivers (Pohl, 2014). However, although the Treaty has been fundamental to manage the distribution of water, it needs to be rethought due to the consequences of climate change, which have especially affected both countries.

Furthermore, there is increasing awareness of the history of cooperation in the context of transboundary waters, the factors that lead to disputes and the importance of institutions for political resilience, as well as more training initiatives around the world to train students and professionals in water diplomacy. Monitoring tools are getting better, especially satellite - based, as are the frameworks for dialogue, treaty-making and negotiations. In short, there is a large and growing community dedicated to water cooperation and diplomacy (Lynn & Wolf, A. 2023).

Cooperation on Transboundary Waters

Stories of war over water periodically make headlines, fueling fears that tensions over water resources are a potential source of conflict. Historical analysis refutes this idea: since 1814, 300 treaties have been negotiated to resolve conflicts; but the past is no guide to the future. Tensions over water resources are increasing, exacerbated by disruptions to the water cycle caused by climate change, population growth and unsustainable patterns of consumption and production.

Peacefully overcoming the water crisis by harnessing the potential for cooperation requires great effort. To get the most out of concerted management of transboundary rivers and lakes, resources must be significantly increased to create and strengthen transboundary basin organizations. International water law (enshrined in the 1992 Helsinki Convention and the 1997 New York Convention) are important aspects to discuss.

Also, the need to balance the principle of no harm with the principle of equitable and reasonable use, which structures it to reinforce cooperation and create upstream-downstream solidarity between coastal States. The regional legal frameworks (with the WFD for the European Union and the SADC protocol in Southern Africa) and those specific to groundwater (draft articles of the International Law Commission on the law of transboundary aquifers in 2008) this imperative of cooperation has become frequent. The Sustainable Development Goals (SDGs) also provide a policy framework to improve the integrated management of transboundary water resources, with a specific target (6.5) and an indicator (6.5.2) to monitor progress (UNESCO, 2021).

Currently, many states opt for cooperation, because it is more effective; from there, they obtain more benefits from the coordinated exercise of their respective sovereignty than from unilateral management (UN Water, 2024). Cooperating in the allocation of water volumes and the benefits derived from the use of the resource, sharing experiences and innovative governance, and investing in joint projects, especially in infrastructure, are an excellent way to prevent conflicts.

Indeed, cross-border cooperation continues to focus too often, and exclusively, on the quantitative management of resources. We must not forget the application of Integrated Water Resources Management (IWRM) to water quality, but it is a particularly difficult challenge to face in transboundary basins. This interstate configuration is even more complex than the already considerable challenge of managing water quality in a national watershed. States concerned about preserving their water security and their supply of water in sufficient quantity and quality make mutual concessions. And not only between economic sectors and administrative levels.

The stakes are high: 40% of the world's population lives in a transboundary basin. Therefore, it is vital to strengthen cooperation between States to manage water quality and preserve the health of aquatic ecosystems. International water law (in particular the 1992 Helsinki Convention and the 1997 New York Convention) contributes by establishing the obligation to take all appropriate measures to prevent, control and reduce water pollution caused or may have a cross-border impact (GWP, 2014).

In this context, efforts must be joined and synergies between these issues promoted to increase the importance of cooperation on transboundary waters on the political agenda at the national, regional and global levels. Sharing good practices and integrating transboundary water issues into the actions, policies and programs of related sectors is one way to accelerate progress. Furthermore, at the 2023 United Nations Water Conference, countries had the opportunity to reinvigorate their commitment to transboundary water cooperation by defining roadmaps to ensure that all their basins are subject to some operational arrangement by 2030. By referring to progress and commitments against SDG indicator 6.5.2 in the voluntary national reviews of the High-Level Political Forum on Sustainable Development, it is also possible to incorporate and strengthen the commitment political with cooperation on transboundary waters; Likewise, integrate cross-border cooperation into national Integrated Water Resources Management and other sectoral measures (UNESCO, 2021).

Transboundary Waters in Latin America and the Caribbean

Of the 33 countries that make up the Latin American and Caribbean region, 22 share transboundary waterways, lakes and aquifers. The region contains a diverse sample of transboundary waters, from large rivers shared by several countries, such as the Amazon (Bolivia, Brazil, Colombia, Ecuador, Guyana and Peru), the Orinoco (Colombia, Guyana and Venezuela) and the Río de la Plata (Argentina, Bolivia, Brazil and Paraguay), to the Guaraní Aquifer System (Argentina, Brazil, Paraguay and Uruguay), and the numerous smaller basins shared between two or more countries. The Amazon Basin covers an area of 850,371 km2, Argentina and Chile share 17 hydrographic basins that represent a total area of 131,221 km2. The 22 countries in the Latin American and Caribbean region that share transboundary waters share river and lake basins. Despite the greater availability of studies and data, these show that most countries have low coverage of operational arrangements. In 10 countries, the area of transboundary river and lake basins covered by operational arrangements does not reach 10%, and only in 4 countries (Argentina, Brazil, Ecuador and Paraguay), at least 90% of the area of river and lake basins cross-border transactions are the subject of this type of arrangement. Among the main hydrographic basins of the region, the Amazon River basin, the Río de la Plata basin and the Lake Titicaca Poopó basin have operational arrangements. However, major efforts will still be required to ensure that all transboundary river and lake basins in the region are covered by operational arrangements by 2030 (UNESCO, 2021).

Recently, between September and November 2022, in the context of the Preparatory Framework for the 2023 World Water Conference, the Regional Group of Experts on Water Resources of Latin America and the Caribbean was formed; that will promote activities that contribute to the fulfillment of SDG 6, will promote the exchange on evolution, current situation and projections of water and water resources management in the member countries of the Andean Community of Nations - CAN, which make up Bolivia, Colombia, Ecuador, Peru), Likewise, common elements were identified and progress was made in the generation of a roadmap to promote regional cooperation in compliance with the Andean Environmental Charter (ECLAC, 2024).

STRATEGIC ACTIONS AND PROSPECTIVE SCIENTIFIC HYDRODIPLOMACY IN THE PUYANGO-TUMBES BINATIONAL BASIN

Peru Ecuador Border Region

Ecuador and Peru share a 1,529 km long border, which runs through a diverse region that extends from the Pacific coast to the Amazon rainforest. This Ecuadorian-Peruvian border region is characterized by a heterogeneity of ecosystems and great socio-cultural wealth. The geographical, cultural, economic, social and demographic characteristics of this area have facilitated integration between neighboring populations.

There are 10 transboundary hydrographic basins of the rivers: Napo, Pastaza, Tigre, Putumayo, Morona, Santiago, Mayo, Catamayo- Chira, Puyango- Tumbes and Zarumilla; distributed in two large slopes, the Amazon and the Pacific, see figure 1.

Figure 1. Transboundary Basins Perú-Ecuador

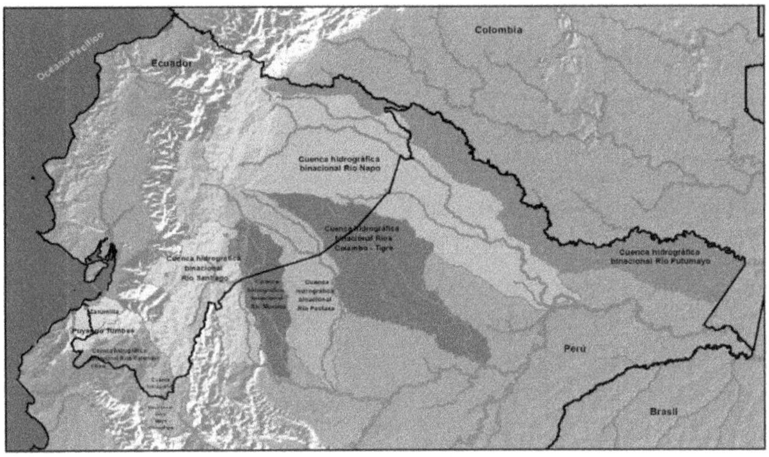

Source: IUCN

Binational Plan Peru-Ecuador

The Peace Agreement between Ecuador and Peru was signed to put an end to the differences that for decades distanced the two countries and give way to the birth of consolidation and integration for development between both nations.

In the city of Brasilia, on October 26, 1998, Their Excellencies Jamil Mahuad Witt, President of the Republic of Ecuador and Alberto Fujimori Fujimori, President of the Republic of Peru, signed the Presidential Act of Brasilia, which contains a Broad Agreement on Border Integration, Development and Neighborhood, which essentially constitute a redefinition of the nature and relations between both border countries, becoming strategic partners that face together, through cooperation and border integration, common challenges of social and order. economic.

The political will and decision of the Governments and Presidents of both countries have made it possible for the Binational Plan, Peru -Ecuador Chapters, to continue its management until 2024; through Presidential Decree No. 1465 of December 2, 2008, the National Secretary of Planning and Development - SENPLADES, has been designated as the institution responsible for the application of the Binational Plan, Ecuador Chapter; subsequently, in the city of Quito, on January 25, 2010, Lautaro Pozo Malo, Ministry of Foreign Affairs, Trade and Integration (e), together with Marcelo Torres Paz, Executive Director (e) of the Binational Development Plan of the Region border Ecuador Chapter, signed the Agreement, through which in its sixteenth article the city of Machala was established as the new headquarters of the Binational Plan. (Izquierdo, 2021).

Puyango-Tumbes Binational Project

The Puyango-Tumbes Binational Project - PEBPT is a joint effort to execute the project that aims to irrigate crop areas of both States, which will boost agricultural activity in the border region. Peru and Ecuador have been working on this project since 1971 and it was part of the infrastructure projects included in the Brasilia Peace Agreements of 1998. In compliance with the commitments assumed in the X Binational Cabinet held in Macas, on March 22, 2017, in the city of Quito, the Water Secretariat - SENAGUA of Ecuador and the Ministry of Agriculture and Irrigation - MINAGRI of Peru, celebrated the Specific Agreement for the Development of the Puyango-Tumbes Binational Project and the Execution of its Works Commons in the Framework of the Integrated Management of Transboundary Water Resources. Through this agreement, the Binational Coordination Unit of the Puyango - Tumbes Binational Project (UCB - PBPT) was created, whose function is to prepare the feasibility study, definitive design and execution of common works and supervision/supervision of the project, in compliance with the objectives of the Agreement for the use of the Puyango - Tumbes and Catamayo - Chira Binational Hydrographic Basins, of 1971, and of the Broad Peruvian - Ecuadorian Agreement on Border Integration, Development and Neighborhood, of 1998.

Puyango Tumbes Transboundary Hydrographic Basin

The Puyango - Tumbes River hydrographic basin, made up of the bordering territories of the department of Tumbes, in northern Peru, and the provinces of Loja and El Oro, in southeastern Ecuador, has a total area of 4,850 km^2, of which 1806 km^2 belong to Peru. In Ecuadorian territory it is called the Puyango River, starting from the confluence of the Pindo River and the Yaguachi River, from Linda Chara to the confluence with Quebrada Cazaderos, it constitutes the border limit between Peru and Ecuador. From this point it enters Peru and changes direction from south to north, being called the Tumbes River. This last section has a length of 94 km, at the end of which it empties into the Pacific Ocean, near Punta Mal Pelo (see figure 2).

Figure 2. Map of the Puyango-Tumbes Binatonal Basin

Source: Binational Plan Perú Ecuador

It constitutes one of the most important binational basins in the South American Pacific, basically due to the important flows of water it collects, the large number of inhabitants it supports and the extension of land that allows irrigation, especially in the valleys on the right and left banks. of the Tumbes River (PNUD - GEF, 2020). Despite this, this basin is not immune to the problem of environmental pollution, with potential sources of pollutants that are frequently discharged as a result of agricultural and domestic activities (discharge of wastewater and urban solid waste).

Transboundary Water Pollution Problems in the Basin

The most critical situation began in the 1980s due to mining activity in the area of Portovelo and Zaruma, with the presence of informal companies that were later relocated to the Calera and Amarillo rivers, the main tributaries of the Puyango - Tumbes river., dumping mining effluents there, focusing the pollution problem, compromising the populations settled downstream of the basin. In research studies carried out by the General Directorate of Environmental Health - Peru and PREDESUR - Ecuador, important findings have been found such as the presence of high levels of mercury and lead in suspended solids, with the presence of Pb being more noticeable in the basin., which exceeds by 0.044 mg/L the limit value established in the General Water Law (ANA, 2013).

Examinations carried out on people who live in the area allowed establishing high levels of lead in the blood, while the biological indicators of mercury are low (Hg in blood, urine and hair). The performance in neurobehavioral tests reflects a deterioration in health and notable differences between the inhabitants of the different ecosystems studied. On the other hand, it is indicated that high levels of thiocyanate were found in the urine in a good percentage of the population.

Likewise, the evaluation of heavy metal results from the monitoring of the Puyango-Tumbes River and tributaries, carried out by the public sectors of Peru and Ecuador, recorded levels that exceeded the maximum permissible limits in some monitoring stations in both the surface waters of the Puyango Tumbes river, as in sediments (Gavilanes, 2016).

The situation becomes more worrying, since the rural and urban populations settled near the river take water for human consumption directly from this source, among them, the city of Tumbes. Although these waters are subjected to a treatment process, the risk is latent considering that this type of contaminants adhere to the colloidal elements of the water, difficult to treat with conventional technologies such as those applied in the water treatment plant. from Tumbes - ATUSA.

On the other hand, the degradation of the Puyango-Tumbes transboundary basin puts the dry forest, one of the most extreme and biodiverse ecosystems, at risk. In addition, a binational irrigation infrastructure project is planned to be built there that further threatens the river.

The solution to the pollution of the tributaries of the Puyango River has been pursued for more than a decade: build an industrial complex next to the El Tablón tailings dam, where the 85 mineral processing plants that currently exist and that will be They distribute mainly along the Calera River. These plants operate with permits from the Mining Regulation and Control Agency (Arcom) and the Ministry of Environment, Water and Ecological Transition of Ecuador, but even so, the basic rule is not respected: minerals cannot be processed less than 200 meters from the river banks.

From this context, it is possible to infer that the waters of the Tumbes River have been and are used, without any management measure that guarantees the maintenance of its quality and the ecological processes that maintain the natural river dynamics; This water source presents a serious environmental problem, on the one hand, it is the common recipient of mining effluents and on the other hand of wastewater, urban solid waste and pollution due to the excessive use of agrochemicals in the agricultural sector (García, B. 2013).

Hydrodiplomatic Agreements Peru Ecuador

-Agreement for the Use of the Puyango - Tumbes and Catamayo - Chira Binational Hydrographic Basins, signed on September 27, 1971.
-The Broad Ecuadorian-Peruvian Agreement on Border Integration, Development and Neighborhood, signed on October 26, 1998.
-Agreement between the Republic of Peru and the Republic of Ecuador for the establishment of the Binational Commission for the Integrated Management of Water Resources of the Transboundary Hydrographic Basin of the Zarumilla River, signed on October 22, 2009.
-Presidential Declaration signed in Loja on October 26, 2010, on the occasion of the Presidential Meeting and IV Meeting of the Binational Cabinet of Ministers, the competent authorities of the respective countries were instructed to adopt urgent joint actions to resolve the problems of contamination in the Puyango - Tumbes and Catamayo - Chira basins.
-X Meeting of the Peruvian-Ecuadorian Neighborhood Commission, held in the city of Piura on October 27, 2011, where the Foreign Ministries of Peru and Ecuador, together with the National Water Authorities of both countries, were commissioned to evaluate the convenience of the formation and operation of a single Binational Commission for the management of all transboundary hydrographic basins (Zarumilla; Puyango -Tumbes; Catamayo - Chira; Mayo - Chinchipe; Morona; Santiago; Pastaza; Conambo-Tigre; and Napo) .
-Presidential Meeting and V Meeting of the Binational Cabinet of Ministers of Ecuador and Peru, held in the city of Chiclayo on February 29, 2012, in which the importance of the installation of the "Binational Commission for the Integrated Management of "Water Resources of the Transboundary Hydrographic Basin of the Zarumilla River" and the competent authorities were ordered to agree on and implement integrated management plans for all transboundary basins, starting with those of Zarumilla, Catamayo - Chira and Puyango Tumbes.
-Joint Presidential Declaration: "Strengthening Integration in the Fight against Poverty", signed in Cuenca on November 23, 2012, and the VI Meeting of the Binational Cabinet of Ministers of Ecuador and Peru, where the installation of the "Binational Commission for the Integrated Management of Water Resources of the Transboundary Hydrographic Basin of the Zarumilla River" and instructed to continue with the development of the institutionality of the Puyango-Tumbes and Catamayo-Chira hydrographic basins.

- Presidential Declaration signed in Piura on November 14, 2013 and the VII Meeting of the Peru-Ecuador Binational Cabinet of Ministers, which provides that the Water Authorities of both Peru and Ecuador advance during the first half of 2014 in the process of the creation of the Binational Commission for the Integrated Management of Water Resources of the transboundary hydrographic basins between the Republic of Peru and the Republic of Ecuador, prioritizing for this purpose the transboundary hydrographic basins of the Puyango-Tumbes rivers; Catamayo-Chira, and gradually incorporating other shared basins of the Mayo-Chinchipe rivers; Santiago; Brunette; Pasta; Conambo-Tigre; and Napo.

-Presidential Declaration signed in Arenillas on October 30, 2014 and in the VIII Peru-Ecuador Binational Cabinet of Ministers, where the Water Authorities of both Peru and Ecuador are ordered to sign the Agreement that allows the creation of the Binational Commission for the Integrated Management of Water Resources of the transboundary basins between Ecuador and Peru, until the second half of 2015.

-Presidential Declaration of Macas, signed on October 7, 2016, which provides for continuing to advance in the negotiation of the Agreement that creates the Binational Commission for the Integrated Management of Water Resources of the nine transboundary hydrographic basins.

-Agreement that establishes the Binational Commission for the integrated management of water resources of the transboundary hydrographic basins between the Republic of Peru and the Republic of Ecuador

On Wednesday, October 25, 2023, the States of Ecuador and Peru signed the Regulations of the Binational Commission for the Integrated Management of Water Resources of the 9 Transboundary Hydrographic Basins between the Republic of Ecuador and the Republic of Peru (Binational GIRH Commission Ecuador -Peru).

This act consolidates a long history of cooperation that dates back to 1971, with the signing of the "Agreement for the Use of the Puyango-Tumbes and Catamayo-Chira Binational Basins by Peru and Ecuador",

The signing of the Agreement Regulations established by the Binational Commission is a fundamental part of this process, since it operationalizes this mechanism as an effective space for bilateral articulation, participation and coordination so that, with an ecosystemic and sustainable vision, it generates a positive impact, in the quality of life of the population, productive activities, the environment and freshwater ecosystems.

Prospective and Strategic Actions

Since the Macas Declaration, in 2016; The Diplomatic Missions of Peru and Ecuador highlight the commitment to carry out the emblematic Puyango-Tumbes Binational Project, highlighting its binational and priority nature; In this sense, it was agreed to continue negotiating a Specific Agreement that allows its development.

Peru and Ecuador have not acceded to the United Nations Convention on International Water Uses, concluded in 1997, nor to the Convention on the Protection and Use of Transboundary Watercourses and International Lakes, concluded in 1992, as they are not Part of these two main international instruments, on the management of shared water resources, hydrodiplomatic actions will have, as a prospective strategy, the alternatives admitted in customary International Law.

It is necessary to specify the inclusion of the right to equitable use of the basin and the duty not to cause damage, taking special consideration to the conservation of the basin ecosystem; regarding the procedural rules that would allow substantive obligations to be put into operation, especially those related to transboundary water pollution; From there, it is essential to establish the duties of cooperation, prior notification, consultation and negotiation related to the exchange of information, and environmental impact assessment.

International cooperation should be mainly aimed at achieving the decontamination of the Puyango - Tumbes River and carrying out control and monitoring activities for the quality of the river's water. In this regard, it is important to indicate that, although it is up to Ecuador to take the necessary actions so that the mining activity that takes place within its territory stops generating negative cross-border impacts on the territory of Peru, the truth is that this country has not managed to fulfill this obligation fully; In this sense, cooperative actions would allow both States to find possible external institutions or agencies that help the river decontamination process.

Although the main solution mechanism is direct negotiation between Peru and Ecuador, other mechanisms could also be included, such as conciliation and the good offices of third parties. In this perspective, it would be important for both States to agree to go to an international court, including the International Court of Justice or arbitral tribunal (Permanent Court of Arbitration) for those disputes that cannot be resolved through hydrodiplomatic means of solution.

CONCLUSION

In this perspective, it is necessary to generate information and knowledge, which leads to the provision of basic data in the context of hydrodiplomacy and scientific and technological development of the water sector, defining priority topics.

Promote and incentivize water scientists, promoting the creation of networks in science, technology and innovation, which link and strengthen spaces to disseminate the scientific and technological knowledge generated, building a Platform to promote the scientific and technological development of the water sector.

Promote the creation of a scientific water security council, national, binational or international, that contributes to the generation of actions through the exchange of ideas and positions, promoting associations and/or proactive alliances between scientists and organizations, mainly in the context integrated management of water resources, pollution and scientific hydrodiplomacy.

It is necessary to continue analyzing the general situation of the Puyango - Tumbes Transboundary Basin, in its various aspects, environmental, organizational; and primarily deal with the legal aspect, taking into account the diplomatic and geopolitical component.

Another perspective lies in carrying out technical-scientific studies, increasingly integrated, that allow establishing a better understanding of the ecosystem and leads to proposing appropriate policies and minimization programs (control and recovery of the Tumbes River).

Generate sustainable practice of Integrated Water Resources Management - IWRM, aimed at the socialization of the Binational Basin, in order to extend the effects of pollution and raise awareness among citizens about this situation, how to confront it and solve it; carry out an ecological flow study, both in the middle and lower basin; Likewise, the medium-term solution to the environmental problem, considering the dangers and threats in the context of environmental biosafety.

Guarantee the diplomatic and geopolitical component, based on the strengthening of hydrodiplomacy, starting from the context of the institutional frameworks for governance in transboundary water courses, taking into consideration the fundamental principles of International Water Law, as well as participatory processes that provide transparency and legitimacy to national and international policies for integrated water resources management, thus consolidating water security.

Undertake consistent hydrological planning strategies by the two States; put at the service of citizens all the data and information available on the state of groundwater or, finally, combine supranational plans, human talent and sufficient budget to preserve water resources.

Likewise, it is necessary to strengthen operational capacities for cooperation in the field of transboundary waters, in particular those of transboundary basin organizations (data collection with networks of control stations dedicated to monitoring water quality and ongoing pollution, development of water information systems and warning systems in case of accidental contamination, upstream-downstream coordination and solidarity in the deployment of hydraulic infrastructures such as wastewater collection and treatment systems.

Finally, specify that, in addition to governance and technology, there is a third factor, which is linked to hydrodiplomacy; and, which is very important: transboundary cooperation for the integrated management of water resources, based on goal 6.5.2, proportion of the surface of transboundary basins subject to operational arrangements for cooperation on water matters; of SDG 6, to guarantee the availability of water and its sustainable management. In this context, through hydrodiplomacy we will not only advance technical-scientific knowledge and innovation, but we will also manage to balance national, cross-border and global interests, building peace to achieve the indicators and goals of Goal 6 of Sustainable Development.

In this context, the formation of the Peru - Ecuador Binational IWRM Commission constitutes a fundamental step for the effective management of transboundary basins that denotes its commitment and understanding of the great responsibility that this shared resource entails, the next strategic step is the creation of the specific Committees for each Basin and the work plans, according to the needs of each body of water.

REFERENCES

Asociación de Derecho Internacional. (2008). Proyecto de artículos sobre el derecho de los acuíferos transfronterizos. Disponible en https://undocs.org/es/A/RES/63/124

Autoridad Nacional del Agua - ANA. (2013). Resultado del Primer Monitoreo Participativo de la calidad del agua superficial en las cuencas de los ríos Tumbes y Zarumilla, lado peruano. Proyecto de modernización de la gestión de los recursos hídricos, Lima- Perú.

Banco Mundial (2021) El agua importa: crecimiento resiliente, inclusivo y verde a través de la seguridad del agua en América Latina.

Baraka, N. (2022) La gestión de agua requiere una nueva estrategia, en Cumbre ONU - Agua sobre Aguas Subterráneas Cap-Net y LA-WETnet (2013). Derecho relacionado con el agua. Manual de capacitación y guía para moderadores sobre la legislación relacionada con el agua y la reforma legal para la gestión integrada de recursos hídricos.

Comisión Económica de las Naciones Unidas para Europa (CEPE/ONU) (2015). Nota de Orientación sobre Políticas: Los Beneficios de la Cooperación en Materia de Aguas Transfronterizas: Identificación, Evaluación y Comunicación Nueva York y Ginebra: Naciones Unidas.

Comisión Económica para América Latina - CEPAL. (2024) Aguas Transfronterizas: Gestión, Cooperación y Paz - Diálogos Regionales del Agua para América Latina y el Caribe 2024, Chile.

García, B. (2013). Contaminación transfronteriza en cuenca binacional Puyango Tumbes. http://genderandwater.org/es/biblioteca/enlaces

Garrido, R., & Hernández, J. (2008). *Agua: recurso único. Conclusiones y propuestas. Expo Agua Zaragoza 2008.* Caja Azul de la Tribuna del Agua.

Gavilanes (2016). Estudio de la concentración del plomo en el agua del río Tumbes, periodo 2012 - 2015 como causa de la minería aurífera y su relación con la salud de los pobladores del caserío de Rica Playa - Tumbes.

Global Water Partnership (2000) Integrated Water Resources Management. Technical Committee Background Paper No. 4. GWP, Stockholm, Sweden. Gual, M. Diplomacia Científica en América Latina y El Caribe: Estrategias, mecanismos y perspectivas para fortalecer la diplomacia de la ciencia, tecnología e innovación-Foro CILAC UNESCO.

Global Water Partnership -GWP. (2014). *Water Security: Putting the Concept into Practice. Technical Committee Background Paper No. 20.* GWP.

International Unión for Conservation for Nature - IUCN. (2017). Gobernanza del Agua desde las bases: un aporte a la diplomacia y a la paz. Gland. Disponible en: https://www.iucn.org/node/27781

Islam, S., & Madani, K. (2017). *Water Diplomacy in Action: Contingent Approaches to Managing Complex Water Problems.* Anthem Environment and Sustainability. 10.2307/j.ctt1jktqgh

Iza, A. (2021). *Implementing Nature Based Solutions in Transboundary Basins.* Centro de Derecho Ambiental de la UICN.

Izquierdo, S. (2021). "Binational Commission for Integrated Water Resources Management in the Ecuadorian-Peruvian transboundary basins – Creation process, experience and mandate". Disponible en: https://www.gwp.org/contentassets/092e5da4a6b0454ab98526442b143961/moocevent-16-march-ppts.pdf

Kumar, S. (2022) Agua: un recurso que puede acelerar el crecimiento verde, inclusivo y resiliente.

Lynn, E., & Wolf, A. (2023). Water, Culture and International Institutions Blue Papers2023 (Vol. 2 No. 2), pp. 16-2510.58981/bluepapers.2023.2.01.

ONU. Yearbook of the International Law Commission (2001). Proyecto de artículos sobre prevención del daño transfronterizo resultante de actividades peligrosas, elaborado por la Comisión de Derecho Internacional y adoptado en 2001.

ONU-Agua. (2021). The Sustainable Development Goal 6 Global Acceleration Framework. Ginebra- Suiza. https://sustainabledevelopment.un.org/hlpf/2020/sdg6accelerationframework

PNUD - GEF. (2020). *Análisis de Diagnóstico Transfronterizo y Programa de Acción Estratégica de las Cuencas y Acuíferos Transfronterizos de Puyango - Tumbes, Catamayo - Chira y Zarumilla.* Primera Edición.

Pohl, B. (2014). Hydro - Diplomacy Can Build Peace Over Shared Waters, But Needs More Support. Wilson Center. Disponible en:https://www.newsecuritybeat.org/2014/09/hydro-diplomacy-builds-peace-shared-waters-support/

Pohl, B. (2014) The Rise of Hydro-Diplomacy - Strengthening foreign policy for transboundary Waters. Climate Diplomacy, Adelphi, Federal Foreign Office. Berlin - Alemania. Disponible en https://climate-diplomacy.org/sites/default/files/2020-10/the_rise_of_hydro-diplomacy_adelphi.pdf

Programa de las Naciones Unidas para el desarrollo (2014). Gestión Integrada de Recursos Hídricos de las Cuecas Transfronterizas y Acuíferos de Puyango - Tumbes, Catamayo - Chira y Zarumilla.

Schmeier, S. (2018). What is water diplomacy and why should you care? Global Water Forum. Disponible en https://www.globalwaterforum.org/2018/08/31/what-is-water-diplomacy-and-why-should-you-care/

UNESCO (2021). Avances en la cooperación en materia de aguas transfronterizas. Situación mundial del indicador 6.5.2 de los ODS y necesidades de aceleración.

Water, U. N. (2024). *The United Nations world water development report 2024: Water for prosperity and peace*. UNESCO.

Wilder, M. (2019). *U. S. Mexico Hydrodiplomacy: Foundations*. Change, and Future Challenges.

Wolf, A. (2005). Gestionando conflictos por el agua y cooperación. Centro de Investigación para la Paz, 2005. Barcelona - España.

Chapter 9
Traditional Knowledge Diplomacy:
A Latin American Case Study on FILAC

Martha Isabel Gómez-Lee
Universidad Externado de Colombia, Colombia

Santiago I. Gómez R.
Pontificia Universidad Javeriana, Colombia

ABSTRACT

The purpose of this chapter was to present a balance of the state of FILAC's diplomatic activities since COP 15 to analyze them in the light of a novel concept, traditional knowledge diplomacy (TKD). A case study is designed that seeks to test the hypothesis that TKD has been used by FILAC as a key asset to integrate Indigenous Peoples' perspectives and voices in the implementation of FILAC's diplomatic practices. Semi-structured interviews and a documentary review of news items were used. The findings were analyzed in three different ways: (a) traditional knowledge (TK) in diplomacy, (b) TK for diplomacy, and (c) diplomacy for TK. The findings were also discussed in light of the characteristics of a TKD. The FILAC case study provided a common understanding of TKD for international relations and an example of how it can be methodologically investigated. In this way new avenues of research, complementary to scientific diplomacy (SD), were opened.

DOI: 10.4018/979-8-3693-2746-3.ch009

INTRODUCTION

The World Conference on Science (WCS) was organized by UNESCO in co-operation with ICSU and took place in Budapest, Hungary, from 26 June to 1 July 1999. The Declaration on Science and the Use of Scientific knowledge, paragraph 26 observes:

that traditional and local knowledge systems as dynamic expressions of perceiving and understanding the world, can make and historically have made, a valuable contribution to science and technology, and that there is a need to preserve, protect, research and promote this cultural heritage and empirical knowledge. (International Council for Science, 2002 p. 1)

In the 21st century the traditional knowledge systems of Indigenous peoples as local users and stewards have become strategic in the contemporary debate on global environmental change. There is an extensive literature that refers to the need for humanity today to work on "developing innovative ways and processes to connect knowledge systems that lead to sustainable learning and recognition of the socio-ecological complexities of systems and the challenges of the Anthropocene" (Berkes et al. 2003; Folke et al. 2011; Ta'bara and Chabay 2012; cited by Tengö et al, 2014, p.579).

This is how the United Nations Intergovernmental Platform on Biodiversity and Ecosystem Services (IPBES), since its creation, has recognized the importance of indigenous and local knowledge for the conservation and sustainable use of ecosystems. For this reason, it has included them among its work products and objectives. Consequently, the IPBES conceptual framework has explicitly considered various knowledge systems and types of values of nature and its benefits (Pascual, et al 2022; Termansen et al (2022).

This recognition is very important in terms of science diplomacy, since the main objective of IPBES is to create a platform between the scientific community and decision-makers to build capacity and strengthen the use of science in the formulation of public policy. Thus, today, a larger role in this field is envisioned for Indigenous peoples.

IPBES is a paradigmatic example of science diplomacy understood as that which emerged in the early years of the 21st century as a new vocabulary and a new concept in international relations, that's refers to professional practices at the intersection of the world of science and that of diplomacy, addressing global challenges (Krasnyak &Ruffini, 2020). As Piñeros et al (2022) point out, SD "facilitates the understanding of how a set of actors, including the state, are involved in "two worlds" that tend to be considered separate: science and political-diplomatic decision-making" (p.7).

The importance of harnessing the potential of indigenous peoples' science for their development and that of society as a whole has also been palpable in other international arenas. In this sense, the First Virtual Forum on Indigenous Peoples, Science and Technology, organized by the IDB and the National Secretariat of Science and Technology under the title "Recognizing, applying and promoting indigenous knowledge" (Senacyt) of the Government of Guatemala (Us and Martin, August 2, 2021 and IDB, August 5, 2021).

In 2023, the Conference Indigenous Diplomacy in the Contemporary World: Indigenous Knowledge as a Pivotal Asset for the Achievement of an Effective Global Sustainable Development recognized that "Indigenous representatives have played a vital role in international sustainability discussions, notably in the development of the 2030 Agenda, highlighting the significance of indigenous diplomacy in global negotiations" (University of Luxembourg, November 14, 2023). Stressing that "Indigenous Knowledge is a pivotal asset for the effective achievement of sustainable development goals and in building more peaceful and inclusive societies" (University of Luxembourg, 14 November 2023).

Traditional knowledge is understood in the sense that the report of the International Council for Science (ICSU) Study Group on Science and Traditional Knowledge, which characterizes it as:

a cumulative body of knowledge, know-how, practices and representations maintained and developed by peoples with extended histories of interaction with the natural environment. These sophisticated sets of understandings, interpretations and meanings are part and parcel of a cultural complex that encompasses language, naming and classification systems, resource use practices, ritual, spirituality and worldview. (International Council for Science, 2002 p.3)

Lanzano (2013) states the concept was developed in the 1980s by anthropology and ethnoscience in an effort to recognize the eurocentric problematic in natural sciences. Later, Traditional Knowledge became part of the political debate, with development agencies and NGOs including traditional knowledge as part of their programs. Although Lanzano (2013) recognizes the concept's success with "scientists, practitioners, and decision-makers" (p.3), the author claims it has rarely been the center of the social sciences and humanities' epistemological reflection. Particularly, Lanzano (2013) questions the lack of "specific ethnographies of the (p.3).

There are some characteristics of the meaning of Traditional Knowledge that are worth highlighting in order to better understand the enormous importance of the topic for science diplomacy. On the one hand, as Tengo (2014) says: "Indigenous and local knowledge systems as well as practitioners' knowledge can provide valid and useful knowledge to enhance our understanding of governance of biodiversity and ecosystems for human wellbeing" (p. 579). On the other hand, it should be noted that the term 'Indigenous knowledge', highlights "the complex associations that Indigenous peoples have with innovations and uses 'associated with genetic resources'" (Robinson & Lumsden, 2024, p.459). It is also important that "Indigenous peoples have customary laws and rules which outline relationships between people, plants, animals and 'country'" (Robinson & Lumsden, 2024, p.459).

In this paper, the term "Traditional Knowledge" is used, in accordance with common usage of the term in the literature. It only refers to "Indigenous Knowledge" and not to the knowledge of other ethnic groups or local communities. Additionally, the paper concludes that the fact of using the word traditional knowledge does not intend to disassociate it from its scientific and validated character, as Luis Enrique López-Hurtado warns (Personal Communication May 10, 2024).

As Vera and López (2023) point out regarding SD, several studies conclude that the possibilities of influencing the production of knowledge and obtaining benefits from international cooperation in the midst of the plurality of private, transnational, subnational, foundations, NGOs or interest groups "are unequally distributed among countries in favor of those with greater relative development" (p. 156). This problem is even greater when it comes to studying Indigenous Peoples who have not been recognized as subjects of international law and have more limitations to influence this production of knowledge. Thus, Indigenous Peoples have difficulties in sharing and transmitting their values and reproducing their own cultural models.

According to Roig (2020), science diplomacy acts as an instrument of "soft power" (Nye, 2008) to improve the influence and position of certain international actors. In this context, it is worth asking to what extent can we speak of a diplomacy of traditional knowledge TKD that acts as an instrument of "soft power" (Nye, 2008) to improve the influence and position of Indigenous peoples, as SD does for other international actors (Roig, 2020)? To what extent does TKD serve to share and transmit the values of Indigenous peoples and reproduce their own cultural models?

The time has come to examine a case for TKD in Latin America and the Caribbean. Following the path of science diplomacy (SD), this article explores a concept of TKD that integrates Indigenous Peoples' perspectives and voices in environmental and climate policy making and implementation.

Based on a Latin American case study, it is argued that international relations research can be enriched by cases that allow for a common understanding of traditional knowledge diplomacy (TKD) and how it should be methodologically investigated.

The particular case of the Fund for the Development of Indigenous Peoples of Latin America and the Caribbean (FILAC) is discussed. FILAC is an important international public law organization created in 1992 by the II Ibero-American Summit of Heads of State and Government held in Madrid, whose governing and operating bodies are based on parity relations between States and Indigenous

Peoples[1]. FILAC works in the processes of self-development and promotion of the rights of indigenous peoples, communities and organizations in Latin America and the Caribbean? This case is interesting because in 2006, the Heads of State and Government attending the XVI Ibero-American Summit in Montevideo, Uruguay, decided to consolidate FILAC as the only international organization specialized in the promotion of Development with Identity focused on the Good Living-Living Well of Indigenous Peoples, as well as the recognition of their individual and collective rights (FILAC, n.d.).

To answer this question the article is divided into four parts: In the first, the research is formulated based on a historical precursor of DTK in contrast to the main driving forces of the SD approach. In the second, an overview of FILAC's activities with respect to the need to integrate Indigenous Peoples' traditional knowledge into environmental and climate policy development and implementation is presented. In the third, it moves on to a more general perspective on TKD as an analytical concept, taking stock of the current status of FILAC's activities under the umbrella of this term and describing the prevailing challenges for the concept. Finally, the limitations of the findings and generalizability, as well as future research directions, are cautiously discussed.

Research questions, frameworks, methods and design

In an effort to conceptualize TKD, this article looks back to the founding report Royal Society and American Association for the Advancement of Science (2010) since, as Krasnyak &Ruffini point out (2020) this publication can be considered an official start for science diplomacy as a new discipline, that has enlarged the scope of the science diplomacy concept by pointing out the interactions between the world of science and that of diplomacy. SD "helped to pave the way toward a better understanding of the role of nonstate actors, such as scientific epistemic communities, in engaging in, and even initiating, science diplomacy actions" (Krasnyak &Ruffini, 2020, p.1).

Considering the influential conceptual framework for understanding science diplomacy de Royal Society & AAAS (2010) it is suggested that TKD is one that occurs in three ways:

(a) traditional knowledge in diplomacy (traditional knowledge experts and data relevant to diplomatic action);
(b) traditional knowledge for diplomacy (using traditional knowledge to foster diplomatic initiatives and relations); and
(c) diplomacy for traditional knowledge (diplomacy and diplomatic practices deployed to encourage international traditional knowledge cooperation).

Observers have noted shortcomings on the proposal of the Royal Society & AAAS (2010). For example, Smith (2014); Copeland (2016); Penca (2018); Flink & Rüffin (2019), point out that actual science diplomacy activities usually have multiple purposes. The Royal Society & AAAS's (2010) definition itself is problematic because it does not involve decisions on funding and organization, as well as official and informal relations between different agencies and institutions (Gluckman, Turekian, Grimes, & Kishi, 2017).

In the absence of new definitions that are broader and more inclusive, it is proposed to work on TKD. We started by doing a literature review in SCOPUS, WOS and Google Scholar and found that only Foyer, Jean & Dumoulin Kervran (2022) have referred to TKD in a case study. Based on ethnographic observations during the 21st Conference of the Parties (COP 21) in Paris in 2015, focuses on the narratives mobilized around the articulation between Indigenous traditional knowledge and the fight against climate change.

In this article an analysis is made based on four characteristics of TKD, according to Foyer& Dumoulin Kervran (2022). TKD is characterized by:

1) Being alternative to the official scientific diplomacy of climate change part of an Indigenous diplomacy whose main objective remains the recognition of Indigenous rights and territories inside nation-states" (p.201).
2) Being primarily a prerogative of indigenous peoples that does not claim the classically organized political form of diplomacy, but rather "indigenous peoples defend their cultural and territorial sovereignty on the basis of the principles of self-determination, and informed consent" (p. 207).
3) Take place "beyond nation-states or rather, in their interstices" (p.207). [2]
4) Although largely subject to the codes of classical Western diplomacy, it also "deviates these codes by introducing other ways of speaking, of presenting oneself, other less technocratic and more embodied formats of discussion, as well as other forms of sanctity and rituality" (p. 207).

In this context, the research question is: To what extent does FILAC's diplomacy enable the engagement of TKD?

A case study was conducted with the purpose of assessing the hypothesis that TKD has been used by FILAC as a key asset to integrate Indigenous Peoples' perspectives and voices in the implementation of FILAC's diplomatic practices.

The objective of the case study was to analyze the way in which FILAC's diplomacy constitutes a TKD. To achieve this general objective, three specific objectives that gave structure to the work were proposed: 1) To identify the TKD experts and relevant data that FILAC has used for diplomatic action. 2) To determine the TKD that FILAC has used to foster diplomatic initiatives and relations. 3) To establish the diplomatic practices that FILAC has deployed to foster international cooperation on traditional knowledge.

With the fulfillment of these objectives, the aim was to contribute to a better understanding of the international processes of knowledge production. In summary, research was designed, from the discipline of international relations, to offer a model that would explain instruments of "soft power" (Nye, 2008) to improve the influence and position of Indigenous peoples in the production of knowledge, just as SD does for other international actors (Roig, 2020).

After an initial evaluation of the article, it was deemed important to clarify that although TK deserves the same treatment as western scientific knowledge, its particularities should always be recognized. As the article's peer evaluator suggested, TK is a convergence between the people's various epistemic collections, ontologies, and languages. This plurality makes TK a dynamic and specific asset in international settings.

Three main stages were followed: (i) data collection through the following techniques: semi-structured interviews and documentary analysis of FILAC news, (ii) measurement and (ii) data analysis. The method applied for the analysis of the data collected was the conceptual framework for understanding science diplomacy of Royal Society & AAAS (2010) and the characteristics of TKD according to Foyer, Jean & Dumoulin Kervran (2022).

Data related to the diplomacy carried out by FILAC was collected from the 25th UN Climate Change Conference COP 25 (2 – 13 December 2019), to the UN Climate Change Conference COP 28 (30 November-12 December 2023). The data were collected from FILAC's official news. To collect the information, a sample of actors who have worked in FILAC diplomacy in the last five years was used.

Accordingly, the interviewees were: Gabriel Muyuy Jacanamejoy, FILAC technical secretary, from February 1, 2021 to February 1, 2023 (May 7, 2024), Luis Enrique López-Hurtado, in charge of the Ibero-American Institute of Indigenous Languages (IIALI) (May 10, 2024), Luis Alfredo Rojas, coordinator of FILAC's Multilateral Area and Ernesto Marconi Ripa, coordinator of FILAC's Initiative for Indigenous Cooperation Program - ICI, (May 22, 2024).

FINDINGS

Since 2017, FILAC has been part of the intergovernmental organizations that have received a standing invitation to participate as observers in the sessions and work of the United Nations General Assembly. Gabriel Muyuy points out this is a status that allows it to participate in different areas of the United Nations system with a voice and also, in some aspects, with a vote. This allows FILAC to have seats in the main headquarters in New York, also in the Geneva headquarters of human rights and in the different spaces of the United Nations system. Within this framework, FILAC has been participating very actively in different spaces related to the processes of implementation of the Framework Convention on Climate Change, the Convention on Biological Diversity, also in the whole process of implementation of the SDGs and in short, there are many other events (personal communication, May 7, 2024). This recognition as a permanent observer, as an international, intergovernmental and peer body, has been important and has made all this participation possible.

FILAC's main purpose is to favor "the exercise of indigenous peoples' rights" (personal communication, 2024). The indigenous peoples lack an active participation on the international system's decision making. Since Gabriel Muyuy's secretariat, efforts have been made to replace this logic by advocating for indigenous participation in public instances (Personal Communication, May 22, 2024). FILAC has worked hard to replace the perception of indigenous peoples as victims with that of indigenous peoples as important assets that can provide alternatives to development (Personal Communication, May 22, 2024). Luís Enrique López-Hurtado (Personal Communication, Mau 10, 2024) recalls the expression "Nothing for us, without us", referring to the fact that indigenous peoples are not interested in aid projects that omit their agency.

FILAC'S diplomatic strategy in the COPs becomes that of seizing "windows of opportunity" when different specific projects that can satisfy the indigenous interests arise (Personal Communication, May 22, 2024). Because of its international organization nature, FILAC's role in COPs has always been that of an observer. This makes the institution dependent on its state members to transmit indigenous peoples' visions. Thus, the Conference of The Parties is not the most important instance of the forums for FILAC's direct diplomatic activities.

COP 25, 27 and 28's approach has been constant: how can indigenous people contribute to tackle climate change? However, at first glance, FILAC'S approach in Dubai suddenly became more specific. During COP 28, FILAC focused its attention on Climate investment funds and especially on the Loss and Damages mechanisms. FILAC argued that the "Mechanisms for accessing funds should be much easier and that resources should be accessed directly (by indigenous people)" (Personal Communication, May 22, 2024).

In this context, one of FILAC's functionaries explained that COPs topics have a slow process of growth and traces the Fund claim back to COP 25 (Personal Communication, May 22, 2024). Here the main purpose was to achieve the implementation of the Paris agreement and thus, a greater participation of the indigenous populations. This claim was developed in a thorough manner during Glasgow a year later. The message on COP 26 was that participation is directly linked to indigenous fund management capacity as "less than 1% of the total of funds destined to indigenous populations gets directly to the indigenous organizations and populations" (Personal Communication, May 22, 2024). In this context, a suggestion framework was created with an intercultural dialogue between the regional indigenous platform, the EU, and the Chilean Government (Personal Communication, May 22, 2024).

In FILAC's diplomacy, bilateral and multilateral allies become crucial for the fulfillment of its interests because as a FILAC member points out, "FILAC calls itself a fund but does not have many financial resources of its own" (2024). Regarding its bilateral relations, the member states are the core. However, some members have a more active interest in participating with FILAC. As López-Hurtado mentions, FILAC has worked "as another arm of Bolivia's diplomacy" (Personal Communication, May 10, 2024). Mostly because of a confluence of agendas in the last years where Bolivia has promoted a "diplomacy of the peoples" (Personal Communication, May 10, 2024).

FILAC's relationship with its members seems to be based on a project-based approach in which concrete programs are selected to be worked upon. For example, currently FILAC and Bolivia have been working on a "strengthening of indigenous territories governance with the elaboration of life plan documents" (Personal Communication, May 22, 2024). Another case is the environmental justice programs promoted throughout the Amazonic countries (Personal Communication, May 22, 2024).

FILAC's most important bilateral ally is the Spanish Agency for International Development Cooperation. The SAIDC is FILAC's biggest sponsor as it not only finances projects but also part of the inner structure of the organization (Personal Communication, May 22, 2024). Another contributor state is the Netherlands, which recently collaborated with FILAC to produce a proposal presented at the UN 2023 Water Conference (Personal Communication, May 22, 2024).

As for its multilateral relations, FILAC displays an important articulation with various United Nations institutions. In recent years, the organization has worked with UNESCO, FAO, UNICEF, IOM, CEPAL and the World Bank in various projects and dialogues. For example, in the year 2021 FILAC worked jointly with UNICEF, FAO, FIAY (Foro Indígena del Abya Yala), and the Red de Jovenes Indígenas in a parallel event to promote indigenous participation "in the processes of dialogue with governments to generate recommendations for the UN Food Systems Summit, as well as to advocate for public policies that guarantee food security in Latin America" (FILAC, 2021a). In addition, in the year 2023 FILAC worked with IOM to promote a Dialogue on Migration, Climate Change and Indigenous Peoples which indigenous peoples' resilience becomes a strategy to tackle the subsequent problems of climate change migration (FILAC, 2023b). On the other hand, collaboration with CEPAL has been seen on diverse instances, for example, on the creation of a monitoring mechanism with indigenous participation for SDG 6 (Personal Communication, May

22, 2024). Various research papers have been produced with this UN institutions, among them FAO's "White/Wiphala Paper on Indigenous Peoples' food systems" (FAO, 2021).

Multilateral relations with indigenous organizations also play a very important role in FILAC's diplomatic practices. The biggest indigenous international organization in the region is the FIAY (Foro Indígena del Abya Yala) whose purpose is to articulate all the indigenous organizations of the continent. On a lower level, FILAC deals with subregional entities such as COICA (Coordinadora de las Organizaciones Indígenas de la Cuenca Amazónica) and CAOI (Coordinadora Andina de Organizaciones Indígenas). On a national level, each country has one or various indigenous organizations, such as Colombia's ONIC. This involves a challenge for FILAC, because in countries in which the indigenous movement is not powerful or has low density, power atomizes and FILAC must deal with various subnational entities at the same time (Personal Communication, May 22, 2024).

FILAC's multilateral programs also include relations with regional state organizations. For instance, OTCA (Organización del Tratado de Cooperación Amazónica) has been an important ally in the development of programs, workshops, and discussions. In 2022, in partnership with OTCA and COICA, FILAC organized the "International Meeting to develop and design an instrument that incorporates traditional knowledge for the good management of biological diversity and good living in the Amazon Region" (COICA, 2022). In 2023, FILAC and the latter discussed possible indigenous participation mechanisms in the context of the meeting of member state leaders in the Cumbre Amazónica held the same year (OTCA, 2023).

In terms of interregional dynamics, FILAC has a very strong bond with the European Union (as mentioned on their collaboration in COP 26). In 2023, for example, FILAC participated in the "Nutritional Food Security and Agri-Food Systems: Agroecology for a green transition" forum organized by the SAIDC as a previous event for dialogue between Latin America and the Caribbean and the European Union. In his intervention, Alberto Pizarro, Technical Secretary for FILAC, "emphasized on the importance of disseminating and expanding indigenous knowledge and best practices that have the potential to transform food systems worldwide" (FILAC, 2023a) and insisted on the importance of diverse alliances that that enable traditional knowledge to contribute to the achievement of the SDGs (FILAC, 2023a).

Networks are also an important asset of FILAC's diplomatic agenda. Private networks of philanthropic organizations are an important source of funding for FILAC's activities (Personal Communication, May 22, 2024). These networks work as a "guild" for global foundations where experiences are shared and possible project contributors are articulated (Personal Communication, May 22, 2024). Civil society networks are one of FILAC's principal vehicles for action. Women's and youth's indigenous associations across the region benefit from FILAC's contributions and workshops, such as the Indigenous Youth of Abya Yala workshop: "Weaving Stories and Resilience for the Implementation of the United Nations Declaration on the Rights of Peoples" in which "the importance of continuing to work for the protection of ancestral knowledge and wisdom, as well as for the preservation of biodiversity, with the objective of guaranteeing equitable access to culturally appropriate health services, was recognized" (FILAC, 2024).

FILAC plays an important role in the development of indigenous individual's abilities and competences. This task is carried out through FILAC's area of training in collaboration with the Universidad Indígena Intercultural. Additionally, various courses and diplomates have been created with renowned universities like the UNAM (Universidad Autónoma de México) and the C3M (Carlos Tercero de Madrid). Financing for individual's projects is also one of FILAC's aid mechanism, as seen on the Kolom No´ooj program, which seeks to "empower young indigenous women, from an economic as well as a social and cultural perspective, with the objective of strengthening their capacities and promoting the recovery of traditional knowledge" (FILAC, n.d.).

DISCUSSION

The first method of analysis of this article consists of discussing the balance of the status of FILAC activities under the umbrella of TKD, from the three classic SD approaches:

a) traditional knowledge in diplomacy (traditional knowledge experts and data relevant to diplomatic action)

By identifying the different sets of experts and information used by FILAC for the development of its diplomatic objectives, it can be affirmed that FILAC uses traditional knowledge experts and data relevant to diplomatic action for diplomatic action.

In the first instance, there are members within the organization who, due to their indigenous status and/or their training, have a broad knowledge of traditional knowledge. This is the case of several members of its diplomatic corps such as Gabriel Muyuy (former secretary), Alberto Pizarro (current secretary), Myrna Cunningham (vice-president of the Board of Directors) and the linguist Luis Enrique López-Hurtado (in charge of IIALI). They are seen as experts, knowledgeable about alternatives for development.

In this context, FILAC makes an important effort to integrate indigenous experts from throughout the region in its diplomatic actions. For example, the event "Indigenous Women in the preservation, reactivation and transmission of traditional knowledge" integrated the experiences of experts from Venezuela, Guatemala, Colombia, Bolivia, and Chile.

FILAC not only recruits experts with existing knowledge, but also plays a key role in training them. Through collaboration with different academic institutions, the organization is responsible for promoting the training of indigenous individuals in different fields and specific skills. An example of this is the eighty scholarships granted to indigenous youth from the Amazon (Personal Communication, May 10, 2024). Additionally, the intergenerational logic of indigenous knowledge adopted by the institution in its practices (Personal Communication, May 22, 2024) creates a favorable space for the older ones to transmit their knowledge to the younger ones and produce new experts. This provides FILAC with a valuable network of specialists in traditional knowledge.

The common pattern in officials and the extended network is that they present themselves as knowledgeable about development alternatives, as mentioned by one of the FILAC officials (Personal Communication, May 22, 2024).

Regarding the data used, FILAC is very emphatic about the fact that 80% of the world's conservation dynamics are in the hands of indigenous communities.

(b) traditional knowledge for diplomacy (using traditional knowledge to foster diplomatic initiatives and relations).

In determining the traditional knowledge that FILAC has used to promote initiatives and diplomatic relations, it can be concluded that FILAC's logic is that it presents traditional knowledge as an asset to solve different problems that compromise the international system. As mentioned by one of its officials, the purpose of traditional knowledge in FILAC goes beyond the logic of conservation of an object, as in old anthropology. Traditional knowledge is a "living and fresh knowledge that acts and transforms reality" (Personal Communication, May 22, 2024). As Luis Enrique López-Hurtado states, traditional knowledge goes from being an endogenous practice and a right of indigenous peoples and becomes an "ethical-political horizon addressed by indigenous people for all, in a Buen Vivir" (Personal Communication, May 10, 2024). In this logic, FILAC has been responsible for operationalizing traditional knowledge in order to improve the current state of the world on different fronts.

First, there is the issue of conservation and climate change. On this front, FILAC has been responsible for demonstrating the potential of traditional knowledge in caring for the environment. This has been reflected in interventions on issues such as water (UN Water Conference) and forest governance (as evidenced in the text produced with FAO "Indigenous and tribal peoples and forest governance"). Likewise, indigenous knowledge is also presented as a tool for prevention and mitigation of environmental risks (FILAC, 2022c). For example, in the dialogue "Application of Indigenous Knowledge in the Prevention and Management of Environmental Risks" developed together with UNESCO, the experience of various Nicaraguan indigenous communities was presented, where different bio-indicators provide early warning of hurricane formation and other possible natural risks (FILAC, 2021c; FILAC, 2021d).

Secondly, indigenous knowledge is presented as an important contributor to the fulfillment of the SDGs. With respect to SDG 2, zero hunger, FILAC has worked to demonstrate the usefulness of TK for global food security. In collaboration with CANEUS, UNOOSA and ECOSOC, FILAC held the webinar "IKRI: A tool to achieve the SDGs" (FILAC, 2023c), where FILAC president Freddy Mamani explains that indigenous food developments "add diversity of scientific knowledge, improve nutrition, the immune system and health; in addition, their practices are in harmony with mother earth and are resilient to natural risks and disasters" (FILAC, 2023c).

Thirdly, FILAC emphasizes the resilience of this knowledge and its importance in migration dynamics. Alberto Pizarro mentioned this during the dialogue "Human Mobility and Climate Change in Latin America and the Caribbean" (FILAC, 2023b) in the framework of COP 28, where he established the need to join efforts from all fields of knowledge to face migration and displacement caused by climate change. This resilience has also been developed by Myrna Cunningham, who exalted the capacity of indigenous women to face problems such as that of COVID 19 through the application of traditional medicine and communal vigilance (FILAC, 2022a; FILAC, 2022b).

Fourth, FILAC positions the gender approach as something characteristic of traditional knowledge. Since 2008, with the Emblematic Program for Indigenous Women, FILAC has been working to mainstream its actions (FILAC, 2022a; FILAC, 2022b). In the intergenerational dialogue with indigenous women "Preservation and transmission of knowledge of Indigenous Peoples" FILAC states: "indigenous women have proven to be essential for the revitalization of traditional knowledge, for the conservation and transmission of indigenous languages; as well as for the coordination of individual and collective self-care actions, and the application of traditional ancestral medicine" (FILAC, 2022b).

In his speech, the president of FILAC highlighted the unbreakable relationship between women and knowledge: "To speak of traditional knowledge is to talk about the millenary and scientific knowledge of indigenous women, in it is deposited the culture, an accumulation of knowledge, experiences and practices, it is the cultural heritage of the peoples" (FILAC, 2022b). The dialogue concluded that "Indigenous women are the pillars of indigenous peoples and play a crucial role in the conservation and transmission of ancestral traditional knowledge, which is a central element for the Buen vivir/Vivir bien of the Peoples and of society in general" (FILAC, 2022b). This leads to the conclusion that traditional knowledge that aims to transform the world needs indigenous women to cope.

(c) diplomacy for traditional knowledge (diplomacy and diplomatic practices deployed to encourage international traditional knowledge cooperation).

By establishing the diplomatic practices that FILAC has deployed to promote international cooperation on traditional knowledge, the findings demonstrate various strategies used by FILAC. Prior dialogue spaces and document redaction and correction becomes the institution's main instrument to influence in the COP system and other international instances. Parallel events are also of utmost importance, as they allow counterparts, cooperation agencies, regional entities, and heads of government missions to receive further information on themes of interest that are not included in the different forum's agenda for that year (Personal Communication, May 22, 2024). Also, to position traditional knowledge FILAC uses the bulk of forums, workshops and webinars mentioned above, the different discussion spaces become FILAC's main space to promote cooperation.

The second method of analysis of the present article consists of discussing the findings in the light of the characteristics to consider regarding TKD, according to Foyer& Dumoulin Kervran (2022) mentioned in the first part of the article. In this sense it is argued that:

Regarding the first characteristic, FILAC's diplomacy does propose a change in traditional indigenous diplomacy and transcends the recognition of rights and territories. This new approach seeks to position traditional knowledge and its practitioners as international actors with alternative development solutions to global dynamics. In this novel perspective, the epistemic communities of official scientific diplomacy are questioned and new communities with living, transformative and validated knowledge are presented (Personal Communication, May 22, 2024). These new epistemic communities present new ways of producing science with diverse and intergenerational spaces for dialogue, where the dynamics of respect and verticality are different from those of the Western academic ladder.

Regarding the second characteristic, FILAC advocates for the full participation of indigenous peoples in the decision-making spaces of the international system. For example, since COP 25 and after the Paris agreement, FILAC's position at COPs has been to achieve greater autonomy in indigenous decision-making. In the interviews conducted, FILAC officials brought up informed consent and the principles of self-determination.

Regarding the third characteristic, FILAC's TKD seeks to project Indigenous peoples beyond national borders by having traditional knowledge that plays a fundamental role in the solution of transnational problems, such as food, climate and environmental problems.

Luis Enrique Lopez-Hurtado identifies FILAC as an arm of Bolivian diplomacy (personal communication, May 10, 2024). While it is true that there is a confluence of agendas between Bolivia and FILAC, it is not a matter of one diplomacy being relegated to the other.

This allows us to emphasize the new characteristic of the TKD and is the underlying structural factor that makes Bolivia's diplomacy and FILAC's TKD different. Returning to Foyer & Dumoulin Kervran (2022), Bolivian diplomacy is still based on a logic of internal national sovereignty. Thus, even if Bolivia tries to expand its ideas outside its borders, they always respond to a national center. On the contrary, FILAC's TKD not only encompasses the diplomatic interests of several countries, but also the interests of various indigenous peoples through representatives who are part of the Board of Directors (FILAC, n.d.) and who seek, through their traditional knowledge, to go beyond national borders and provide transnational solutions.

Regarding the fourth characteristic, concerning the codes of the TKD, when analyzing the minute by minute of a FILAC event, "Participatory Dialogue with Indigenous Peoples' Organizations: Towards the Summit of the Future and COP 28", (FILAC & Secretaria General Iberoamericana, 2023), an alternative diplomatic protocol that modifies the official codes of Western diplomacy is evident. As mentioned by Foyer & Dumoulin Kervran (2022) this dialogue acquires a ritual character by beginning with a Mayan invocation ceremony by Rosalina Tuyuc, who remembers the grandparents, the god Tohil, Mother Earth, the Mayan calendar, and oxiajuj toj is spoken (FILAC & Secretaria General Iberoamericana, 2023).

The interventions in this space also present a more horizontal character, where officials address each other as "brothers and sisters" and exchange words of fraternity in their respective indigenous languages, "Jallalla brothers and sisters" (FILAC & Secretaria General Iberoamericana, 2023), concludes the president of the FILAC Board of Directors. In addition, when referring to the other, they do not use their position, but their name, which evokes a greater proximity in the dialogue.

CONCLUSION

Considering that FILAC's diplomacy complies with the three classic ways of SD and with the characteristics of TKD according to Foyer & Dumoulin Kervran (2022) FILAC the answer to the research question is that FILAC, through the involvement of traditional knowledge as an alternative diplomacy, has managed to expand the sets of institutions and representatives among which the interactions of its diplomacy take place. In this way, the research hypothesis that traditional knowledge has served FILAC to share and transmit across borders the values of Indigenous peoples and to position its own cultural models has been validated. FILAC's TKD aims to legitimize forms of knowledge other than western scientific knowledge.

While it is true that Indigenous peoples do not have the capacity of a State to use economic incentives or military force to influence the behavior of other actors, through TKD they can influence the actions or interests of other actors using their traditional knowledge, complemented by diplomatic means. This, in the same way as SD does for other international actors.

TKD in the FILAC case study has taken hold in climate change, food and environmental policy, as well as other international issues, to interconnect the fields of policy and traditional knowledge. The concept of TKD, like the concept of SD can serve to build bridges between the drivers of two different fields, such as diplomacy and, in this case, traditional knowledge. As with SD, TKD enables actors to take hold of an understanding of traditional knowledge that allows them to address problems and/or solutions to transnational issues such as environmental and climate change. The relevant elements that coincide in SD and TKD is that both use knowledge to achieve a diplomatic end.

This case study confirms the theme of the Conference Indigenous Diplomacy in the Contemporary World: Indigenous Knowledge as a Pivotal Asset for the Achievement of an Effective Global Sustainable Development. It is worth exploring this TKD concept further to confirm whether the motto could become: there is no more diplomacy for global development issues without traditional knowledge, nor is there traditional knowledge without diplomacy on these issues.

In this way TKD is a concept of diplomacy that is more inclusive of Indigenous peoples by going beyond the scientific knowledge of SD. While it is certainly a concept that is consistent with the three classical ways of understanding SD, it brings new insights that offer an alternative vision.

The FILAC case study has provided a common understanding of a TKD for international relations and a sample of how it can be methodologically investigated. Specifically, with regard to traditional knowledge for diplomacy, it has provided 4 new features of TKD that had not previously been contemplated. First, it is a preventive diplomacy that anticipates potential environmental risks and prepares for their mitigation from its knowledge base. Second, it is a diplomacy that contributes to the fulfillment of the SDGs. Thirdly, it is a diplomacy that offers resilient solutions based on the resilience of traditional knowledge. Fourth, it is a cross-cutting diplomacy with a gender approach because it is based on traditional knowledge in which women play an intrinsic role.

In this way, new avenues of research are opened to explore the practical implications of TKD, its challenges and the ways in which the article can be improved by future researchers. For example, it is worthwhile to inquire into the relationship that has been raised in this article between the gender approach with TKD and what would be in this case the differences with SD.

FILAC's analysis also raises new questions that enrich SD, such as: What conflicts or challenges does the organization face in these areas, and how does it address power disparities related to Western knowledge and those who hold this legitimized knowledge? What diplomatic tools or strategies does it use? How does FILAC address power disparities between Western and traditional knowledge, as well as between diverse knowledge holders?

REFERENCES

Banco Interamericano de Desarrollo. (2021, 5 de agosto). Primer foro virtual de ciencia, tecnología y pueblos indígenas. [Archivo de video]. YouTube. https://www.youtube.com/watch?v=L53yTJBgU-s&t=1s

COICA. (2022, April 8). *Encuentro internacional entre COICA, OTCA y FILAC, por la protección de la biodiversidad y el buen vivir en la Región Amazónica*. COICA. https://coicamazonia.org/encuentro-internacional-entre-coica-otca-y-filac-por-la-proteccion-de-biodiversidad-y-el-buen-vivir-en-la-region-amazonica/

Copeland, D. (2016). Science diplomacy. In Constantinou, C. M., Kerr, P., & Sharp, P. (Eds.), *The SAGE handbook of diplomacy* (pp. 628–641). SAGE. 10.4135/9781473957930.n52

FAO. (2021). Los pueblos indígenas y tribales y la gobernanza de los bosques - Una oportunidad para la acción climática en Latina América y el Caribe. In *FAO eBooks*. 10.4060/cb2953es

FILAC. (2021a). *Evento Paralelo: Conocimientos tradicionales de los Pueblos Indígenas fundamentales en la transformación de los Sistemas Alimentarios | Fondo para el desarrollo de pueblos indígenas de América Latina y el Caribe*. https://www.filac.org/evento-paralelo-conocimientos-tradicionales-de-los-pueblos-indigenas-fundamentales-en-la-transformacion-de-los-sistemas-alimentarios/

FILAC. (2021c). *Presentaron experiencias de cómo los conocimientos indígenas ayudan en la prevención de riesgos ambientales | Fondo para el desarrollo de pueblos indígenas de América Latina y el Caribe*. https://www.filac.org/presentaron-experiencias-de-como-los-conocimientos-indigenas-ayudan-en-la-prevencion-de-riesgos-ambientales/

FILAC. (2021d). *Taller Global: Aplicación de los Conocimientos Indígenas en la Prevención y Gestión de Riesgos Ambientales | Fondo para el desarrollo de pueblos indígenas de América Latina y el Caribe*. https://www.filac.org/taller-global-aplicacion-de-los-conocimientos-indigenas-en-la-prevencion-y-gestion-de-riesgos-ambientales/

FILAC. (2022a). *FILAC realiza diálogo con mujeres indígenas para la preservación de los conocimientos tradicionales | Fondo para el desarrollo de pueblos indígenas de América Latina y el Caribe*. https://www.filac.org/filac-realiza-dialogo-con-mujeres-indigenas-para-preservar-los-conocimientos-tradicionales/

FILAC. (2022b). *Mujeres indígenas juegan un papel crucial en la transmisión de conocimientos tradicionales | Fondo para el Desarrollo de Pueblos Indígenas de América Latina y el Caribe.* https://www.filac.org/mujeres-indigenas-juegan-un-papel-crucial-en-la-transmision-de-conocimientos-tradicionales/

FILAC. (2022c). *Pueblos Indígenas contribuyen con sus conocimientos a la reducción de riesgos ambientales | Fondo para el desarrollo de pueblos indígenas de América Latina y el Caribe.* https://www.filac.org/pueblos-indigenas-buscan-contribuir-con-sus-conocimientos-a-la-reduccion-de-riesgos/

FILAC. (2023). *Infraestructura de Investigación del Conocimiento Indígena, un aporte valioso para proteger los conocimientos y Sistemas Alimentarios de los Pueblos Indígenas | Fondo para el desarrollo de pueblos indígenas de América Latina y el Caribe.* https://www.filac.org/nfraestructura-de-investigacion-del-conocimiento-indigena-un-aporte-valioso-para-proteger-los-conocimientos-y-sistemas-alimentarios-de-los-pueblos-indigenas/

FILAC. (2023a). *Conocimientos tradicionales de los Pueblos Indígenas para la transformación de los sistemas alimentarios | Fondo para el desarrollo de pueblos indígenas de América Latina y el Caribe.* https://www.filac.org/conocimientos-tradicionales-de-los-pueblos-indigenas-para-la-transformacion-de-los-sistemas-alimentarios/

FILAC. (2023b). *COP 28: FILAC destaca la realidad que enfrentan los Pueblos Indígenas frente a urgencia de la Crisis Climática en América Latina | Fondo para el desarrollo de pueblos indígenas de América Latina y el Caribe.* https://www.filac.org/cop-28-filac-destaca-la-realidad-que-enfrentan-los-pueblos-indigenas-frente-a-urgencia-de-la-crisis-climatica-en-america-latina/

FILAC. (n.d.). *Consejo Directivo | Fondo para el desarrollo de pueblos indígenas de América Latina y el Caribee.* https://www.filac.org/consejo-directivo/

FILAC. (n.d.). *Fondo para el desarrollo de pueblos indígenas de América Latina y el Caribee | FILAC.* https://www.filac.org/

FILAC. (n.d.). *Kolom No'ooj | Fondo para el desarrollo de pueblos indígenas de América Latina y el Caribe.* https://www.filac.org/iniciativas/kolom-noooj/

FILAC & Secretaria General Iberoamericana. (2023). *Diálogo participativo con Organizaciones de Pueblos Indígenas: Rumbo a la Cumbre del Futuro y la COP 28.*

Flink, T., & Rüffin, N. (2019). The current state of the art of science diplomacy. In Simon, D., Kuhlmann, S., Stamm, J., & Canzler, W. (Eds.), *Handbook on science and public policy* (pp. 104–121). Edward Elgar Publishing. 10.4337/9781784715946.00015

Foyer, J., & Kervran, D. David (2022) Traditional knowledge in the global climate change regime: Narratives for an alternative science diplomacy. In Mays C, Laborie L, Griset P (eds) Inventing a shared science diplomacy for Europe: Interdisciplinary case studies to think with history. Zenodo. 10.5281/zenodo.6582702

Gluckman, P. D., Turekian, V. C., Grimes, R. W., & Kishi, T. (2017). Science diplomacy: A pragmatic perspective from the inside. *Science & Diplomacy*, 6(4). https://www.sciencediplomacy.org/article/2018/pragmatic-perspective

International Council for Science. (2002). Report from ICSU Study Group on Science and Traditional Knowledge. SCIENCE AND TRADITIONAL KNOWLEDGE. https://council.science/wp-content/uploads/2017/05/Science-traditional-knowledge.pdf

Krasnyak, O., & Ruffini, P. (2020). *Science Diplomacy*. Oxford Universisty Press., 10.1093/obo/9780199743292-0277

Lanzano, C. (2013). What kind of knowledge is 'indigenous knowledge'? Critical insights from a case study in Burkina Faso. *Transcience*, 4(2), 3–18.

Nye, J.Jr. (2008, March). Public Diplomacy and Soft Power. *The Annals of the American Academy of Political and Social Science*, 616(1), 94–109. 10.1177/0002716207311699

Pascual, U., Balvanera, P., Christie, M., Baptiste, B., González-Jiménez, D., Anderson, C., Athayde, S., Barton, David., Chaplin-Kramer, R., Jacobs, S., Kelemen, E., Kumar, R., Lazos, E., Martin, A., Mwampamba, T., Nakangu, B., O´Farrel, P., Raymond, C., Subramanian, S., Termansen, M., van Noordwijk, M., & Vatn, A. (2022). Resumen para los encargados de la formulación de políticas de la evaluación metodológica relativa a la conceptualización diversa de los valores múltiples de la naturaleza y sus beneficios, incluidos la diversidad biológica y las funciones y los servicios de los ecosistemas. Plataforma Intergubernamental Científico-Normativa sobre Diversidad Biológica y Servicios de los Ecosistemas (IPBES).

Penca, J. (2018). The rhetoric of "science diplomacy": Innovation for the EU's scientific cooperation? (Working Paper EL-CSID Project). https://aei.pitt.edu/102624/

Piñeros Ayala, R., Echeverría-King, L. F., & Andrade-Sastoque, E. (2022). Política exterior y de CTI en Colombia: la diplomacia científica como frontera. Desafíos, *34*(Especial), 1-36. https://doi.org/10.12804/revistas.urosario.edu.co/desafios/a.11881

Robinson, D. F., Raven, M., & Lumsden, S. (2024). Epistemic injustice: Intellectual property, biodiversity and traditional knowledge. In Amani, B., Ncube, C. B., & Rimmer, M. (Eds.), *The Elgar Companion to Intellectual Property and the Sustainable Development Goals* (pp. 458–480). Edward Elgar Publishing Limited. 10.4337/9781803925233.00027

Roig, A. (2020). *Diplomacia científica y tecnológica.* Secretaría de Relaciones Exteriores de México.

Royal Society & AAAS. (2010). New frontiers in science diplomacy. London, England: Royal Society. Retrieved from https://royalsociety.org/topics-policy/publications/2010/new-frontiers-science-diplomacy

Smith, F. L.III. (2014). Advancing science diplomacy: Indonesia and the US Naval Medical Research Unit. *Social Studies of Science*, 44(6), 825–847. 10.1177/030631271453586425608440

Tengö, M., Brondizio, E. S., Elmqvist, T., Malmer, P., & Spierenburg, M. (2014). Connecting diverse knowledge systems for enhanced ecosystem governance: The multiple evidence base approach. *Ambio*, 43(5), 579–591. 10.1007/s13280-014-0501-324659474

Termansen, M., Jacobs, S., Mwampamba, T. H., Ahn, S., Castro, A., Dendoncker, N., Ghazi, H., Gundimeda, H., Huambachano, M., Lee, H., Mukherjee, N., Nemogá, G. R., Palomo, I., Pandit, R., Schaafsma, M., Ngouhouo, J., Choi, A., Filyushkina, A., Hernández-Blanco, M., . . . González-Jiménez, D. (2022). Chapter 3: The potential of valuation. En: Balvanera, P., Pascual, U., Christie, M., Baptiste, B., and González-Jiménez D. (eds). *Methodological Assessment Report on the Diverse Values and Valuation of Nature of the Intergovernmental Science-Policy Platform on Biodiversity and Ecosystem Services.* IPBES secretariat, Bonn, Germany. DOI: 10.5281/zenodo.6521298

Universidad de Luxemburgo. (2023). Conference Indigenous Diplomacy in the Contemporary World: Indigenous Knowledge as a Pivotal Asset for the Achievement of an Effective Global Sustainable Development. https://www.uni.lu/wp-content/uploads/sites 2/2023/11/20231114_AFF_DSOC_Dyplomacy_VF_WEB.pdf

Us, H., & Martin, C. (2021, 2 de agosto). *Pueblos indígenas, ciencia y tecnología. BID Blog.* https://blogs.iadb.org/igualdad/es/los-pueblos-indigenas-la-ciencia-y-tecnologia/

Vera, N., & López, M. P. (2024). Diplomacia Científica en la Argentina contemporánea. Un mapeo de iniciativas en los Ministerios de Ciencia, Tecnología e Innovación y de Relaciones Exteriores (2019-2022). *Estudios Internacionales (Santiago)*, 55(204). Advance online publication. 10.5354/0719-3769.2023.67792

ENDNOTES

1 The subscribing States to this Agreement were Argentina, Bolivia, Brazil, Colombia, Costa Rica, Cuba, Chile, Ecuador, El Salvador, Spain, Guatemala, Honduras, Mexico, Nicaragua, Panama, Paraguay, Portugal and the Dominican Republic.

2 The authors recognize the fact that indigenous peoples live in territories that do not align with the geographic boundaries imposed by western countries.

Chapter 10

Vaccine and Health Diplomacy in Latin America and the Caribbean in the Pre- and Post-COVID-19 Context:

Past and Present Learned Lessons as a Contribution to the Sustainable Development in Our Region

Danev Ricardo Perez Valerino
Tokyo International University, Japan & Finlay Vaccine Institute, Japan

ABSTRACT

The COVID-19 pandemic caused an important global change in political, financial, and health sectors to face the generated crisis. Therefore, new strategies have been developed, assumed, and implemented, with the support of governments and international organizations, on innovative capabilities that make it sustainable to invest in science to advance human development improvement of our nations. The objective of this chapter is to present in three different sections, before, during, and after the pandemic, a comprehensive overview of this topic in LATAM, which has generated close links among vaccine development and foreign policy strategies considering health and vaccine diplomacy as paramount branches included on science diplomacy actions in this region. Addressing the challenges and benefits of those

DOI: 10.4018/979-8-3693-2746-3.ch010

strategies constitutes an element to be considered in foreign policy by governments interested in allocating efforts and resources into science, health, and sustainable development.

I. LATIN AMERICA AND THE CARIBBEAN FACING THE GREAT PANDEMIC CHALLENGE OF THE 21st CENTURY

When in December 2019, the World Health Organization (WHO) received information about patients detected in Wuhan province, China, with a clinical picture apparently linked to a viral pneumonia (WHO, 2020), for the Latin American and Caribbean countries it did not seem to be an event that would affect the region, let alone the planet, in the way that the virus later recognized as COVID-19 finally did[1].

The extraordinary speed in which the virus spread through the environment, the initial lack of knowledge about its characteristics, as well as the high degree of infection of the disease caused by the SARS-CoV-2 virus, led the WHO to declare a state of public health emergency of global importance on January 30th, 2020, and later on March 11th it was declared as a pandemic situation (WHO, 2020). By the end of February, the Brazilian Ministry of Health had already identified and made public, in the city of Sao Paolo, the first positive case in the region, but it was not yet possible to foresee the great economic and social challenges that the world and the region would have to face (BBC News, 2020).

As the pandemic progressed at an accelerated pace, Latin American and Caribbean (LAC) nations adopted and decreed strict measures such as confinement and mobility restrictions to slow down the level of infection among citizens (Cid & Marinho, ECLAC, 2022). In addition to the regrettable and harsh situation of the loss of human lives, since by mid-2022 the pandemic reached very worrying figures as more than 73 million people had been infected by the virus and more than 1.7 million were dead, we should also consider the impact of the restrictions that at times lead to the total paralysis of non-essential economic and social activities (Cid & Marinho, ECLAC, 2022).

In July of that first pandemic year, Mexican diplomat Alicia Bárcenas Ibarra, then executive secretary of the Economic Commission for Latin America and the Caribbean (ECLAC)[2], clearly stated: "The pandemic is a wake-up call to address the region's inequalities and rethink historical patterns of urban development. Rebuilding society in LAC needs to mean rebuilding with equality and sustainability"[3].

These words, joined with a pragmatic and concrete analysis of what happened, show that it was no coincidence that at times countries in the region became the main generator of cases at global level. From the social point of view, the region is characterized by a high level of inequality, as well as informal urbanization, where one in five urban residents lives in slums with limited access to water and sanitation, which, together with overcrowding, leads to a significant increase in the risk of spreading COVID-19 coronavirus and other diseases (Cid & Marinho, ECLAC, 2022).

Additionally, and this is the main point of analysis of this essay, the LAC region has been specifically characterized by the lack promotion of actions to support science and human health as essential elements to generate development, equity, and sustainability in their inhabitants by governments and national or regional political authorities, with some exceptions. COVID-19 undoubtedly took advantage from this reality, and so would be the case of any other health crisis of similar impact (Batthyany, 2021).

The author considers that although most public health systems in LAC are used to dealing with seasonal epidemics such as dengue, Zika virus or Yellow Fever in the past, we were not prepared, neither in the region nor in the world, for a global health crisis such as COVID-19. In this sense, it is necessary and essential to promote strategies and mechanisms aimed at increasing social protection and generic support for new development paradigms that can be linked to strategies established internationally as part of the so-called 2030 Agenda[4].

II. SOME HISTORICAL BACKGROUND OF HEALTH DIPLOMACY

The speed with which the virus spread across the continent and the high degree of contagion and deaths described a challenging scenario for Latin American nations, just as it was happening in the rest of the world. In this context needs to be a priority to generate policies and strategies that start from theory and lead to practical actions, so during the pandemic programs linked to Science Diplomacy were recognized and re-launched at the international level, which already had precedents, but not enough recognition, attention, and support in the region: Health Diplomacy and Vaccine Diplomacy.

In the case of Health Diplomacy, it has achieved important advances in international practice based on the conviction and acceptance by political authorities that public health occupies an enormous space in the global agenda beyond the WHO competencies[5].

Historically, international discussions related to providing solutions to issues related with diseases or pandemics were considered to concern the monitoring and attention of health authorities and health specialists. Hence, as far as foreign policy of countries was concerned, or at the multilateral level, these issues were not high priority when compared to others such as geopolitical issues related to peace, international security, economic development, as well as ideological aspects promoted by the power blocs at the time (Kevany, 2014).

This perspective began to change as Global Health, an academic discipline promoted internationally in important universities and research centers involved in the training of personnel to investigate and take action to face transnational health problems and conflicts, providing solutions for the improvement of health and equity at the global level (Kirckbush & Buss, 2011), begins to forcibly establish close links with the set of public decisions made by States in accordance with national interests and directly or indirectly linked with the rest of the actors in a country's international system. In other words, Health Diplomacy transitions from an academic action to a group of practical actions to the extent that Global Health and the Foreign Policy of the states converge in a political strategy with established plans and objectives (Kirckbush, 2011).

In this sense, the International Sanitary Conference of 1851 is considered the initial moment of this process when a group of countries met in the French capital to find a collaborative solution, beyond borders, to the spread of cholera, plague and yellow fever, threats that were growing rapidly due to the acceleration of international trade and migration (Fidler, 2001). More than a century later, in 1978 in Alma Ata, around 130 nations and 65 international organizations gathered in a new global conference organized by the WHO and UNICEF and defined, through the final declaration, the importance of primary health care as a strategy to achieve a better level of health of nations, as well as the priority as a socioeconomic issue that should be assumed directly by the governments present in the Congress (WHO, 1978)[6].

Other turning points at the political level were, on the one hand, the launch of the Global Health and Foreign Policy Initiative (GHFP) signed in March 2007 as the Oslo Declaration by the Ministers of Foreign Affairs of Brazil, France, Indonesia, Norway, Senegal, South Africa, and Thailand. On the other hand, the respective United Nations resolutions approved such as 63/33 of November 2008 and, at the regional level, the CD48/15 of the Governing Council of the Pan American Health Organization (PAHO). In the case of the aforementioned initiative, one of its objectives was to promote a foreign policy in the signatory countries that would make health care and its problems a priority beyond global crises and health emergencies (Amorin et al, 2007). On the other hand, the WHO and PAHO resolutions, both documents approved recognized globally and regionally the increasingly close link between health and the foreign policy of the states and the need to use more resources to include these issues as part of the analysis and objectives of comprehensive development by the countries (WHO, 2008; PAHO, 2008).

III. VACCINES: FROM PRODUCTS OF SCIENCE TO SOLUTIONS IN THE POLITICAL CONTEXT DURING HEALTH CRISES

In the case of Vaccine Diplomacy, although received this name due to the pandemic scenario and its consequences, it is a phenomenon that had been occurring and developing long before the pandemic, even in the Latin American region[7]. Vaccine Diplomacy is a concrete example of soft power that requires, on the one hand, human resources with preparation and training both in scientific issues linked to the development of vaccines, as well as in diplomatic and international negotiation skills, for its implementation by the nation that assumes it. Furthermore, it can be politically perceived as a strategy to be followed due to economic or financial interests or to enhance political ties and alliances between countries[8].

In the world we are living in, the Vaccine Diplomacy is only permissible for very few nations or international organization and could be conducted in specific scenarios such as outbreaks diseases as the recent COVID-19 pandemic, where the production, distribution and control of vaccines have generated concrete opportunities to implement soft power actions in international relations, as well as the introduction of new actors in the global political scenario (Hotez, 2021). It was in this context (COVID-19 pandemic) that visibly came to the fore strong negotiations between international coalitions with flagrant internal contradictions, the participation of government alliances, biopharmaceutical transnationals, universities, military study centers and think tanks, in a race whose objective was to achieve multi-million profits, as well as international prestige for achieving the registration of effective vaccines against the virus[9]. This race, called "vaccine nationalism" by the WHO top management, generated a competition among major powers, linked to comparative sales, measuring the power of their scientific capabilities, and investing not only financial, but also human and technological resources to find effective and safe vaccines against the virus[10]. Contradictorily, such a position, which foresaw the accumulation of doses of possible vaccines to satisfy certain nations or population sectors, marked an unknown scenario with a high probability of leading to structural failure in view of the high degree of global spread of the virus and the prevailing globalization, which is very difficult to control (Hafner and col, 2020 and Adeel Riaz and col, 2021).

IV. MODELING THE INTERNATIONAL CONTEXT IN WHICH HEALTH DIPLOMACY AND VACCINE DIPLOMACY CONVERGE

An attempt to model clearly and individually each of the actors involved in the complex, yet fascinating and necessary scenario of international relations in which Health and Vaccine Diplomacy converge is represented in Figure 1.

Figure 1. Modeling of the International Scenario of Vaccine and Health Diplomacy Connected to Science Diplomacy

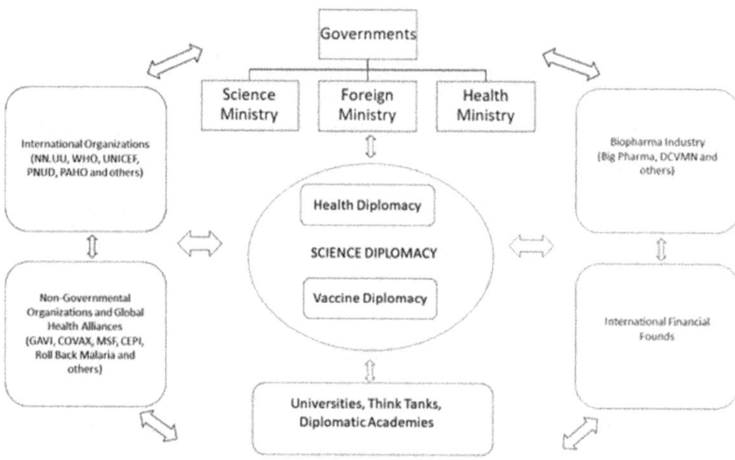

Both, Health and Vaccine Diplomacy have specific characteristics and their own implementation mechanisms. However, they both share common ground such as the generation of investment policies from the public and private sectors, the use of negotiating skills to reach agreements between the interested parties in finding solutions to health conflicts, the direct connection between local or regional scientific development with public policies, the foreign policy profile of the states through the diplomatic corps and the respective links with regional and international organizations that promote international collaboration[11].

The modeling presented shows specific characteristics for each geographic area taking into account the regional context. Carrying out an analysis to show how both diplomatic tools have been used in the Latin American and Caribbean region is the objective to be developed next from the practical perspective generated before, during and after the COVID-19 pandemic.

V. HEALTH DIPLOMACY IN LATIN AMERICA: SOME EXPERIENCES FROM THE PAST

Considering the words of the co-developer of the theory of international relations known as *neoliberalism*, the American politician and academic Dr. Joseph Nye[12], soft power in international politics assumes the ability of a country to influence others through non-violent mechanisms and strategies such as culture, diplomacy, human values, science, education, technology, among others (Nye, 2004). Faced with the perspective that reduces power to clearly material, quantifiable elements, the vast majority of which come from the political initiative and are subject to its direct control, Nye highlights the existence of another series of immaterial factors. These are not always controllable by the State and are nonetheless capable of contributing to the achievement of the objectives set by a nation as much or even more than military pressure and economic coercion (Nye, 2004 and Gallarotti, 2011).

The soft power, it is an indirect way of exercising power where: "*A country can obtain the results it wants because other countries want to follow in its wake, admiring its values, emulating its example, aspiring to its level of prosperity and openness*"[13] (Nye, 2003)In other words, under this concept it is no longer enough to measure the power of an international actor based on its tangible elements such as its armed forces, economy, territory and population, among others. In this new globalized world, there are factors such as scientific development, culture, leadership, identity, values and language, which are equally important to consider in international relations and which become attractive to others when it comes to influencing their decisions. This capacity generally constitutes an asset for countries that have a high level of development that allows them to deploy, in crisis scenarios, strategies that benefit to a certain extent their prominence at a regional or even international level.

In the current international scenario, it is increasingly visible that throughout history it has been possible to build bridges and reduce political tensions between nations through the universal language promoted by science and its positive results. Scientific collaboration between experts from different nations does not have geographical boundaries even when there are barriers that try to prevent it from a political point of view. Tense diplomatic relations have been overcome, generating trust between the parties and mutually beneficial relationships based on the positive influence of international scientific collaboration networks[14].

But the global impact of Covid-19 has generated an important turning point in this scenario. The international context generated by the COVID19 was a favorable moment for nations that traditionally implement soft power strategies in their international relations such as persuasion and influence instead of coercion. They identified the situation as an opportunity to carry out projects, links, and procedures directly connected to the real capabilities of producing, distributing, and controlling the market for resources, medicines, and vaccines against the pandemic. This prominence linked to purely commercial actions or through collaborations or even donations reinforced the strategic interest in promoting health policies and scientific and technological development as useful tools to face the challenge that the deadly virus presented to humanity since the end of 2019.

However, although political strategies connected to Vaccine Diplomacy and Health Diplomacy, on which we will later focus this chapter, became popular globally and specifically in LAC as a result of the recent pandemic scenario, it would be a mistake to consider this situation as a new scenario in the geographical area of interest of our work. The date of February 18th, 1881, could be considered the embryo of Health Diplomacy actions in our region and that would later allow actions of international scope for the currently recognized relationship between Science and Diplomacy. On that day, as part of the academic program of the "V International Health Conference", held in Washington DC, the eminent Cuban scientist Carlos J. Finlay proclaimed to the world his revolutionary hypothesis, which would subsequently transform medical and scientific thinking related with communicable diseases. During his conference, Dr. Finlay presented to the academic and scientific world attending the event the hypothesis about the existence of an "intermediary agent" that acted as a transmitter among human beings affected by yellow fever, a disease that had had numerous Epidemic outbreaks mainly in Central Caribbean countries at that time[15]. Six months later, on August 14th of the same year, in a new conference given at the Academy of Sciences in Havana, the Cuban scientist identified the agent as the mosquito currently known as *Aedes aegypti* (Damiani, 2016). These claims were the result of research conducted by Dr. Finlay that revolutionized the theories, until then unproven by the international scientific community, which considered the transmission of the disease through the air or by direct contact between people. (Chavez Carballo, 2005, Staples and Monath, 2008, and Lodola and Gois, 2015.)For almost two decades, international scientific commissions that visited the island rejected Finlay's hypothesis as well as his conclusions and prophylactic proposals to close the path to the disease through vector control. At the end of the war between Cuba and the United States against the Spanish colonial domination, North American troops were suffering as many casualties from yellow fever as from combat. Then a commission of key American doctors and specialists, considering the favorable echo of the results previously presented by Finlay in a sector of the Hispanic scientific community, decided to verify the veracity and relevance of the Cuban doctor's hypothesis. As a result of a sanitary campaign conducted by American specialists, with the support of the military government on the island, following the precise indications of Dr. Finlay and his collaborators, it was possible to eradicate the disease[16].

This was a collaboration between Cuban and American scientists. On the Cuban side there was Dr. Finlay as well as other eminent doctors such as Claudio Delgado Amestoy, who was his inseparable collaborator, and Dr. Arístides Agramonte, a graduate of Columbia University. On the American side, there were specialists led by the army surgeon, pathologist, bacteriologist, and Robert Koch's disciple, Dr. Walter Reed, as well as his renowned colleagues the bacteriologists James Carroll, Jesse W. Lazear and William C. Gorgas[17].

These results would be recognized by British doctors from the Liverpool School of Tropical Medicine and shortly after, they would be replicated in the port of Veracruz in Mexico, as well as by experts from the São Paolo Health Service in Brazil, who reached the same result. Even from the Pasteur Institute in Paris, scientists who collaborated with their Brazilian delegate colleagues in Rio de Janeiro and São Paulo tested the Finlay thesis coordinated with North American experts, demonstrating its veracity and scientific accuracy (Damiani, 2016).

This collaboration would reach its climax when its results were used, by some of the researchers mentioned above, as a solution to the complex situation that had arisen for several years in the construction works of the Panama Canal. During the construction of this demanding engineering work, thousands of lives were lost due to the impact of tropical diseases, mainly yellow fever, and malaria (Lechuga and Castro, 2017). Currently, a statue of Dr. Finlay is erected in Panama City and part of their research and experiences are in the historic building of the Gorgas Memorial Institute of health Studies in recognition of this scientific collaboration, without which it would have been much more difficult to complete the construction works and without a doubt saved thousands of lives. A park in Tegucigalpa, the Honduran capital, recognizes this legacy as well. And at an international level, to settle debates about the authorship of the thesis that occurred at the time and subsequently, the scientific and academic community recognized in 1954 during the XIV International Congress of the History of Medicine held in Rome and Salerno, the authorship of Dr. Finlay of the discovery of the transmission of diseases through vectors (Damiani, 2016). In addition to this, it was also recognized the corresponding relevance that these results meant for the research and development of vaccines and other mechanisms to confront epidemic outbreaks caused by this type of diseases[18]. On the other hand, every December 3, since 1933, Latin American Medicine Day has been celebrated and the United Nations Commission for Education, Science and Culture (UNESCO), years later, would name Dr. Finlay along with Edward Jenner[19], Robert Koch[20], Alexander Fleming[21] and Louis Pasteur[22], among the five great microbiologists in History, and would issue a biannual international award with his name to recognize the outstanding work of scientists at a global level in the branch of microbiology[23].

These and other awards are very relevant, but for the objective of this paper the fundamental thing is to recognize that, in the author's opinion, Dr. Finlay's collaboration with his North American colleagues and other countries in the region, could be the first great example of Health Diplomacy in LAC region in the modern era.

Interestingly, to show what is in our consideration the first example of Vaccine Diplomacy in LAC according to the concepts presented above, we would have to go back several decades before 1881 when Dr. Finlay presented in public his revolutionary results and when the nations of the region still remained colonies of the Kingdom of Spain.

In the first years of the Spanish conquest, smallpox had already been introduced in the region decimating the native population and, by the 19th century the old organizational structure of public health transferred by the colonizers and gradually implemented in cities such as Mexico, Lima, and Havana, based on the institution of the Royal Protomedical Court, was completely insufficient to stop the multiple epidemic outbreaks in the colonized lands[24].

These were times when the British rural doctor Edward Jenner had already shown that, if a person became infected with a harmless disease, it was possible for them to acquire immunity to a similar pathogen. This principle had been proven precisely by using the cowpox virus to protect human beings from the variant of the disease that affected them (Rusnock, 2016).

The results in Europe were so prominent and the situation in the American colonies so worrying that the Spanish monarch Charles IV followed the suggestions of his advisors, mainly his military doctor and honorary royal surgeon Dr. Francisco Xavier de Balmis, whom he instructed to organize and lead what would be called the Royal Philanthropic Vaccine Expedition that would allow this health tool to be introduced in the colonies (Mark and Rigau Perez, 2009).

The expedition led by Dr. Balmis and his assistant Dr. José Salvany Lleopart left with the vaccine at the end of November 1803 towards the Canary Islands and from there to the lands currently occupied by nations such as Venezuela, Mexico, Puerto Rico, Cuba, Guatemala, Colombia, Ecuador, Peru, Bolivia, Chile, and Argentina (Mark and Rigau Perez, 2009).

It was a very complex time in which there were no cold chains, as there are today, to keep the temperature of the vaccine or culture media that would maintain the microorganisms available for reproduction and development. For this reason, experts had to use their inventiveness and come up with a procedure that could currently be seen as questionable from an ethical point of view but that, given the circumstances of the early 19th century, it could also be considered to have been an innovative, novel, and peculiar solution. Dr Balmis and his colleague's solution was to select about twenty healthy orphan children who were inoculated with the vaccine fluid, becoming live containers of the vaccine. This way, upon arriving at the colonies the vaccine maintained its immunogenic properties and it was possible to progressively use it[25].

The Vaccine Expedition not only allowed controlling the epidemiological situation in the Hispanic colonies but also administratively established Vaccination Boards that, in addition to organizing and developing local vaccination campaigns, generated training actions for health experts to conduct these processes and additionally collect statistical and epidemiological data that would be useful for making political decisions in situations of health crises in the LAC region.

VI. CAPACITIES TO IMPLEMENT PROJECTS LINKED TO HEALTH DIPLOMACY AND VACCINE DIPLOMACY IN LATAM

As it has been mentioned before, throughout history the fields of health and international relations, although they are not completely detached, they have not had a prominent relationship either, since health has traditionally been considered a matter of low priority by the foreign ministries. In this sense, the general tendency has always been to consider issues of war and peace as well as economy and trade as "high politics." However, it must be taken into account that since ancient times health and illness (as an effect and as a weapon) were a matter of high relevance in wars, particularly in the great wars of the 19th and 20th centuries.

From the perspective of health policies, it is recognized by several observers that international trade and finance, population mobility, environmental change, international conflicts, disasters, as well as international security issues, among other factors, have clear and demonstrable repercussions on health. It is also acknowledged that the strongest links between health and foreign policy are found in the fields of transportation, commerce, tourism, and emigration[26].

In LAC, as throughout the entire planet, Health Diplomacy and Vaccine Diplomacy programs and strategies, assumed by different governments, accelerated their visibility and execution especially after the World War II. However, the trend in the region has been for it to occur in an unequal manner, depending fundamentally on the actions and decisions implemented by the governments in power and the procedures and strategies assumed by regional organizations[27].

At this point, it is important to establish the practice of Health Diplomacy and Vaccine Diplomacy strategies in two perfectly differentiated capacities that depend significantly on resources, whether natural, technological, financial, and political decisions expressed by governments and national and regional organizations, with the intention of using them jointly for the development of both strategies or focusing on at least one of them. The first capacity is defined as a practical variant and is identified through concrete actions of Health Diplomacy or Vaccine Diplomacy by a country or a group of countries in collaboration, in situations of disasters or health crises/issues that occur in the region. The second capacity could be defined as an *academic or human resource training variant* to execute specific actions of one strategy or another following a program of academic actions that range from the organization and conduct of seminars, workshops, scientific events or social communication actions, to the training of experts through postgraduate studies aimed at health, scientific, diplomatic and other specialties personnel in academic centers such as universities, think tanks and other high level institutions in the field of education.

There are several factors that influence the possibility of executing these two capabilities mentioned before in LAC. On one hand, the low degree of investment in health policies and the development of science, technology and innovation applied to health. On the other hand, the repeated economic crises in the countries of the region that have generated repeated crises of inequality and increased poverty in Latin American societies in recent decades, and that in many cases forced scientists, experts, and academic personnel to emigrate to more developed nations[28]. In addition to this, there is a third factor, that has been already addressed to a certain extent, which is the disconnection, in most countries in the region, between scientific development and the foreign policy of governments, which generally and unfortunately depends on ideology, political leaning and strategic projection of the governments in power[29].

Although it is a fact that prior to COVID-19 the region had not suffered major natural disasters that affected the entire region, on several occasions phenomena such as hurricanes and earthquakes, especially in the Caribbean area, had been combined with the precarious economic situation in some countries to generate national crises that rarely affected multiple states (UNDRR, 2023). However, some regional organizations such as the Pan American Health Organization (PAHO), the Organization of American States (OAS), the Southern Common Market (MERCOSUR), the Caribbean Community of Nations (CARICOM), the Inter-American Development Bank (IDB), the Economic Commission for Latin America and the Caribbean (ECLAC) among others had issued resolutions and directives promoting investment in health and scientific policies among member nations (World Bank Report, 2023). The reality is that the scope of government decisions or regional organizations such as those mentioned above was not sufficient to consider the possibility of establishing these options as strategic policies of governments, leaving everything in non-binding agreements.

This situation lead to the consequence that, when the pandemic hit the region in early 2020, very few countries had scientific development strategies, not to mention efficient productive capacities to generate their own vaccines and medicines. Only nations such as Mexico, Brazil, Argentina, the particular case of Cuba and to a lesser extent Venezuela and Colombia had certain capabilities and previous experiences to face the greatest epidemiological challenge in the history of the region.

VII. LATIN AMERICAN EXPERIENCES IN HEALTH DIPLOMACY DURING THE CURRENT CENTURY PRIOR TO THE COVID-19 PANDEMIC.

The concept of international collaboration as a possible solution to face the health challenges that arise in times of epidemics or from natural disasters, as well as its practical execution through programs between governments, has been achieving success since the second half of the 19th century. (Bynum, 1993). These threats have been exponentially increasing as fast as other issues such as intercontinental trade, legal and illegal migrations, and mainly social inequality between countries and even among different provinces or regions of the same country. Challenges such as pandemics, chronic non-communicable diseases, antimicrobial resistance, climate change and environmental pollution and social inequality have influenced the development of a system of rules and norms by the international community that attempt to control and conduct actions of states in the face of natural or man-made events that may become international threats.

According to a report from the United Nations Office for Humanitarian Affairs released at the beginning of 2020, just before the pandemic was declared, LAC was considered the second region most prone to natural disasters (OCHA, 2020). During the first two decades of this century, more than 152 million Latin Americans and Caribbeans had been affected by more than 1,200 disasters, including floods, hurricanes and storms, earthquakes, droughts, avalanches, fires, extreme temperatures, and volcanic events (OCHA, 2020). In LAC, as in other regions of the world, the negotiating roles between government authorities, decision-making figures in international politics, the private business sector, as well as international organizations, have been essential to achieve alternative solutions to these threats through actions linked to Health Diplomacy and Vaccine Diplomacy.

In general, there are three essential aspects that must be considered and that are common to the dynamics in LAC. The first of them is the important challenge of governments to internally manage to conceptualize health problems that may affect outside national borders as an issue to be included in the foreign policy strategy. The second one is the objective reality of promoting that these problems can be coordinated with other sectors of the economy and local politics, having the government as a regulator and organizer. A third aspect of great importance, is the priority that governments assign to the need to collaborate with the countries of the region and that generally depends on the state of existing bilateral or multilateral relations.

Under this perspective, several strengths were developed, and decisions were made during pre-pandemic years by some of the Latin American governments, which undoubtedly had a positive impact in recent times as part of the capabilities to execute Health Diplomacy and Vaccine Diplomacy actions in LAC. In a way, we have already mentioned the two main ones: having an effective synergy that promotes health policies based on the foreign policy of governments (and vice versa) as well as human resources with full capacity and skills related to this work both in the diplomatic corps and in the scientific and academic community.

From a practical point of view, this has translated into relevant actions such as the creation in Mexico of a Global Health Program coordinated by the National Institute of Public Health from which academic training in global health issues has been promoted since 2003 (Solimano and Valdivia, 2014). The results obtained with this program are based on a concept that defines global health as a global public good, as an issue of social justice and as a universal right. Additionally, the training perceives the need to focus analysis and political decisions around equity, ethics and human rights that deserve to be respected from all social sectors, emphasizing the work with socially disadvantaged population groups, common challenges with local depth and global impact, citizen participation, as well as interdisciplinarity and inter-sectorality (Almeida, 2020).

Similar experiences were implemented at the School of Public Health of the University of Chile, at the National Faculty of Public Health of the University of Antioquia in Colombia, and at the Peruvian University Cayetano Heredia (Franco, 2015). In each of these cases, teaching and scientific research on national and regional health issues highlighted the possibility of having a critical mass of experts in the region who have assumed at certain times political or administrative responsibilities in national governments, regional and international organizations, promoting public health policies and expanding the strategic vision from national health to global health, taking into account precisely the epidemiological warnings that had already materialized during the first years of the 21st century.

A special mention must be made to the Brazilian institutions, mainly to Sergio Arouca National School of Public Health, from Oswaldo Cruz Foundation in Rio de Janeiro and to the Faculty of Public Health of the University of São Paulo in which a training approach for experts in global health and health diplomacy is implemented, considering international collaboration and coordination with international organizations and global health investment funds as a bridge of connection (Solimano and Valdivia, 2014). Under this premise, international experts from several countries shared their experiences and knowledge in the different study programs from diplomas, seminars, master's, and doctoral programs that include ethical and bio-ethical analyses and address the impact of globalization on health and international cooperation in local and regional policies.

From the connection established between these and other academic institutions with similar interests, the possibility of establishing the Latin American Alliance for Global Health (ALASAG) arose (Buss and Tovar, 2017). An institution like this whose declared purpose is precisely the strengthening of global health on the continent by providing a Latin American perspective to issues on the global agenda on health and development in spaces for academic discussion, research, and strategic policies on these issues, has undoubtedly generated a positive scenario in the region that has been recognized by international organizations such as the WHO and PAHO. It is important to highlight in these coordination actions the role of relevant academics such as Dr. Paulo Buss, Dr. Sebastian Tovar, Dr. Celia Almeida, and others who have been the protagonists in generating a circle of advice and counseling to different governments in the region and have participated in the training of human resources with suitable capabilities for these projects and strategies. Proof of this has been the possibility that in a joined collaboration with other Latin American experts, they have managed to insert their publications, articles and reflections in important international journals that address topics related to Global Health and Health Diplomacy (Buss and Tovar, 2017).

This advanced training of professionals from the region in Latin American institutions, which has also had the support and collaboration of European and American experts, has also been useful in achieving important progress during the last two decades between the Ministries of Health, the national regulatory agencies[30], non-governmental organizations, as well as international sources to finance development devoted to the health sector. It has been key in this case the political connection between the highest health authorities of the countries with the PAHO's Directing Council where the strategy for training experts in Health Diplomacy has been recognized and endorsed (Buss and Tovar, 2018). Based on this dynamic, the Program to Strengthen Cooperation for Health Development, whose main objective is the training of experts in Health Diplomacy with the professional tools to promote diplomacy and cooperation in health in the economic, commercial and environmental scenarios, was strengthened and implemented since 2015A clear example of this commitment by the highest regional health authority, PAHO, has been to maintain and expand a Leader Training Program under the name Edmundo Granda Ugaldesince 2009, from which more than 660 experts from around 40 nations on the continent had graduated until the first year of the pandemic[31].

At a practical level, it is also worth mentioning the multinational health response to natural disasters or to requests from governments interested in improving the health care of their populations in several countries in the region over the last 30 years. In the first case, the impact of hurricanes Mitch, Irma and Maria in Central American countries, or the earthquakes in Peru, Haiti and Chile are striking.

Likewise, the response to requests from the respective governments of countries such as Mexico, Ecuador, Guatemala, Nicaragua, Venezuela, Bolivia, Uruguay as well as Caribbean Island nations for medical brigades provided by Cuba within the framework of international collaboration protocols should be mentioned. This specific case generated a strategy by the Cuban government to employ thousands of health specialists in these countries in geographic locations where primary health care was deficient or simply did not exist (Huish and Kirk, 2007). Despite this being a program that has received criticism and questioning from some academic and political sectors at a regional and international level for not agreeing with some of its premises and considering them negative at a political level, it is impossible not to recognize the actions of these specialists who were an important link during the first two decades of this century in improving health indicators in the areas where they carried out their work.

Additionally, these brigades implemented specific health care programs such as Operation Miracle, which allowed thousands of Latin Americans to improve or recover their vision, or the implementation of academic programs with Cuban and other teaching staff for the creation of Medical Universities in areas with a low presence of trained human resources for health care.

These examples reflect the essential interaction between the public and private sectors to, on one hand, invest resources in the creation of training sources in academic institutions and, on the other, invest in projects, strategies and initiatives both in health and in scientific and technological development who understand how to take advantage in the best possible way of the human resources that acquire this knowledge and skills.

VIII. LATIN AMERICAN EXPERIENCES IN VACCINE DIPLOMACY DURING THE CURRENT CENTURY PRIOR TO THE COVID-19 PANDEMIC.

To appreciate the role of LAC nations in Vaccine Diplomacy prior to COVID-19, it is important to analyze three main factors. The first is the region's health strategy regarding the implementation of National Immunization Programs (NIP). Secondly, the economic-financial aspects involved in the development of a vaccine production industry. And third, the practical role assumed by the countries that had the capabilities and opportunities to execute actions linked to this strategy in pre-pandemic times.

With the discovery of the first vaccines starting at the end of the 18th century, and their continuous progress during the 19th and 20th centuries, the present 21st century arrived with the general consideration at an international level of the need to introduce this type of products within the global reach of science, for the benefit of humanity as a result of its own intelligence. Initially conceived to face infectious diseases, their accelerated advancement has allowed in recent decades to investigate and develop candidates against addictions, metabolic or mental diseases, cardio-vascular conditions, and against different variants of cancer (Greenwood, 2014) which allowed us to perceive, with the advent of the 21st century, the possible availability in a near future of a greater menu of vaccines accessible to the health authorities of countries (Delany, Rappuoli and de Gregorio, 2014). However, it was possible at the same time to perceive the high level of elitism and the extraordinary degree of investment by public or private sectors to achieve the development of scientific projects of this type that also included productive capacities under the quality and safety requirements demanded by international and national regulatory organizations[32]. Before the pandemic expanded, the vaccine industry was already recognized internationally for the high added value of its products, which included a combination of knowledge, technology, innovation and quality culture. Both in universities, private companies, and government research institutions, the personnel dedicated to this industry of research, development, production and marketing of vaccines had global recognition by generating necessary and demanded products in the international market to combat different diseases (Greenwood, 2014).

The economies of developing countries such as LAC, generally did not have the necessary capabilities to create national industries that would guarantee a series of variables that affect the technological and productive competences of an industry of this type (Guzman-Holst et al, 2020). These include the recurrent economic weaknesses caused by international or internal crises, the institutional instabilities of governments that hinder the effectiveness of public policies, the low private investment in R&D, the notable increase in dynamic industrial sectors dominated by transnational companies whose development and growth strategies are largely independent of the local economy, the complex direct influence of international speculative finance on national economies, as well as the geopolitical disadvantages generated by the sum of the above factors in negotiations between developing countries with multinationals to assume technological transfer processes of projects or production chains linked to the development of vaccines and the bio-pharmaceutical industry in general (Guzman-Holst at al, 2020, and Diaz, Aruvito and Gefner, 2023)

These circumstances, with specific exceptions, have been recurrent in LAC even considering the relevance that vaccines have had through the NIPs that have managed to eradicate, eliminate, and control various diseases preventable by vaccination. The region has thus been considered to have one of the highest levels of vaccination coverage in the world (UNICEF, 2018).

The countries in the LAC region with the best Immunization Programs covering vaccination from an early age with products supplied by multinationals or through PAHO's Revolving Fund or UNICEF supply programs, were Chile and Panama. At a second level were the EIP of Argentina, Uruguay, and Costa Rica, with the vaccination of other age groups in addition to childhood, and specific vaccination campaigns such as stationary influenza (Rombini et al, 2019). Then, at a third level, Brazil, Colombia, and Mexico showed more outdated NIP, with programs not implemented for various reasons, as well as low national vaccination coverage, largely due to the high territorial extension of these countries and the difficulties in delivering coverage equitably. Finally, Paraguay and Peru closed the ranking of this study, showing delayed vaccination schemes, low coverage, and information gaps (Rombini et al, 2019).

Consequently, despite being one of the regional markets with the highest levels of immunization from expanded programs supported by PAHO and overseen by local health authorities, vaccine development and production in the region has been concentrated in a few countries and at levels that do not cover real demand. This occurs, largely due to economic and social inequalities, and the combination of the above-mentioned variables between countries and within the same country (Cortes, Cardoso, Fitzgerald and Di Fabio, 2012).Some initiatives allowed at the time to have productive capacities to generate vaccines against rabies, yellow fever, and other cases in nations such as Mexico, Argentina and to a lesser extent Chile, Venezuela and Colombia, but their projects did not achieve international recognition to allow the implementation of Vaccine Diplomacy strategies. Significantly, in LAC there are two examples that should be specifically highlightedThe first of them is Brazil where two prestigious scientific institutions, with the support of important investments from the State, have become hallmarks of national identity as vaccine producers in this century: Butantam Institute (founded in 1901) in Sao Paolo, and Oswaldo Cruz Foundation, "FIOCRUZ" (officially inaugurated in 1898).

Both institutions supply a high percentage of the vaccines used in the EPI such as hepatitis A, hepatitis B, HPV (Human Papillomavirus), dTpa (Diphtheria + Tetanus for adults) and rabies (in the case of Butantam) as well as DTP (Diphtheria + Tetanus + Pertussis), Yellow Fever, Haemophilus influenzae B, Meningitis A and C, 10-valent Pneumococcus, Inactivated Poliomyelitis, Oral Poliomyelitis, Tetravalent Viral Human Rotavirus, Triple Viral and Measles and Rubella (attenuated) (in the case of FIOCRUZ)[33]. This productive capacity, also implemented to obtain Active Pharmaceutical Ingredients (API), generated mainly in recent decades, is supported by a solid investment in R&D and a recognized quality control system. This has allowed several of these products, obtained mainly at FIOCRUZ, to be received in more than 70 countries, through agreements negotiated with international entities such as PAHO, UNICEF and the Global Alliance for Vaccine and Immunization

(GAVI)[34]. Under this scheme, the vaccine against yellow fever and immunizations against meningococcal diseases have been used to a greater extent to face epidemic outbreaks in developing countries (Grabois Gadelha et al, 2020).

The other productive potential in terms of vaccines in the region prior to the pandemic is Cuba where, starting in the 1980s, the government decided to accelerate scientific progress specifically in the complicated field of the bio-pharmaceutical industry, to guarantee Cuba's incorporation into the global trend of promoting an industry of high value-added products. The Biological Front was then inaugurated as the seed of what later became the first productive Scientific Cluster in the country based in the west of Havana, created with the aim of promoting the advancement of biotechnology and the Pharmaceutical Industry and which has been part since 2012 of the cluster called BioCubaFarma (Arencibia et al, 2016).

The development and production of human vaccines constitutes a key link in BioCubaFarma, highlighting a group of scientific institutions such as the Finlay Vaccine Institute (FVI), the Center for Genetic Engineering and Biotechnology (CIGB), the Center for Molecular Immunology (CMI) and the Center for Biopreparations (BioCen) among others. In a strategy called “closed cycle,” these companies have achieved high national recognition by supplying several of the vaccines included in the NPI. Also, internationally this industry has reached agreements with entities such as UNICEF, UNDP, and other United Nations agencies, as well as NGOs to use the vaccines they produced in developing countries (Thorstendottir and Saez, 2012, Sierra-Gonzalez, 2019, and Sheridan, 2022)

Precisely, based on the productive strengths of both countries Brazil and Cuba, it was possible to implement a joint collaboration to supply vaccines against meningitis to African countries through UNICEF in order to face the supply crisis generated between 2006 and 2010 of annual epidemics that affected Sub-Saharan African countries[35]. Both Latin American governments joined their efforts to, with the support of their respective regulatory bodies and their diplomatic corps, implement a “fast-track” process that allowed them to receive the corresponding authorizations from the WHO, UNICEF, and the African authorities in just 9 months to immunize millions of children and teenagers (Thorstendottir and Saez, 2012).

Millionaire investments both from a structural point of view in the construction of productive facilities as well as in the purchase of equipment and resources were executed by both countries to achieve the high standards of quality and technological efficiency that a project like this demanded. The success achieved in the rigorous pre-qualification process granted by the WHO, with the support of local regulatory authorities, allowed millions of doses of this vaccine to be supplied to the African population for several years, managing to reduce the negative impact of the epidemic in some of the affected countries (Thorstendottir and Saez, 2012). Additionally, this process allowed to increase direct relations with the international organizations involved in the project (WHO, UNICEF, UNDP) as well as visualize new collaboration projects from Southern countries to contribute to the development of other developing countries[36].

This result has been recognized as the first successful South-South cooperation in the bio-pharmaceutical field. Besides, this experience demonstrated that the capacities implemented in LAC countries can be used in response to a health crisis based on the exchange and connection between scientific and political factors for the benefit of improving health conditions in both the region and faraway countries.

IX. HEALTH AND VACCINE DIPLOMACY IN THE LAC REGION DURING THE PANDEMIC.

The data provided allows us to perceive the technological gap and vulnerability of health systems in the specific case of LAC region based on the impact of the COVID-19 pandemic. As globally, the health systems of most countries in the region were overwhelmed by the high level of infection caused by the virus. In the first moments, practically each country had to sustain the outbreak with its own resources, which resulted in a high degree of difficulty and complexity in being able to execute health diplomacy actions in the region..

In the pandemic context, were interesting to highlight the speed with which some countries, which initially carried out health donations, logistical support with material or human resources based on negotiations between governments, went in a matter of weeks from being donors to becoming receivers in need of all types of resources necessary to face the impact of the disease. On the other hand, and closely linked to the previous factor, due to the extent of the crisis, many states assumed nationalist positions, gathering all types of resources to face the situation within their borders and even diverting shipments of contracted supplies from their destination to pass through national airports to use them in their own population. (Riaz et al, 2021, Murhula and Singh, 2022, and Zhou, 2022).

The LAC region was particularly affected, also considering the aforementioned aspects of poverty, social inequality and health vulnerability in the vast majority of countries. However, an example to be highlighted was that of countries such as Mexico, Venezuela, Chile, and Brazil which shared some resources with nations in the area who were in a worse situation, as well as the traditional response of the Cuban government by sending brigades of doctors and health staff to collaborate with local health forces to take care of patients in the most affected areas.(Gorry, 2020; Cyr, Bianchi, Gonzalez and Perini, 2021; and San Martin and Bortz, 2022). In each of these cases, the exchange between governments and agreements between the parties were crucial to enhance the positive effect of these collaborations.

From a scientific point of view, the political decision of most countries in the region to invest resources to face the epidemic with results from local science was relevant (Garcia et al, 2020). Both public and private sectors from nations that had certain capabilities or facilitated multilateral agreements with other countries or even with transnational companies in the bio-pharmaceutical industry participated in this intervention through the execution of clinical trials of developing products, the direct introduction of vaccines, diagnosis, innovative treatments, and partial technological transfer to produce capabilities developed abroad.

In this context, the COVAX initiative, a platform to support the development, manufacturing and distribution of vaccines in a fair and equitable manner, in coordination with United Nations entities such as WHO, UNICEF, PAHO and global initiatives such as GAVI or CEPI[37] played a significant role in the distribution of vaccines in much of the continent starting in the second quarter of 2021. However, the high demand for the necessary doses was impossible to be covered through this mechanism alone. Based on this limitation, several countries in the region, due to the lack of local productive capacities, carried out strong negotiations with transnational companies or production companies from countries such as China and India, and managed to reach agreements to carry out clinical studies of vaccine candidates in development even before these products were approved by international regulatory bodies, with the aim of achieving advantages in access to vaccines.

Based on this, Brazil, and Mexico and Argentina jointly, signed important agreements for the transfer and production of the vaccine developed by the British laboratory AstraZeneca and the University of Oxford. Likewise, vaccines developed by Chinese companies Sinopharm, Sinovac and to a lesser extent Cansino, were introduced on a large scale. Additionally, agreements were signed in several countries for the introduction and joint production in the region of the Sputnik vaccine from the Gamaleya Institute of Russia. Besides, based on donations from the European Union, the United States government as well as direct financing from some of the developed countries, millions of doses of vaccines developed by the international laboratories Pfizer, Moderna and Johnson & Johnson were introduced. It is important to highlight the decisions made by the governments of Mexico, Brazil, Argentina and Cuba that, even in difficult times in their own nations, donated vaccines to other countries in the region with fewer possibilities and resources.

Another very interesting case was the close relationship between Taiwan and Paraguay, the only country in Latin America that has an embassy of the Asian island, taking into account the political conflict with the People's Republic of China[38]. From this relationship, the Paraguayan nation received donations of vaccines, various supplies in the most complex moments of the epidemic and even its regulatory authority authorized the Emergency Use of a candidate developed by an important laboratory on the Asian island. This constitutes an interesting case study within Vaccine Diplomacy and of which several academics have written valuable contributions (Nolte, 2023).

Unfortunately, in this maelstrom surrounded by the context of the pandemic, cases of corruption multiplied in several LAC nations, political interference to prioritize some products over others, internal discussions and debates between governments and their respective political positions on commercial strategies and/or of vaccination and even cases of breach of contracts that generated diplomatic tensions at high levels of governments that, although not the object of analysis in this paper, must also be recognized as a significantly negative aspect of this complex scenario.

Despite these circumstances, several countries in the region such as Brazil, Argentina, Mexico, Chile, Ecuador, and Peru did not stop investing and tried to advance local projects to develop their own vaccines. Among these national projects, the case of Cuba should be particularly mentioned, since its vaccine industry managed to advance 5 vaccine candidates, 3 of which obtained the pharmaceutical registration and were used in the national population and subsequently in several countries that requested doses despite not having obtained the WHO pre-qualification. Additionally, two of these products, Soberana02 and Abdala, were the subject of important technology transfer processes to Iran and Venezuela during the pandemic, which is a unique result in the region and in the Global South.

X. CONCLUSION

In the author's opinion, it would be risky to talk about the final impact of the pandemic in the region since its effects are still being measured in various health sectors and it remains to be seen what may happen in the future. But without a doubt, the experience acquired has generated a different vision of health in an important sector of global politics and especially in LAC.

Lessons of the past showed that in the face of health crises that overflow borders and surpass political interests, economic-financial potential, or even military dominance, it is a human responsibility to keep health within the political agenda of governments. And this can be achieved by increasing the quality of health systems, promoting regional and multilateral integration, investing comprehensively in science, in new productive capacities, in new joint strategies where several actors from societies and the international scene could participate.

Countries such as Brazil, Argentina, Mexico, Cuba, among others, continue to expand their academic training capacities of experts with skills to face these scenarios. New agreements between governments for scientific projects and collaborations have multiplied in the region since 2023, with prestigious international laboratories and with the financial support of investment funds and specialized organizations like UNICEF, WHO, PAHO and others. Vaccines are today a viral topic both in social networks and in national political strategies, electoral campaigns, and in the economic and commercial link between nations. Although there is still political interference that does not support the adoption of cooperative strategies, a more positive scenario is currently seen to understand the importance of the connection between science and health with the diplomatic strategies of governments.

Science and Health Diplomacy, as well as its collateral branch, Vaccine Diplomacy, must be sustained under principles of multilateral collaboration, with solid technical support, programmed financing and led by experts trained at both the scientific and diplomatic levels. We have intended to demonstrate that, given the structural and development differences between developed countries and those with fewer resources, it is not possible to implement these tools and processes in the same way. However, it is possible, with their results, to be better prepared to face natural crises that negatively impact human beings without differentiating them by gender, race, wealth, religion, or political ideology. Hence the need and priority, in our opinion, to continue promoting and professionalizing these tools globally and specifically in our geographic region.

REFERENCES

Amorim, C., Douste-Blazy, P., Wirayuda, H., Store, J.G., Gadio, C.T., Diamini-Zuma, N., & Pibulsonggram, N. (2007) Oslo Ministerial Declaration--global health: a pressing foreign policy issue of our time. *The Lancet* ; Vol 369 Issue 9570: p1373-1378, Apr. 21, DOI: .10.1016/S0140-6736(07)60498-X

Arencibia-Jorge, R., Corera-Alvarez, E., Chinchilla-Rodríguez, Z., & de Moya-Anegón, F. (2016). Cuban biopharmaceutical industry: A scientometric approach. *Scientometrics*, 108(3), 1621–1636. 10.1007/s11192-016-2023-1

Battyany, K. (2021). Latin America and the Caribbean, the SDGs, and the pandemic. *Online Idees Review No. 55*. Available at https://revistaidees.cat/en/monografics/rethinking-development-cooperation-to-meet-the-challenges-of-the-21st-century/

BBC News World. (2020) Coronavirus: Brazil confirms first case in Latin America. Posted online on Feb 26th, 2020. Available at https://www.bbc.com/mundo/noticias-america-latina-51641436#:~:text=El%20Ministerio%20de%20Salud%20de,detectado%20oficialmente%20en%20Am%C3%A9rica%20Latina

Brazil, A. N. V. I. S. A. (n.d.). https://www.gov.br/anvisa/en

Buss, P., & Tobar, S. (2018) Health Diplomacy in the Political Process of Integration in Latin America and the Caribbean. Edited and published by Oxford Research Encyclopedia of Public Health, UK. DOI: 10.1093/acrefore/9780190632366.013.24

Butantan Institute. Official website https://en.butantan.gov.br/portfolio.php

Bynum, W. F. (1993) Policing Hearts of Darkness: Aspects of the International Sanitary Conferences. *History and Philosophy of the Life Sciences.* Vol. 15, No. 3 published by: Stazione Zoologica Anton Dohrn – Napoli

Carrillo-Roa, A., & Santana, J. P. (2012). Regional integration and south–south cooperation in health in Latin America and the Caribbean. *Revista Panamericana de Salud Pública*, 32(5), 368–375. 10.1590/S1020-4989201200110000723338694

. Chaves-Carballo E. (2005) Carlos Finlay and yellow fever: triumph over adversity. Mil Med. Oct;170(10):881-5.

Cid, C., & Marinho, M. L. (2022) Dos años de pandemia de COVID-19 en América Latina y el Caribe. ECLAC- Report in Spanish. Available at https://repositorio.cepal.org/server/api/core/bitstreams/3a821906-213d-4cfc-bd26-66167328fb9d/content

Ciocca, D. R., & Delgado, G. (2017). The reality of scientific research in Latin America; an insider's perspective. *Cell Stress & Chaperones*, 22(6), 847–852. 10.1007/s12192-017-0815-828584930

Cortes, M. A., Cardoso, D., Fitzgerald, J., & Jose Luis DiFabio, J. L. (2012). Public vaccine manufacturing capacity in the Latin American and Caribbean region: Current status and perspectives. *Biologicals*, 40(1), 3–14. 10.1016/j.biologicals.2011.09.01322033155

Cyr, J., Bianchi, M., González, L., & Perini, A. (2021). Governing a Pandemic: Assessing the Role of Collaboration on Latin American Responses to the COVID-19 Crisis. *Journal of Politics in Latin America*, 13(3), 290–32. 10.1177/1866802X211049250

. Damiani, A. (2016). Carlos J Finlay (1833-1915) Su impronta en el Canal de Panamá en el centenario de su Muerte. Article in spanish Available at *Revista de la Asociación Médica Argentina*, Vol. 129, Número 1 de 2016

de Almeida, C. M. (2020). Gobernanza global en el sector de la salud: Cambios en el "orden mundial", la arena internacional y los impactos en la salud. *Astrolabio*, (24), 80–124. 10.55441/1668.7515.n24.25394

. Delany, I., Rappuoli R., De Gregorio, E. (2014) Vaccines for the 21st century. EMBO Mol Med. Jun;6(6):708-20.

Díaz, F. E., Arruvito, L., & Geffner, J. (2023). The future of vaccination in Latin America: Learning from the COVID-19 pandemic. *Current Opinion in Immunology*, 85(December), 102390. 10.1016/j.coi.2023.10239037806096

Elmgren, L., Li, X., Wilson, C., Ball, R., Wang, J., Cichutek, K., Pfleiderer, K., Cavaleri, M., Southern, J., Jivapaisarnpong, T., Minor, P., Griffiths, E., Sohn, Y., & Wood, D. (2013, April 18). A global regulatory science agenda for vaccines. *Vaccine*, 31(Suppl 2), B163–B175. 10.1016/j.vaccine.2012.10.11723598478

Fidler, D. P. (2001). "The Globalization of Public Health: The First 100 Years of International Health Diplomacy" Articles by Maurer Faculty. 599. https://www.repository.law.indiana.edu/facpub/599

Fiocruz Foundation. Brazil. Official website https://portal.fiocruz.br/en/production-and-innovation

Franco Giraldo, A. (2015) Academic review of global health approaches: an analytical framework. *Cad. Saúde Pública* 31 (9) • Set 2015 https://doi.org/10.1590/0102-311X00027115

Gallarotti, G. M. (2011). Soft Power: What it is, Why it's Important, and the Conditions Under Which it Can Be Effectively Used. Division II Faculty Publications. Paper 57.

Gorry, C. (2020) Global Collaboration in Times of COVID-19: Cuba's Emergency Medical Contingent. MEDICC Review.

Grabois Gadelha, C. A., Seixas da Costa Braga, P., Mattoso Montenegro, K. B., & Bahia Cesário, B. (2020). Access to vaccines in Brazil and the global dynamics of the Health Economic-Industrial *Complex.Cadernos de Saude Publica*, 36(Suppl 2), e00154519. 10.1590/0102-311x0015451932876097

. Greenwood, B. (2014) The contribution of vaccination to global health: past, present and future. Philos Trans R Soc Lond B Biol Sci., 369(1645):20130433.

Guzman-Holst, A., Rodrigo De Antonio, A., Prado-Cohrs, D., Patricia Juliao, P. (2020) Barriers to vaccination in Latin America: A systematic literature review. Vaccine. Volume 38, Issue 3, 16 January, Pages 470-481

Hafner, M., Yerushalmi, E., Fays, C., Dufresne, E., & van Stolk, Ch. (2020). COVID-19 and the cost of vaccine nationalism. Published by Rand Corporation. e RAND Corporation, Santa Monica, Calif., and Cambridge, UKHotez, P.J. (2014) "Vaccine Diplomacy": Historical Perspectives and Future Directions. *PLoS Neglected Tropical Diseases*, 8(6), e2808. 10.1371/journal.pntd.0002808

Hotez, P. J. (2021) Preventing the Next Pandemic: Vaccine Diplomacy in a Time of Anti-science. Edited by Johns Hopkins University Press. DOI??

Huish, R., & Kirk, J. M. (2007) Cuban Medical Internationalism and the Development of the Latin American School of Medicine. *Latino American perspectives* Vol. 34, No. 6, Aggressive Capital and Democratic Resistance (Nov, 2007), pp. 77-92 Published By: Sage Publications, Inc.

International Labour Organization. (2021). Executive Summary "Overview of social protection systems in Latin America and the Caribbean: Progress and setbacks in the face of the pandemic". Available at https://www.ilo.org/wcmsp5/groups/public/---americas/---ro-lima/---sro-port_of_spain/documents/publication/wcms_824905.pdf

Kevany, S. (2014). Global health diplomacy, 'smart power', and the new world order. *Global Public Health: An International Journal for Research, Policy and Practice*, 9(7), 787–807. 10.1080/17441692.2014.92121924953683

Kickbusch, I. (2011, June 10). Global health diplomacy: How foreign policy can influence health. *BMJ (Clinical Research Ed.)*, 342(jun10 1), d3154. 10.1136/bmj.d315421665931

. Kickbusch, I., & Buss, P. (2011). Global health diplomacy and peace. *Infect Dis Clin North Am.* 2011 Sep;25(3):601-10, ix. .10.1016/j.idc.2011.05.006

Lechuga, R. I., & Castro, A. C. (2017, March). Dramatic effects of control measures on deaths from yellow fever in Havana, Cuba, in the early 1900s. *Journal of the Royal Society of Medicine*, 110(3), 118–120. 10.1177/014107681769458328278397

Lisboa, V. (2023). La producción nacional de vacunas refuerza soberanía y planificación. Article in spanish. Available at https://agenciabrasil.ebc.com.br/es/saude/noticia/2023-09/la-produccion-nacional-de-vacunas-refuerza-soberania-y-planificacion

. Lódola S, and Góis Junior E. (2015) Theories about the propagation of yellow fever: the scientific debate in the São Paulo press between 1895 and 1903. Historia, Ciencias, Saude--manguinhos. Jul-Sep;22(3):687-704.

Marck, C., & Rigau-Perez, J. (2009). The World's First Immunization Campaign: The Spanish Smallpox Vaccine Expedition, 1803–1813. *Bulletin of History of Medicine*. Vol. 83, No. 1, Special Issue: Reassessing Smallpox Vaccination, 1789–1900 (Spring 2009)

Murhula, P. B. B., & Singh, S. B. (2022). The Impact of COVID-19 Vaccine Nationalism on Global Health and Human Rights to Health Standards. *The Oriental Anthropologist*, 22(1), 93–101. 10.1177/0972558X221096263

Neghme, A. (1962) La vida y obra de Carlos J Finlay. Edited by Centro de Investigaciones de Historia de la Medicina. Universidad de Chile. Available at https://obtienearchivo.bcn.cl/obtienearchivo?id=documentos/10221.1/55353/1/203253.pdf

Nolte, D. (2023). China's mask and vaccine diplomacy in Latin America: A success story? *International Politics*. Advance online publication. 10.1057/s41311-023-00525-w

Nye, J. S. (2003). *La paradoja del poder norteamericano*. Editorial Taurus Madrid.

Nye, J. S. (2004). *Soft Power: The Means To Success In World Politics*. New York: PublicAffairs. p. ix, xi. ISBN 9781586482251

PAHO. (2008). Res CD48/15 Health and international relations: linkages with national health development. Available at https://www3.paho.org/english/gov/cd/cd48-15-e.pdf

Perez-Perez, A., & Vallejo, J. R. (2023, June 5). The Smallpox Vaccine in Latin America: A New Approach (1801-1804). *Medicina (Kaunas, Lithuania)*, 59(6), 1093. 10.3390/medicina5906109337374296

Perez Valerino, D. R. (2020) Health, Science and Diplomacy: Necessity in Epidemic Times – A Cuban Perspective. *Science Diplomacy Review*, Vol. 2, No. 3,November 2020.

Riaz, M. M. A., Unaiza, A., Anmol, M., dos Santos-Costa, A. C., Khan, H., Maryam Salma Babar, M. S., Hasan, M. M., & Essar, M. Y. (2021). Global impact of vaccine nationalism during COVID-19 pandemic. *Tropical Medicine and Health*, 49(1), 101. 10.1186/s41182-021-00394-034963494

Risnock, A. (2016) Historical context and the roots of Jenner's discovery. *Hum Vaccin Immunother*. Aug; 12(8): 2025–2028.

Rojas-Ochoa F., et al. (2011) Vacunas. Cuba. 1959-2008. Editorial Ciencias Médicas (in Spanish)

Rombini, M.F., Mauas, R.P., & Urueña, A. (2019) Ranking de los programas de inmunización en América Latina. *Rev Panam Salud Publica.* 2022;;46:1. Article in spanish

Ruckert, A., Almeida, C., Ramírez, J., Guerra, G., Salgado de Snyder, V. N., Orozco, E., Alvarenga, A. A., Nassim de Saboya, A., Martins dos Santos, R., Sepúlveda, D., Rivera Vivian, E. C., Santos, M. D. S., Doriani, K., & Labonté, R. (2022). Global Health Diplomacy (GHD) and the integration of health into foreign policy: Towards a conceptual approach. *Global Public Health: An International Journal for Research, Policy and Practice*, 17(6), 1041–1054. 10.1080/17441692.2021.190031833736572

Sanmartin, A. M., & Bortz, G. (2023) Between regional convergence, and locally rooted technological and cultural aspects: lessons from the case of COVID-19 vaccines in Latin America. UN Report available at https://sdgs.un.org/sites/default/files/2023-05/B8%20-%20Cecilia%20Sanmartin%20-%20Lessons%20from%20the%20COVID-19%20pandemic%20in%20Latin%20America.pdf

Sharma, J., Pérez-Valerino, D. R., Widmaier, C. N., Lima, R., Gupta, N., & Varshney, S. K. (2022). Science Diplomacy and COVID-19: Future Perspectives for South–South Cooperation, *Global Policy*, London School of Economics and Political Science, vol. 13(2), pages 294-299

Sheridan, M. B. (2022) How Cuba became a pioneer in covid-19 vaccines for kids. The Washington Post. Available at https://www.washingtonpost.com/world/2022/06/18/cuba-coronavirus-vaccine-abdala-soberana/

. Sierra Gonzalez, G.V. (2019) Cuban Meningococcal Vaccine VA-MENGOC-BC:30 Years of Use and Future Potential. MEDICC Rev. Oct;21(4):19-27.

Solimano, G., & Valdivia, L. (2015). Global Health in Latin-American academic institutions: towards a distinctive development and identity. *Dossie Saude Global* • *Saude* soc. 23 (2) • https://doi.org/10.1590/S0104-12902014000200001

. Staples JE, Monath TP. (2008) Yellow fever: 100 years of discovery. JAMA. Aug 27;300(8):960-2. 10.1001/jama.300.8.960

United Nation Office for the Coordination of Humanitarian Affairs (OCHA). (2020) Desastres Naturales en America Latina y el Caribe 2000-2019. Report in Spanish Available at https://reliefweb.int/report/world/desastres-naturales-en-am-rica-latina-y-el-caribe-2000-2019

United Nations Children's Fund. (2018). Immunization Regional Analytics. Latin America and the Caribbean. Available at https://www.unicef.org/timorleste/media/3251/file/Immunization-regional-snapshots-LAC-2020.pdf.pdf

United Nations General Assembly. (2008) Resolution A/63/33 Global Health and Foreign Policy. Available at http://www.worldlii.org/int/other/UNGA/2008/71.pdf

Valderrama, B., Islas, P., Tibaduiza, Y., & Gual-Soler, M. (2021) Vaccine Diplomacy and Political Diversification in Mexico's Science and Technology Stakeholders. Available at https://www.sciencediplomacy.org/article/2021/vaccine-diplomacy-and-political-diversification-in-mexicos-science-and-technology

. Vianna-Franco, M.P (2021) Geopolitics, culture, and the scientific imaginary in Latin America, *Latin American Science, Technology and Society*, 4:1, 1919417, DOI: 10.1080/25729861.2021.1919417

WHO. (1978) Declaration of Alma Ata. Available at https://www.who.int/docs/default-source/documents/almaata-declaration-en.pdf

WHO. (2020a) Report of the WHO-China Joint Mission on Coronavirus Disease 2019 (COVID-19). Available at https://www.who.int/publications/i/item/report-of-the-who-china-joint-mission-on-coronavirus-disease-2019-(covid-19)

WHO. (2020b) Report. Origin of the SARS-Cov-2 Available at https://iris.who.int/bitstream/handle/10665/332197/WHO-2019-nCoV-FAQ-Virus_origin-2020.1-eng.pdf

WHO. (2020c) Timeline COVID-19. Available at https://www.who.int/news/item/27-04-2020-who-timeline---covid-19

World Bank Report (2023) Building resilience against disasters in Latin America and the Caribbean. Available at https://www.worldbank.org/en/results/2023/11/19/building-resilience-against-disasters-in-latin-america-and-the-caribbean

Zhou, Y. R. (2022) Vaccine nationalism: contested relationships between COVID-19 and globalization. GLOBALIZATION, VOL. 19, NO. 3, 450–465https://<ALIGNMENT.qj></ALIGNMENT>10.1080/14747731.2021.196320

ENDNOTES

1 In this regard, official details can be found in the documents issued between December 2019 and March 2020 Available at the WHO website. https://www.who.int/emergencies/diseases/novel-coronavirus-2019/

2 Mexican diplomat with an extensive and successful career in several functions at the United Nations. Between 2008 and 2022 Executive Secretary of ECLAC. In September 2022, began her duties as Secretary of Foreign Affairs of the Mexican government appointed by President Lopez Obrador.

3 Phrase stated within the framework of the virtual event "Local Action for Global Commitments: Inclusive, Resilient and Green Covid-19 Response and Recovery at The Local Level" organized by the United Nations in July 2020, as part of the High-Level Forum 2020

4 This consideration is even prior to the pandemic stage. Batthyany (2021) established once UNGA approved in Sept 2015, the declaration on the Sustainable Development Goals (SDGs), 2030 Agenda was confirmed as an ambitious work agenda that the signatory states committed to accomplishing by this year. The outcome of these 8 objectives, predecessors of the Millennium Development Goals (MDGs), during their last years of validity was marked by the economic and social upheavals unleashed by the international financial crisis that began in 2008 and the way their regulatory agenda contrasted with the progress of other issues, such as climate change and the persistence of inequalities.

5 As will be mentioned later, capital importance in this sense was the United Nations resolution established in 2008 in which the relevance of health policies in the international context is recognized. This circumstance was obviously promoted by the epidemic outbreaks at the beginning of the century that occurred mainly in Asian and African countries, which would come to be controlled at that moment.

6 It should be noted that after the Alma Ata Declaration, the global health agenda was guided more by the epidemic outbreaks and diseases that emerged, instead of health promotion as should have occurred with the emergence of the term health security. This was fundamentally due to the ambiguity of the language

presented in the declaration that allowed many countries the possibility of not complying with what was initially agreed.

7 As will be demonstrated later, in the Latin American and Caribbean region, some countries had implemented diplomatic strategies based on capabilities in their territory for the research, production and marketing of human vaccines.

8 One of the authors with the most contributions on this topic is the US scientist Dr. Peter J. Hotez who, in his publications, long before the pandemic scenario, has highlighted on several occasions political strategies implemented at the international level by large transnational companies as well as governments through the scientific results provided by these biopharmaceutical products. In this regard, it is recommended to read his article "Vaccine Diplomacy": Historical Perspectives and Future Directions (Reference nº 17)

9 An interesting analysis is carried out by Valderrama et al (2021), authors of the article "Vaccine Diplomacy and Political Diversification in Mexico's Science and Technology Stakeholders" published in the online magazine ScienceDiplomacy.org. Likewise, case studies have been published concerning nations such as China, Russia, and India that enable a comprehensive scenario in which this type of negotiations take place and that had a recognized role in the field of international relations during the pandemic.

10 The term Vaccine Nationalism has been defined as: "*A situation in which countries push to get first access to a supply of vaccines, potentially hoarding key components for vaccine production*." Although there are antecedents prior to COVID19, the reality is that in the past the situation has never been so tense on a political level. Hence, the WHO Director General of the Dr Tedros Ghebreyesus was forced to use it in a letter sent in August 2020 to world leaders to warn about the danger that this circumstance would bring to stop the advance of the pandemic. In parallel, the aforementioned scientist Dr Hotez has used the term in several of his articles, conferences and essays.

11 Authors such as Kickbursh (2011), Kevany (2014), Almeida (2020), Buss and Tovar (2018), Hotez (2021), and Valderrama (2021) maintain criteria in their respective publications that support this statement.

12 Dr Nye is a political scientist, considered one of the most influential scholars in the field of international relations in the past 20 years. In a join collaboration with Robert Keohane published in 1977 their book Power and Interdependence where Theory of Neoliberalism was presented by both authors. Former Dean of the John F. Kennedy School of Government at Harvard University where he currently holds the position of University Distinguished Service Professor,

Emeritus. In 2014, he was appointed by Secretary of State John Kerry to the Foreign Affairs Policy Board and later also to become a member of the US Defense Policy Board. In the author's opinion, his numerous books and articles are of relevant importance for understanding the complexities of international politics in the current century.

13 Nye J (2003), La paradoja del poder norteamericano, Taurus Ed, Madrid.

14 Practical examples of these statements are addressed in other chapters of this book. (Which chapters?). Include the full reference. OK. But I have no idea about the makeup of the chapters. It would depend on it since I am sure that several colleagues will address the topic in their respective chapters.

15 A detailed document in chronological order on the impact of this disease in Central America was prepared by the United States-CDC based in Atlanta. It is Available at the website https://www.cdc.gov/yellowfever/resources/training/mod-1/files-b/History_Yellow_Fever_document.pdf.

16 For more than 20 years, since he made his theory known, it was difficult for Dr. Finlay to prove it as he did not have the support of the Spanish colony or his own resources to convince the international scientific community. During that time, multiple investigations addressing different theories had been unsuccessful. Only once the Health Commission of the North American Army approached the Cuban specialists and an exchange of opinion was generated, and procedures were planned following suggestions and instructions from Dr. Finlay was it possible to achieve concrete and true results. The article titled *Notable Scientific Achievement of Latin Americans*, authored by journalist Danna Bell, published on December 19, 1915, in The Sunday Star of Washington D.C, highlights the beginnings of this collaboration and the multiple opinions into the scientific community about the research of the Cuban specialist and the support of his theory subsequently extended to practice through this collaboration.

17 A relationship of respect and consideration as scientists had developed between Dr. Finlay and Dr. Gorgas. Despite the opinions against the theory promoted by the Cuban scientist, the renowned North American specialist would manage, together with his colleague Dr Lazear (who would die from the effects of the disease to demonstrate the transmission theory proposed by Finlay) to convince a large part of health section of the US army to follow up on the results previously presented by the Cuban doctor.

18 In addition, the prestigious Thomas Jefferson University, from which degreed in 1855, would recognize the Cuban doctor as one of its notable students in the history of the high center of academic studies.

19 Dr. Edward Jenner (1749 –1823). English physician and scientist who pioneered the concept of vaccines and created the smallpox vaccine, the world's first vaccine.

20 Dr. Heinrich Hermann Robert Koch (1843 –1910), German physician and microbiologist. One of the main founders of modern bacteriology and father of microbiology (with L. Pasteur). Discoverer of the specific causative agents of deadly infectious diseases including tuberculosis, cholera, and anthrax.

21 Sir Alexander Fleming (1881 –1955), Scottish physician and microbiologist, best known for discovering the world's first broadly effective antibiotic substance, named penicillin. Nobel Prize in Physiology or Medicine (shared) in 1945.

22 Dr. Louis Pasteur (1822 –1895), French chemist, pharmacist, and microbiologist renowned for his discoveries of the principles of vaccination, microbial fermentation, and pasteurization, the last of which was named after him. His research in chemistry led to remarkable breakthroughs in the understanding of the causes and preventions of diseases, which laid down the foundations of hygiene, public health and much of modern medicine. Pasteur's works are credited with saving millions of lives through the developments of vaccines for rabies and anthrax. Considered as one of the founders of modern bacteriology and has been honored as one of the "father of bacteriology" and the "father of microbiology".

23 This award was established by UNESCO since 1980 with the sponsorship of the Cuban government, a total of 28 scientists (from countries such as Argentina, Belgium, France, Sweden, Mexico, Brazil, Spain, Malaysia, Pakistan, and others) have been awarded this important award. The most recent was Dr. Dilfuza Egamberdieva, Head of the research group at the Faculty of Biology, National University of Uzbekistan

24 Established in Spain in the fourteenth century, the protomedicato evolved very slowly in the Spanish colonies in the American continent. A board of physicians were responsible for regulating medical practitioners and inspecting apothecary shops in the main cities of the colonies being the roots of the current Health Ministries.

25 Considering the high incidence of the disease there was a possibility that any adult alive probably had already lived through the disease, and anyone with existing immunity would fail to develop the blisters needed to harvest more pus and propagate the vaccine. That's why Dr Joseph Flores, a personal physician of the Spanish crown, suggested the recruitment of the children as the only subpopulation that could keep the vaccine alive. A group of specialists (physicians, nurses, assistants, secretary, and the director of the orphanage as main care) joined the Expedition during the long crossing. One of the most recent articles written by Perez and Vallejo (2023) addresses additional details of this expedition.

26 Since its founding in 1948, the United Nations General Assembly (UNGA) has promoted the recognition of the importance of health to maintain peace and social stability at a global level through resolutions approved mostly by the plenary session of the member countries. The role of the WHO in leading global health policies from the organization has been key. Already during the first year of the pandemic, a resolution was passed emphasizing the importance of affordable healthcare for all. It also urged Member States to strengthen national health systems and improve their resilience, especially in the face of emergencies such as the COVID-19 pandemic.

27 Dr Paolo Buss, a Brazilian specialist internationally recognized as an expert in Health Diplomacy, is one of the academics who has written and contributed the most on this topic. In several publications Dr Buss highlights the role of multilateralism and the need to expand international collaboration between governments and regional and international organizations as an essential step for social development. His book "Health Diplomacy and Global Health: Perspectives from Latin America" (Edited by Editora FioCruz, 2017) is a main work whose reading and analysis must be prioritized to be updated on these issues in the Latin American region.

28 Migration is a historical phenomenon that characterizes the history of human society and even more so in the current globalized world. However, when this migration is linked to knowledge, to the scientific capabilities of specialists who seek better economic conditions to develop their intellect, it brings with it an imbalance that is difficult to complement in societies with fewer resources. Most countries in the LAC region have this issue within the group of recognized circumstances that slow down or prevent development in the general theme of Science. In the present century, the scientific diaspora, a term used in the academic setting, has been the object of interesting analysis through the characteristics of case studies, by specialists from various sectors

with the aim of implementing actions and strategies that achieve a greater link through scientific collaboration with the origins of these specialists. Although it is surely addressed by my colleagues from DiploCientifica in other chapters of this book, I would like to suggest reading two recently published articles on the subject: Bonilla K, Arrechea S, and Vazquez Perez LG (2022) Connecting Scientists Residing Abroad: A Review of Converciencia as a Practice to Engage the Guatemalan Scientific Diaspora From 2005–2020 available at Front Res Metr Anal. 2022; 7: 898496 and Echevarria King L, Camacho Toro R, Galvis L, Gonzalez A, Suarez VR, Torres Atencio I, and Widmaier Muller CN (2022) Organized Scientific Diaspora and Its Contributions to Science Diplomacy in Emerging Economies: The Case of Latin America and the Caribbean, originally published at Front. Res. Metr. Anal., 19 May 2022. In both cases, reading allows you to acquire an updated and proactive assessment of this phenomenon.

29 This is a topic with such extensive debate that several books can be written. The relationship between the development of science and political ideologies has generated different stereotypes and considerations such as perceiving that politicians with conservative ideas do not approve government investments for the advancement of science and technology based on development. Or governments with more liberal ideas do not have the administrative capabilities to control new technology that they only use for electoral interests. In the author's opinion, science cannot be separated from politics and ideologies and for this it is key to establish a connecting channel between one and the other through a bridge of investments and comprehensive development programs. Discoveries in the past such as nuclear fission or the synthesis of Nerve gases during Nazi time in Germany are examples of the potential of scientific activities also in totalitarian and extremist regimes. For this reason, the current tendency to consider that all sciences are rooted in political ideology and to deny their privileged status as generators of knowledge. It is complex and dangerous. In LAC throughout history there have been positive and negative examples at both extremes: from denialist governments that do not give importance to the unconditional support of innovative scientific projects to generate a development matrix through science, to governments that have relied on science and technological innovation as a significant pillar of their scientific projects but without enough scope or the control of resources necessary to use the results for the open benefit of society. In this regard, an interesting analysis is made by Ciocca and Delgado (2017) as well as Viana Franco in his book Geopolitics, culture, and the scientific imaginary in Latin America (2022) edited by the Florida University

30 It is a relevant and internationally recognized result that the LAC region has up to 6 National Reference Regulatory Authorities. They are the National Health Surveillance Agency (ANVISA, Brazil), the National Institute for Food and Drug Surveillance (INVIMA, Colombia), the Center for State Control of Medicines, Equipment and Medical Devices (CECMED, Cuba), the Institute of Public Health (ISP, Chile) and the Federal Commission for the Protection against Sanitary Risks (COFEPRIS, Mexico). This category is granted by WHO/PAHO and allows these institutions to serve as a reference to other National Authorities with a view to supporting WHO/PAHO in the regulatory strengthening of medicines in the region. Additionally, they provide an exemplary profile in improving good pharmaceutical practices and other innovations related to regulation, as well as strengthening the use of the decisions of other regulatory authorities through guidance and advice.

31 Dr Edmundo Grande Ugalda (1946-2008) Outstanding Ecuadorian doctor during his professional life as a trainer of specialists in social medicine and public health in LAC. For several years permanent advisor to the Pan American Health Organization (PAHO), the Latin American Association of Social Medicine (ALAMES), the Latin American Association of Schools of Public Health (ALAESP) and the International Center for Research and Development (C11D) of Canada, in more than a dozen countries in America in fields related to management, research, epidemiology, social health sciences, planning and health education. A well-deserved tribute that his name honors a continental program for the training of health leaders in the region.

32 The extraordinary complexity of this topic can be understood by reading articles where the scientific, technological, and regulatory challenge underlying the development of productive platforms of this type under established and regulated quality conditions is based. An example we suggest is Elmgren et al (2013).

33 Data obtained from the official website of Butantam Institute (https://en.butantan.gov.br/portfolio.php), FIOCRUZ Foundation (https://portal.fiocruz.br/en/production-and-innovation) and ANVISA (https://www.gov.br/anvisa/en), Brazil.

34 Global Alliance for Vaccine Inmunization (GAVI). Since its creation in 2000, GAVI is a global public and private health association whose main objective is to increase access to vaccines in countries with fewer economic resources through donations made by governments of developed countries and technical and collaboration. logistics of institutions such as WHO, UNICEF, the World Bank the vaccine industry in both industrialized and developing countries,

research and technical agencies, civil society, the Bill & Melinda Gates Foundation, and other private philanthropists.

35 This crisis was generated from two fundamental events. On the one hand, the decision made by big biopharmaceutical companies to use the facilities dedicated to this vaccine to produce others aimed at higher lucrative markets. On the other hand, the confirmation that other vaccine-producing institutions located in developing countries would not be ready to supply the product in large quantities until after 2010. This combination of events made international regulatory authorities consider requesting a global call requesting for support for this crisis. Cuba and Brazil being the only ones that took on the challenge and risk of carrying out a project of this magnitude in a very short time.

36 Based on the success achieved in this experience, new and successful collaborations were identified for both Brazilian and Cuban scientific institutions with developing countries such as China, Vietnam, Syria, India, Venezuela, Algeria, Angola, Thailand among others.

37 Based on donations from public, private, philanthropic and civil society organizations, the Coalition for Epidemic Preparedness Innovations (CEPI) is a Foundation officially established in 2015 with the objective of financing independent research projects to develop vaccines against infectious diseases. emerging. The Wellcome Trust, Bill & Melinda Gates Foundation, the World Economic Forum and governments such as Norway and India were involved in its creation.

38 More information on this topic can be accessed through a press release from the Ministry of Health of Paraguay available at COVID-19: vacuna taiwanesa Medigen es aprobada para uso de emergencia - Ministerio de Salud Publica y Bienestar Social (mspbs.gov.py) and from the Reuters agency at the link: https://www.reuters.com/article/world/americas/taiwn-dice-que-est-ayudando-a-su-aliado-paraguay-a-comprar-vacunas-covid-19-idUSKBN2B903D/

Compilation of References

. Chaves-Carballo E. (2005) Carlos Finlay and yellow fever: triumph over adversity. Mil Med. Oct;170(10):881-5.

. Damiani, A. (2016). Carlos J Finlay (1833-1915) Su impronta en el Canal de Panamá en el centenario de su Muerte. Article in spanish Available at *Revista de la Asociación Médica Argentina*, Vol. 129, Número 1 de 2016

. Delany, I., Rappuoli R., De Gregorio, E. (2014) Vaccines for the 21st century. EMBO Mol Med. Jun;6(6):708-20.

. Greenwood, B. (2014) The contribution of vaccination to global health: past, present and future. Philos Trans R Soc Lond B Biol Sci., 369(1645):20130433.

. Kickbusch, I., & Buss, P. (2011). Global health diplomacy and peace. *Infect Dis Clin North Am.* 2011 Sep;25(3):601-10, ix. .10.1016/j.idc.2011.05.006

. Lódola S, and Góis Junior E. (2015) Theories about the propagation of yellow fever: the scientific debate in the São Paulo press between 1895 and 1903. Historia, Ciencias, Saude--manguinhos. Jul-Sep;22(3):687-704.

. Sierra Gonzalez, G.V. (2019) Cuban Meningococcal Vaccine VA-MENGOC-BC:30 Years of Use and Future Potential. MEDICC Rev. Oct;21(4):19-27.

. Staples JE, Monath TP. (2008) Yellow fever: 100 years of discovery. JAMA. Aug 27;300(8):960-2. 10.1001/jama.300.8.960

. Vianna-Franco, M.P (2021) Geopolitics, culture, and the scientific imaginary in Latin America, *Latin American Science, Technology and Society*, 4:1, 1919417, DOI: 10.1080/25729861.2021.1919417

Abdullah, H., & Roig, A. (2021). *City-led science diplomacy: building urban sustainability and resilience at the science-policy interface*. Cidob.

Acharya, A., Estevadeordal, A., & Goodman, L. W. (2023). Multipolar or multiplex? Interaction capacity, global cooperation and world order. *International Affairs*, 99(6), 2339–2365. 10.1093/ia/iiad242

Adler, E., & Haas, P. (1992). Conclusion: Epistemic Communities, World Order, and the Creation of a Reflective Research Program. *International Organization*, 46(1), 367–390. 10.1017/S0020818300001533

Adler, J. E., & Rips, L. J. (2008). *Studies of Human Inference and Its Foundations*. Cambridge University Press. 10.1017/CBO9780511814273

Aguirre-Bastos, C. (2023). Internationalization of Science and Diplomacy, Concepts and Practices: Lessons for Developing Countries. In Ittekkot, V., & Baweja, J. K. (Eds.), *Science, Technology and Innovation Diplomacy in Developing Countries. Research for Development*. Springer. 10.1007/978-981-19-6802-0_2

Alavosius, M. P., & Houmanfar, R. A. (2020). Global warming: Analysis of behavior and organizational practices as climate impacts increase. In *Behavior science perspectives on culture and community* (pp. 221-256). Academic Press.

Alexander von Humboldt Foundation. (2024, March 1.). Retrieved from https://www.humboldt-foundation.de/

Alismail, H. A., & McGuire, P. (2015). 21st century standards and curriculum: Current research and practice. *Journal of Education and Practice*, 6(6), 150–154.

Allen, M. M. C. (2015). The national innovation system in Germany. In Narayanan, V. K., & O'Connor, G. (Eds.), *Wiley encyclopedia of management: Technology and innovation management* (Vol. 130047). Wiley. 10.1002/9781118785317.weom130047

Álvarez, I. Natera, J. M. & Castillo, Y. (2019). Generación y transferencia de ciencia, tecnología e innovación como claves de desarrollo sostenible y cooperación internacional en América Latina. Documentos de Trabajo. Fundación Carolina 19/2019.

Álvarez, J. P. (2023). Esto invierten en ciencia los países de Latinoamérica: Brasil cuadriplica a México. Retrieved 1 March, 2024, from Bloomberg línea: https://www.bloomberglinea.com/2023/09/05/esto-invierten-en-ciencia-los-paises-de-latinoamerica-brasil-cuadriplica-a-mexico/

Alvial-Palavicino, C. (2015). The future as practice: A framework to understand anticipation in science and technology. *Tecnoscienza–Italian Journal of Science & Technology Studies*, 6(2), 135–172.

Amorim, C., Douste-Blazy, P., Wirayuda, H., Store, J.G., Gadio, C.T., Diamini-Zuma, N., & Pibulsonggram, N. (2007) Oslo Ministerial Declaration--global health: a pressing foreign policy issue of our time. *The Lancet* ; Vol 369 Issue 9570: p1373-1378, Apr. 21, DOI: .10.1016/S0140-6736(07)60498-X

ANID. (2022). Fondo CAS-ANID 2022. Available at: https://ayuda.anid.cl/hc/es/articles/7973636144532-Fondo-CAS-ANID-2022

ANID. (2023a). Bases del concurso Becas de Doctorado Igualdad de Oportunidades Fulbright–ANID Convocatoria 2023. Available at: https://s3.amazonaws.com/documentos.anid.cl/BecasChile/2023/fullbright/Bases_Doctorado_BIO_Fulbright_2023_LA_28062023.pdf

ANID. (2023b). Se adjudica Concurso Fondo CAS-ANID, propuestas de investigación de posiciones postdoctorales, Programa Nacional para el Desarrollo de la Astronomía y Ciencias Afines. Convocatoria 2022. Resolución Nº: 8903. Available at: https://s3.amazonaws.com/documentos.anid.cl/rec /2022/CasAnid/REX8903-2023_adjudica_Concurso_Fondo_CAS-ANID_2022_sF4LzC6QbVwN1XhT3eR-tY7uIaM5oKvP.pdf

ANID. (2024). Página institucional. Disponible en: https://anid.cl/

Ansell, C., Sørensen, E., & Torfing, J. (2024). The democratic quality of co-creation: A theoretical exploration. *Public Policy and Administration*, 39(2), 149–170. 10.1177/09520767231170715

Arencibia-Jorge, R., Corera-Alvarez, E., Chinchilla-Rodríguez, Z., & de Moya-Anegón, F. (2016). Cuban biopharmaceutical industry: A scientometric approach. *Scientometrics*, 108(3), 1621–1636. 10.1007/s11192-016-2023-1

Arevalo Robles, G. A. (2017). Indigenous diplomacy: A transdiplomatic approach. *SI SOMOS AMERICANOS-REVISTA DE ESTUDIOS TRANSFRONTERIZOS*, 17(1), 141–169. 10.4067/S0719-09482017000100141

Argüello, G., & Rafaly, V. (2023). Science diplomacy and Asian states: Transforming the governance landscape in the Arctic. *The Polar Record*, 59, e41. 10.1017/S0032247423000281

Asociación de Derecho Internacional. (2008). Proyecto de artículos sobre el derecho de los acuíferos transfronterizos. Disponible en https://undocs.org/es/A/RES/63/124

Autoridad Nacional del Agua - ANA. (2013). Resultado del Primer Monitoreo Participativo de la calidad del agua superficial en las cuencas de los ríos Tumbes y Zarumilla, lado peruano. Proyecto de modernización de la gestión de los recursos hídricos, Lima- Perú.

Awandare, G., André, E., Corrales-Aguilar, E., Chen, C., Mostajo-Radji, M. A., Jancoriene, L., & Nemer, M. (2020). Science advisers around the world on 2020. *Nature*, 588(7839), 586–588. 10.1038/d41586-020-03557-x33340028

Babatunde, D. E., Denwigwe, I. H., Babatunde, O. M., Gbadamosi, S. L., Babalola, I. P., & Agboola, O. (2019). Environmental and societal impact of nanotechnology. *IEEE Access : Practical Innovations, Open Solutions*, 8, 4640–4667. 10.1109/ACCESS.2019.2961513

Banco Interamericano de Desarrollo. (2021, 5 de agosto). Primer foro virtual de ciencia, tecnología y pueblos indígenas. [Archivo de video]. YouTube. https://www.youtube.com/watch?v=L53yTJBgU-s&t=1s

Banco Mundial (2021) El agua importa: crecimiento resiliente, inclusivo y verde a través de la seguridad del agua en América Latina.

Baraka, N. (2022) La gestión de agua requiere una nueva estrategia, en Cumbre ONU - Agua sobre Aguas Subterráneas Cap-Net y LA-WETnet (2013). Derecho relacionado con el agua. Manual de capacitación y guía para moderadores sobre la legislación relacionada con el agua y la reforma legal para la gestión integrada de recursos hídricos.

Barcons, X., & de Figueiredo Melo, C. (2022). Observatorio Europeo Austral (ESO): astronomía a través de la cooperación internacional, en Nuevos mapas para la diplomacia. Ciencia, Conocimiento, Tecnología e Innovación. Santiago de Chile: Academia diplomática de Chile.

Barraza García, R. (2014). Escuela Intercultural de Diplomacia Indígena: Academia, posdesarrollo y diálogo de saberes. *Desafíos*, 26(1), 13–56. 10.12804/desafios26.1.2014.01

Barrett, P., Hansen, N.-J., Natal, J.-M., & Noureldin, D. (2021). Why Basic Science Matters for Economic Growth. *IMF Blog*. Retrieved March 1, 2024, from https://www.imf.org/en/Blogs/Articles/2021/10/06/blog-ch3-weo-why-basic-science-matters-for-economic-growth

Barrios, V., & Sancho, C. (2023). Ciberseguridad y Relaciones Internacionales: Interdependencia y Cooperación en la Era Digital. In Nuevos mapas para la diplomacia. Ciencia, conocimiento, tecnología e Innovación. Santiago de Chile: Academia diplomática de Chile.

Barth, K. (2006). Catalysts of Change: Scientists as Transnational Arms Control Advocates in the 1980s. *Osiris*, 21(1), 182–206. 10.1086/507141

Battyany, K. (2021). Latin America and the Caribbean, the SDGs, and the pandemic. *Online Idees Review No. 55*. Available at https://revistaidees.cat/en/monografics/rethinking-development-cooperation-to-meet-the-challenges-of-the-21st-century/

Bayerisches Hochschulzentrum für Lateinamerika (BAYLAT) an der Friedrich-Alexander-Universität Erlangen-Nürnberg (FAU). (2024. March 3). LATinBAY Kooperationsinformationssystem. Retrieved from https://www.baylat.org/latinbay-kooperationsinformationssystem.html?tx_tweprivacy_eprivacydialog%5Baction%5D=dialog&tx_tweprivacy_eprivacydialog%5Bcontroller%5D=Subject&cHash=93bc279ae1fdc7ca7480bc0a1d9c7943

BBC News World. (2020) Coronavirus: Brazil confirms first case in Latin America. Posted online on Feb 26th, 2020. Available at https://www.bbc.com/mundo/noticias-america-latina-51641436#:~:text=El%20Ministerio%20de%20Salud%20de,detectado%20oficialmente%20en%20Am%C3%A9rica%20Latina

Belli, S., & Morín Nenoff, J. (2022). Cooperation in Science and Innovation between Latin America and the European Union. *Journal of Open Innovation*, 8(2), 94. 10.3390/joitmc8020094

Beltran, L. (2024). Latin America is moving towards a sustainable economy. These are the challenges and opportunities' *World Economic Forum Annual Meeting*. Retrieved April 4, 2024, from https://www.weforum.org/agenda/2024/01/latin-america-is-moving-towards-a-sustainable-economy-these-are-the-challenges-and-opportunities/

Berkman, P. A. (2019). Evolution of science diplomacy and its local-global applications. *European Foreign Affairs Review*, 24(Special Issue), 63–80. 10.54648/EERR2019019

Berlin Partner for Business and Technology. (2014). Informationen rund um den Wirtschaftsstandort Berlin. Retrieved February 23, 2024, from www.businesslocationcenter.de

Berlin Partner for Business and Technology. (2024, February 23). Explore the Berlin-Brandenburg Ecosystem | Berlin tech ecosystem map. Retrievedfrom https://startup-map.berlin/intro

Berlin Partner for Business and Technology. (2024, February 24). Science. Business Location Center. https://www.businesslocationcenter.de/en/science

Berlin Science Week. (2024, March 13) Das Internationale Wissenschaftsfestival. Retrieved from https://berlinscienceweek.com/de/

Berlin University Alliance (BUA). (2024, March 2). Retrieved from https://www.berlin-university-alliance.de/

Beskaya, A., Savas, B., & Samiloglu, F. (2010). The impact of education on economic growth in Turkey. *The Journal of Faculty of Economics and Administrative Sciences*, 15(3), 43–62.

Boghossian, P. (2014). What is inference? *Philosophical Studies*, 169(1), 1–18. 10.1007/s11098-012-9903-x

Bompongo Nkombe, A. (2020). Management of Congo Water Basin: Opportunities and Challenges for Science Diplomacy. In Research and Information System for Developing Countries (Coord.), Southern Perspectives on Science Diplomacy. Nueva Delhi: RIS.

Borda, S., & Paz Berger, M. (2012). Relaciones bilaterales China y Colombia: 1990-2010. *Colombia Internacional*, 75(75), 83–129. 10.7440/colombint75.2012.04

Borlovan, P. (2023) Dive into the German Startup Ecosystem Key Players, Trends and Opportunities. SeedBlink Blog. Retrieved from https://seedblink.com/2023-10-19-dive-into-the-german-startup-ecosystem-key-players-trends-and-opportunities

Bound, K. (2016). *'Innovating Together? The Age of Innovation Diplomacy'. Global Innovation Index 2016* (Vol. 1). World Intellectual Property Organization.

Boytchev, H. (2020). How Germany retains one of the world's strongest research reputations. Nature Index. *Nature*, 587(7834), S104–S105. 10.1038/d41586-020-03318-w

Brain City Berlin. (2024, February 24). Berlin's science hub. Retrieved from https://braincity.berlin/en/science-location-berlin-1.

Brandt, M. (2023, June 20). So hat sich Deutschlands Wettbewerbsfähigkeit entwickelt. World Competitiveness Ranking 2023. Statista. Retrieved from https://de.statista.com/infografik/30229/abschneiden-von-deutschland-im-world-competitiveness-ranking/

Brazil, A. N. V. I. S. A. (n.d.). https://www.gov.br/anvisa/en

Brazilian Ministry of Foreign Affairs. (2023). *Relatório Programa de Diplomacia da Inovação*. Retrieved April 5, 2024, from https://www.gov.br/mre/pt-br/assuntos/ciencia-tecnologia-e-inovacao/Relatrio2023ProgramadeDiplomaciadaInovao.pdf

Brigg, M., & Graham, M. (2023). Approaching First Nations diplomacy from the Australian continent. *Australian Journal of International Affairs*, 77(6), 585–589. 10.1080/10357718.2023.2265849

Broekel, T., Rogge, N., & Brenner, T. (2018). The innovation efficiency of German regions – A shared-input DEA approach. *Jahrbuch für Regionalwissenschaftt = Review of Regional Research*, 38(1), 77–109. 10.1007/s10037-017-0112-0

Brown, J. (2016, March 23). Diplomacies, from public to public. *Huffington Post*. Retrieved February 20, 2024, from https://www.huffingtonpost.com/john-brown/diplomacies-from-public-to-public_b_9530412.html

Bueger, C. (2015). Making things known epistemic practices, the United Nations, and the translation of piracy. *International Political Sociology*, 9(1), 1–18. 10.1111/ips.12073

Buhl, C. M., Meier zu Köcker, G., & Sedlmayr, B. (2019). Development of Cluster Initiatives in Germany from 2008 - 2018. VDI/VDE Innovation + Technik GmbH. Retrieved from https://www.researchgate.net/publication/337671892_Development_of_Cluster_Initiatives_in_Germany_from_2008_-_2018

Bundesministerium für Bildung und Forschung (BMBF). (2021) The German research landscape. Deutscher Akademischer Austauschdienst (DAAD). https://www.research-in-germany.org/website/public/epapers/german-research-landscape/#0

Bundesministerium für Bildung und Forschung (BMBF). (2024, March 2) Research. High-tech and Innovation. Research Campus – Public Partnerships for Innovation. Retrieved from https://www.bmbf.de/bmbf/en/research/hightech-and-innovation/research-campus/research-campus_node.html

Bundesministerium für Bildung und Forschung (BMBF)/Federal Ministry of Education and Research. Division General for European and International Cooperation in Education and Research. (2016). Science Diplomacy: Wissenschaftsdiplomatie. Retrieved from https://www.bmbf.de/bmbf/de/europa-und-die-welt/vernetzung-weltweit/bildungs-und-wissenschaftsdiplomatie/bildungs-und-wissenschaftsdiplomatie.html

Bundesministerium für Bildung und Forschung, BMBF/Federal Ministry of Education and Research. Division General for European and International Cooperation in Education and Research. (2016). Internationalisation of Education, Science and Research: Strategy of the Federal Government. Retrieved from https://www.bmbf.de/SharedDocs/Publikationen/de/bmbf/FS/31286_Internationalisierungsstrategie_en.pdf?__blob=publicationFile&v=3

Bundesministerium für Bildung und Forschung. (2022) CALAS: Wissenschaftskooperation in und mit Lateinamerika. Retrieved from https://www.geistes-und-sozialwissenschaften-bmbf.de/de/CALAS-Wissenschaftskooperation-in-und-mit-Lateinamerika-2666.html

Buss, P., & Tobar, S. (2018) Health Diplomacy in the Political Process of Integration in Latin America and the Caribbean. Edited and published by Oxford Research Encyclopedia of Public Health, UK. DOI: 10.1093/acrefore/9780190632366.013.24

Butantan Institute. Official website https://en.butantan.gov.br/portfolio.php

Büyüktanir, D., & Ruffini, P. B. (2023). Science Diplomacy in the Global South – an introduction. *Science & Public Policy*, 50(4), 742–748. 10.1093/scipol/scad028

Bynum, W. F. (1993) Policing Hearts of Darkness: Aspects of the International Sanitary Conferences. *History and Philosophy of the Life Sciences.* Vol. 15, No. 3 published by: Stazione Zoologica Anton Dohrn – Napoli

Camacho Toro, R., Cumba Garcia, L. M., Galvis, L. A., Echeverría-King, L. F., Pantović, B., Alarcón-López, C., Suarez, V. R., Figueroa, P., Torres-Atencio, I., Widmaier, C., Fraga, T. R., & Benavides, S. (2024). The needed link between open science and science diplomacy -A Latin American perspective. [Frontiers.]. *Frontiers in Research Metrics and Analytics*, 9, 1355393. 10.3389/frma.2024.135539338903656

Campbell, C. A. (2014). US Science Diplomacy with Arab Countries. In *Science diplomacy: New day or false dawn.* World Scientific.

Cancilla, R., & Gargano, M. (2011). *Global environmental policies: Impact, management and effects.* Nova Science Publishers, Inc.

Cancilleria Colombiana. (2021). *Unión Europea y Colombia firman acuerdo de colaboración relativo a la cooperación en el ámbito de los sistemas mundiales de navegación por satélite | Cancillería.* Available at: https://www.cancilleria.gov.co/newsroom/news/union-europea-colombia-firman-acuerdo-colaboracion-relativo-cooperacion-ambito

Cancilleria Colombiana. (2023a). *Colombia y Estados Unidos sostendrán el X Diálogo de Alto Nivel en el que abordarán los temas prioritarios de la relación bilateral | Cancillería.* Avaialbale at: https://www.cancilleria.gov.co/newsroom/news/colombia-estados-unidos-sostendran-x-dialogo-alto-nivel-abordaran-temas-prioritarios

Cancilleria Colombiana. (2023b). *Colombia y China suscribieron 12 instrumentos de cooperación | Cancillería.* Available at: https://www.cancilleria.gov.co/newsroom/news/colombia-china-suscribieron-12-instrumentos-cooperacion

Carabajal, M. I. (2022). Coproducción de conocimiento y redes de colaboración en el Sur de América del Sur. *Revista Redes - Revista de Estudios Sociales de la Ciencia y la Tecnología, 53.* https://doi.org/10.48160/18517072re54

Carabajal, M. I., & Hidalgo, C. (2021). Making sense of climate science: From climate knowledge to decision making. In Sillitoe, P. (Ed.), *The Anthroposcene of weather and climate: Ethnographic contributions to the climate change debate* (pp. 120–140). Berghahn Books., https://www.berghahnbooks.com/title/SillitoeAnthroposcene10.2307/j.ctv29sfz69.12

Carpenter, K., & Tsykarev, A. (2021). Indigenous Peoples and Diplomacy on the World Stage. *AJIL UNBOUND*, 115, 118–122. 10.1017/aju.2021.7

Carrier, M. (2022). What Does Good Science-Based Advice to Politics Look Like? *Journal for General Philosophy of Science*, 53(1), 5–21. 10.1007/s10838-021-09574-2

Carrillo-Roa, A., & Santana, J. P. (2012). Regional integration and south–south cooperation in health in Latin America and the Caribbean. *Revista Panamericana de Salud Pública*, 32(5), 368–375. 10.1590/S1020-4989201200110000723338694

Cash, D. W., Clark, W. C., Alcock, F., Dickson, N. M., Eckley, N., Guston, D. H., & Mitchell, R. B. (2003). Knowledge systems for sustainable development. *Proceedings of the National Academy of Sciences of the United States of America*, 100(14), 8086–8091. 10.1073/pnas.123133210012777623

CD4D2. (2023). Through the transfer of knowledge and expertise, CD4D2 engages diaspora to support the development of their countries of origin. Retrieved December 28, 2023, from https://connectingdiaspora.org/about-cd4d2/

CEMARIN. (2024, March 1). Retrieved from https://cemarin.org/

Chudnovsky, D., López, A., & Porta, F. (1997). Market or policy driven? The Foreign direct investment boom in Argentina. *Oxford Development Studies*, 25(2), 173–188. 10.1080/13600819708424128

Cid, C., & Marinho, M. L. (2022) Dos años de pandemia de COVID-19 en América Latina y el Caribe. ECLAC- Report in Spanish. Available at https://repositorio.cepal.org/server/api/core/bitstreams/3a821906-213d-4cfc-bd26-66167328fb9d/content

Ciocca, D. R., & Delgado, G. (2017). The reality of scientific research in Latin America; an insider's perspective. *Cell Stress & Chaperones*, 22(6), 847–852. 10.1007/s12192-017-0815-828584930

Cisneros, G. A. T. (2013). Diplomacia indígena: Transitando del problema a la solución. *Revista Mexicana de Política Exterior*, 98, 197–232.

Cohen, W. M., & Levinthal, D. A. (1990). Absorptive Capacity: A New Perspective on Learning and Innovation. *Administrative Science Quarterly*, 35(1), 128. 10.2307/2393553

COICA. (2022, April 8). *Encuentro internacional entre COICA, OTCA y FILAC, por la protección de la biodiversidad y el buen vivir en la Región Amazónica*. COICA. https://coicamazonia.org/encuentro-internacional-entre-coica-otca-y-filac-por-la-proteccion-de-biodiversidad-y-el-buen-vivir-en-la-region-amazonica/

Colciencias. (2010). *Colombia y China firmaron acuerdo para la cooperación en Ciencia, Tecnología e Innovación | Colciencias*. Available at: https://legadoweb.minciencias.gov.co/noticias/colombia-y-china-firmaron-acuerdo-para-la-cooperaci-n-en-ciencia-tecnolog-e-innovaci-n

Colglazier, E. W., Vafai, H. A., Lansey, K. E., & Bryson, M. D. (2021). *Sustainable Development for the Americas: Science, Health, and Engineering Policy and Diplomacy*. CRC Press. 10.1201/9781003220503

Colombia, D. A. A. D. (2024, March 1). Retrieved from https://www.daad.co.). Retrieved from https://www.daad.co

Colombiana De Ciencias Exactas. Físicas Y Naturales. Retrieved March 16, 2024, from https://accefyn.org.co/libros-mision-de-sabios/Ministerio de Ciencia, Tecnologia e Innovacion. (n.d.-c). *Linea de Tiempo Minciencias*. Retrieved in March 16, 2024, from https://minciencias.gov.co/sites/default/files/ckeditor_files/Linea%20de%20Tiempo%20Minciencias.png

Comisión Económica de las Naciones Unidas para Europa (CEPE/ONU) (2015). Nota de Orientación sobre Políticas: Los Beneficios de la Cooperación en Materia de Aguas Transfronterizas: Identificación, Evaluación y Comunicación Nueva York y Ginebra: Naciones Unidas.

Comisión Económica para América Latina - CEPAL. (2024) Aguas Transfronterizas: Gestión, Cooperación y Paz - Diálogos Regionales del Agua para América Latina y el Caribe 2024, Chile.

CONICET. (2018). Científicos chinos y argentinos intercambiaron conocimientos sobre desertificación, ecología de desiertos y restauración. CONICET (03/10/2018). Available at https://www.conicet.gov.ar/cientificos-chinos -y-argentinos-intercambiaron-conocimientos-sobre-desertificacion-ecologia-de-desiertos-y-restauracion/ Retrieved in December 2023.

CONICET. (2019). Argentina y China fortalecen lazos en materia científica. CONICET (21/06/2019) Available at https://www.conicet.gov.ar/argentina-y-china-fortalecen-lazos-en-materia-cientifica/ Retrieved in December 2023.

CONICET. (2019). El CONICET y la Academia China de Ciencias acuerdan la creación de un Centro Conjunto de Ciencia y Tecnología. CONICET (03/12/2019). Available at https://www.conicet.gov.ar/el-conicet-y-la-academia-china -de-ciencias-acuerdan-la-creacion-de-un-centro-conjunto-de-ciencia-y-tecnologia/ Retrieved in December 2023.

CONICET. (2022). Se presentó el Centro Binacional Chino-Argentino de estudio de políticas para la Innovación y la Tecnología. CONICET (09/05/2022). Available at https://www.conicet.gov.ar/se-presento-el-centro-binacional-chino -argentino-de-estudio-de-politicas-para-la-innovacion-y-la-tecnologia/.Retrieved in December 2023.

CONICET. (2023). Se establece el Comité Directivo y Científico del Centro Conjunto de Ciencia y Tecnología entre el CONICET y la Academia de Ciencias de China. CONICET (27/01/2023). Available at https://www.conicet.gov.ar /se-establece-el-comite-directivo-y-cientifico-del-centro-conjunto-de-ciencia-y-tecnologia-entre-el-conicet-y-la-academia-de-ciencias-de-china/ Retrieved in December 2023.

CONICYT. (2018). Adjudica el Concurso para Proyectos de Investigación Conjunta en el manejo de Desastres Naturales en Chile y China, CONICYT Chile – NSFC China. Convocatoria 2018. Resolución N° at: https://www.conicyt.cl/pci/files/2018/10/REX.100438 .del_.25.10.2018.NSFC_.adjudicacion.pdf10.048. Available

CONICYT. (2019). Aprueba Fallo Y adjudica el Concurso para Proyectos de Investigación Conjunta en Chile y China, CONICYT Chile – NSFC China. Convocatoria 2019. Resolución N° at: https://www.conicyt.cl/pci/files/2019/11/REX.10707.del _.12.11.2019.NSFC_.Fallo_.pdf10.707. Available

CONICYT. (2024). Convenios Internacionales. Available at: https://www.conicyt .cl/sobre-conicyt/convenios/convenios-internacionales/

Copeland, D. (2016). Science diplomacy. In Constantinou, C. M., Kerr, P., & Sharp, P. (Eds.), *The SAGE handbook of diplomacy* (pp. 628–641). SAGE. 10.4135/9781473957930.n52

Cormann, M. (2023). Science, technology and innovation are changing the world: international cooperation to address global challenges. In Nuevos mapas para la diplomacia. Ciencia, conocimiento, tecnología e Innovación. Santiago de Chile: Academia diplomática de Chile.

Cortes, M. A., Cardoso, D., Fitzgerald, J., & Jose Luis DiFabio, J. L. (2012). Public vaccine manufacturing capacity in the Latin American and Caribbean region: Current status and perspectives. *Biologicals*, 40(1), 3–14. 10.1016/j.biologicals.2011.09.01322033155

Council of the European Union. (2021). Council recommendation on a Pact for Research and Innovation in Europe (Interinstitutional File: 2021/0230 (NLE)). Retrieved from https://www.bmbf.de/bmbf/en/international-affairs/the-european-research-area/the-european-research-area_node.html

CPD. (2024). USC Center on Public Diplomacy. Retrieved January 14, 2024, from https://uscpublicdiplomacy.org/page/what-is-pd

Cruz Chavarro, N., & Arevalo Robles, G. A. (2021). Indigenous Summits: Politics and Ancestral Diplomacy in Latin America. *NOVUM JUS*, 15(1), 133–160. 10.14718/NovumJus.2021.15.1.7

CTCI. (2022). *Estrategia Nacional De ciencia, Tecnología, Conocimiento e Innovación para el Desarrollo de Chile—2022*. Consejo Nacional de Ciencia, Tecnología, Conocimiento e Innovación para el Desarrollo.

Cumba García, L. M. (2020, September 15). Reinventing myself in a pandemic: From immunologist to science communicator. *Sister Stem*. Retrieved March 1, 2024, from https://sisterstem.org/2020/09/15/reinventing-myself-in-a-pandemic-from-immunologist-to-science-communicator/

Cumba García, L. M. (2021, October 19). La importancia de la divulgación científica en tiempos de pandemia: Un llamado. *Editorial Politics*. Retrieved from https://editorialpoliticscom.wordpress.com/2021/10/19/la-importancia-de-la-divulgacion-cientifica-en-tiempos-de-pandemia-un-llamado/

Cumba García, L., & Carbajal, M. I. (2024, February 22). "One Health" as a pillar of the Loss and Damage Fund. *Latinoamerica21*. Retrieved March 2, 2024, from https://latinoamerica21.com/en/one-health-as-a-pillar-of-the-loss-and-damage-fund/

Cyr, J., Bianchi, M., González, L., & Perini, A. (2021). Governing a Pandemic: Assessing the Role of Collaboration on Latin American Responses to the COVID-19 Crisis. *Journal of Politics in Latin America*, 13(3), 290–32. 10.1177/1866802X211049250

D'agrosa Okita, S., & Pantovic, B. (2023, October 10). La investigación marítima para la diplomacia científica en Argentina: La Iniciativa Pampa Azul. *La red de diplomacia científica en Latinoamérica y El Caribe*. Retrieved March 1, 2024, from https://diplomaciacientifica.org/la-investigacion-maritima-para-la-diplomacia-cientifica-en-argentina-la-iniciativa-pampa-azul/

da Cruz, A. S., Jr. (2010). Diplomacia da inovação. *Parcerias Estratégicas*, 15, 297–308.

da Silva, R., Ferreira, G., Onuki, J., & Nunes de Oliveira, J. N. (2021). The institutional building of science and innovation diplomacy in Latin America: Toward a comprehensive analytical typology. *Frontiers*, 49–62.33987524

Dangond Gibsone, S. (2022). *Colombia-Estados Unidos—200 años de relaciones diplomáticas*. Gobierno de Colombia.

Das Ibero-Amerikanische Institut (IAI). (2024. March 3). Geschichte des IAI. Retrieved from https://www.spk-berlin.de/iai/institut/geschichte-des-iaiDas Ibero-Amerikanische Institut (IAI). (2024, March 3). Geschichte des IAI. Retrieved from https://www.spk-berlin.de/iai/institut/geschichte-des-iai

Dauchert, H., & Garzik, L. (2022). Berlin Innovation System. In Garzik, L. (Ed.), *Successful Innovation Systems. Future of Business and Finance*. Springer., 10.1007/978-3-030-80639-2_7

de Almeida, C. M. (2020). Gobernanza global en el sector de la salud: Cambios en el "orden mundial", la arena internacional y los impactos en la salud. *Astrolabio*, (24), 80–124. 10.55441/1668.7515.n24.25394

De Greiff, A. (2014). La norteamericanización de la tecnociencia en América Latina: diplomacia científica y hegemonía cultural. In Kreimer, P. (Ed.), *Vessuri, H.; Velho, L. y Arellano, A. (coords.). Perspectivas latinoamericanas en el estudio social de la ciencia, la tecnología y la sociedad* (pp. 194–207). Siglo XXI Editores.

Departamento Nacional De Planeación. (2021). *CONPES. Política Nacional de Ciencia, Tecnología e Innovación 2022-2031*. https://colaboracion.dnp.gov.co/CDT/Conpes/Econ%C3%B3micos/4069.pdf

Department for Business, Energy and Industrial Strategy. (2022). *Partner Country Case Study: China Final Evaluation of The Newton Fund* (Partner Country Case Study, p. 77) [Final evaluation]. Department for Business, Energy and Industrial Strategy.

Department for Science. (2024). *The UK Science and Technology Framework*. Retrieved July 13, 2024, from https://www.gov.uk/government/publications/uk-science-and-technology-framework/the-uk-science-and-technology-framework

Deutsche Forschungsgemeinschaft (DFG). (2014, March 3). Retrieved from https://www.dfg.de/en/research-funding/funding-opportunities/international-cooperation

Deutsche Forschungsgemeinschaft (DFG). (2024, March 2). Retrieved from https://www.dfg.de/resource/blob/172938/5b6fd952251824b1… (PDF file)

Deutsche Gesellschaft für Internationale Zusammenarbeit (GIZ). (2023, March 1). Retrieved from https://www.giz.de/

Deutsche Gesellschaft für Internationale Zusammenarbeit (GIZ). (2024, March 14). Triangular Cooperation with Latin America and the Caribbean: International Competence Building. Retrieved from https://www.giz.de/en/worldwide/13032.html

Deutsche Startup Monitor (DSM). (2022, March 2). Deutscher Startup Monitor. Retrieved from https://deutscherstartupmonitor.de/

Deutsche Welthungerhilfe (WHH). (2024. March 14). Lateinamerika. Retrieved from https://www.welthungerhilfe.de/informieren/laender?tx_igxprojektkarte_projektkarte%5Baction%5D=list&tx_igxprojektkarte_projektkarte%5Bcontroller%5D=Projektkarte&cHash=3f8e08b14fd8714962bad01c8125db48

Deutscher Akademischer Austauschdienst (DAAD). (2024. March 14). Oficina Regional del DAAD en Bogota.Movilidad entre Alemania y Países de la Región Andina (Colombia, Ecuador y Venezuela). Estadistics Actuales Septiembre 2023. Retrieved from https://www.daad.co/files/2023/12/DAAD_Estadisticas_Estudiantes_en_Alemania_ColEcuPerVen_2023_Stand_06_12_23_NM_RB-1.pdf

Deutscher Akademischer Austauschdienst (DAAD). (2024. March 14).The DAAD. Retrieved from https://www.daad.de/en/

Dewitz, P., Carr, E. M., & Patberg, J. P. (1987). Effects of inference training on comprehension and comprehension monitoring. *Reading Research Quarterly*, 22(1), 99–121. 10.2307/747723

Díaz, F. E., Arruvito, L., & Geffner, J. (2023). The future of vaccination in Latin America: Learning from the COVID-19 pandemic. *Current Opinion in Immunology*, 85(December), 102390. 10.1016/j.coi.2023.10239037806096

DiCenzo, C., Menezes, S., Smith, H., Murray-Johnson, K., Azizi, M., & McDuffie, K. (2021). *Inclusive science communication starter kit*. Metcalf Institute, University of Rhode Island., Retrieved from https://inclusivescicomm.org/wp-content/uploads/sites/1568/ISC-Starter-Kit_FINAL.pdf

Dilling, L., & Lemos, M. C. (2011). Creating usable science: Opportunities and constraints for climate knowledge use and their implications for science policy. *Global Environmental Change*, 21(2), 680–689. 10.1016/j.gloenvcha.2010.11.006

DiploCientifica. [@diplomaciacientifica]. (2022). *Hablemos de Diplomacia Científica con Carlos Nobre*. Tomado de Youtube. https://www.youtube.com/watch?v=QPogfuMQvgM (Abril 14 de 2022).

Dolan, B. M. (2012). Science and Technology Agreements as Tools for Science Diplomacy. *AAAS Center for Science and Diplomacy*. https://www.sciencediplomacy.org/article/2012/science-and-technology-agreements-tools-for-science-diplomacy

Domínguez, R. (2013). El pragmatismo en las relaciones UE y Colombia/Perú. In *Después de Santiago: Integración regional y relaciones Unión Europea—América Latina*. Miami-Florida European Union Center of Excellence.

Doong, S. H., & Ho, S. C. (2012). The impact of ICT development on the global digital divide. *Electronic Commerce Research and Applications*, 11(5), 518–533. 10.1016/j.elerap.2012.02.002

Doty, D. H., & Glick, W. H. (1994). Typologies as a unique form of theory building: Toward improved understanding and modeling. *Academy of Management Review*, 19(2), 230–251. 10.2307/258704

Dusdal, J., & Powell, J. J. W. (2021). Benefits, motivations, and challenges of international collaborative research: A sociology of science case study. *Science & Public Policy*, 48(2), 235–245. 10.1093/scipol/scab010

Dwyer, C. P. (2017). Inference. In *Critical Thinking: Conceptual Perspectives and Practical Guidelines* (pp. 117–128). Cambridge University Press. 10.1017/9781316537411.009

Dwyer, C. P., Hogan, M. J., & Stewart, I. (2014). An integrated critical thinking framework for the 21st century. *Thinking Skills and Creativity*, 12, 43–52. 10.1016/j.tsc.2013.12.004

Echeverría King, L. F., Piñeros-Ayala, R., De Luque-Montaño, O., Flores-Zamora, A. F., Pantović, B., & Figueroa, P. (2023). Teaching Science Diplomacy at University Programs for Latin America. In Global Science 's Cooperation Opportunities, Challenges, and Good Practices. Academic Press.

Echeverría King, L. F., Piñeros-Ayala, R., De Luque-Montaño, O., Flores-Zamora, A. F., Pantović, B., & Figueroa, P. (2023). Teaching science diplomacy at university programs for Latin America. In M. Moussaoui, *Global science's cooperation opportunities, challenges, and good practices* (pp. 165-189). IGI Global.

Echeverría King, L. F. (2020). Diplomacia científica e internacionalización de la investigación: Nuevos horizontes de trabajo colaborativo con base en dos experiencias de Colombia. *Revista Educación Global*, 24, 11–21.

Echeverría King, L. F., González, D. A., & Andrade-Sastoque, E. (2021). Science Diplomacy in Emerging Economies: A Phenomenological Analysis of the Colombian Case. *Frontiers in Research Metrics and Analytics*, 6(636538), 636538. Advance online publication. 10.3389/frma.2021.63653833997599

Echeverría-King, L. F., Camacho Toro, R., Figueroa, P., Galvis, L. A., González, A., Suárez, V. R., & Torres Atencio, I. (2022). Organized Scientific Diaspora and Its Contributions to Science Diplomacy in Emerging Economies: The Case of Latin America and the Caribbean. *Frontiers in Research Metrics and Analytics*, 7. 10.3389/frma.2022.893593

Echeverría-King, L., Widmaier Müller, C., & Aquino Valle, K. (2020). La importancia de la Diplomacia Científica y los Objetivos de Desarrollo Sostenible: Desafíos y Oportunidades. Revista Internacional De Cooperación Y Desarrollo, 7(1), 166-179. 10.21500/23825014.4570

Echeverría-King, L. F., Fossati, A., Raja, N. B., Bonilla, K., Urbani, B., Whiffen, R. K., & Vizinová, T. (2023). Scientific collaborations between Latin America and Europe: An approach from science diplomacy towards international engagement. *Science & Public Policy*, 50(4), 794–806. 10.1093/scipol/scad025

Echeverría, L., Aquino, K., & Widmaier, C. (2020). Science Diplomacy and Sustainable Development Goals: A Latin American Perspective. *Science Diplomacy Review*, 2(1), 3–13.

Economic Commission for Latin America and the Caribbean. (n.d.). *The Digital Economy for Latin America and Caribbean Initiative (DE4LAC)*. Retrieved from World Bank.

Ellerbeck, S. (2023). Lithium: Here's why Latin America is key to the global energy transition. *World Economic Forum*. Retrieved April 4, 2024, from https://www.weforum.org/agenda/2023/01/lithium-latin-america-energy-transition/

Elmgren, L., Li, X., Wilson, C., Ball, R., Wang, J., Cichutek, K., Pfleiderer, K., Cavaleri, M., Southern, J., Jivapaisarnpong, T., Minor, P., Griffiths, E., Sohn, Y., & Wood, D. (2013, April 18). A global regulatory science agenda for vaccines. *Vaccine*, 31(Suppl 2), B163–B175. 10.1016/j.vaccine.2012.10.11723598478

Embajada Argentina ante la Unión Europea (2019). 20° Aniversario de la Cooperación en Ciencia, Tecnología e Innovación entre la Argentina y la Unión Europea. Ministerio de Relaciones Exteriores, Comercio Internacional y Culto (26/06/2019). Available at https://eceur.cancilleria.gob.ar/es/20%C2%BA-aniversario-de-la-cooperaci%C3%B3n-en-ciencia-tecnolog%C3%ADa-e-innovaci%C3%B3n-entre-la-argentina-y-la-uni%C3%B3n-0. Retrieved in December 2023.

Embajada de Colombia en China. (2023). *Invest in Colombia.*

Embajada de Estados Unidos en Chile. (2024). Educación y Cultura. Consejo Chile-Estados Unidos para la Ciencia, Tecnología e Innovación (STIC). Disponible en: https://cl.usembassy.gov/es/education-culture-es/consejo-chile-estados-unidos-para-la-ciencia-tecnologia-e-innovacion-stic/

Embajada de Estados Unidos en Chile. (2024). Nuestra relación. Available at: https://cl.usembassy.gov/es/our-relationship-es/policy-history-es/

Embraer. (2024). History of EMBRAER. Retrieved March 22, 2024, from https://historicalcenter.embraer.com/global/en/history

Emprende, A. (2024). Amazonía Emprende-Escuela Bosque. Retrieved April 4, 2024, from https://uplink.weforum.org/uplink/s/uplink-contribution/a012o00001pTVciAAG/A%20knowledge%20and%20incubation%20hub%20for%20nature-based%20solutions%E2%9C%AA

Ennis, R. H., Millman, J., & Tomko, T. N. (2005). *Cornell Critical Thinking Tests.* The Critical Thinking Co.

Ennis, R. H., & Weir, E. (1985). *The Ennis-Weir Critical Thinking Essay Test.* Critical Thinking Press and Software.

Epping, E. (2020). Lifting the smokescreen of science diplomacy: Comparing the political instrumentation of science and innovation centres. *Humanities & Social Sciences Communications*, 7(1), 111. 10.1057/s41599-020-00599-4

European Commission. (2023). *International cooperation with Latin America and Caribbean in research and innovation.* https://research-and-innovation.ec.europa.eu/strategy/strategy-2020-2024/europe-world/international-cooperation/regional-dialogues-and-international-organisations/latin-america-and-caribbean_en

European Commission. (2024, March 2). Research and innovation, international cooperation: Why and how? Strategy on Research and Innovation, Strategy 2020-2024—Europe in the World. Retrieved February 28, 2024, from https://research-and-innovation.ec.europa.eu/strategy/strategy-2020-2024/europe-world/international-cooperation_en

European Commission. (2024, March 2). Science diplomacy in the global approach to research and innovation. Retrieved from https://research-and-innovation.ec.europa.eu/strategy/strategy-2020-2024/europe-world/international-cooperation/science-diplomacy_en

European Commission. (n.d). Research and Innovation. Argentina. Available at https://research-and-innovation.ec.europa.eu/strategy/strategy-2020-2024/europe-world/international-cooperation/bilateral-cooperation-science-and-technology-agreements-non-eu-countries/argentina_en. Retrieved in March 2024.

European Union (2023). La UE y Chile se comprometen a una asociación con visión de futuro. European External Action Service. Available at: https://www.eeas.europa.eu/eeas/la-ue-y-chile-se-comprometen-una-asociaci%C3%B3n-con-visi%C3%B3n-de-futuro-0_es

European Union. (n.d.). *Comisión Europea firma Acuerdos de Cooperación para observación espacial con Colombia, Chile y Brasil | EEAS*. Retrieved March 15, 2024, from https://www.eeas.europa.eu/node/40977_es

Facione, P. (1990). *Critical thinking: A statement of expert consensus for purposes of educational assessment and instruction (The Delphi Report)*. American Philosophical Association.

Facione, P., & Facione, N. (1992). *The California Critical Thinking Dispositions Inventory (CCTDI); and the CCTDI Test Manual*. The California Academic Press.

Fägersten, B. (2022). *Leveraging Science Diplomacy in an Era of Geo-Economic Rivalry* (1; p. 22). The Swedish Institute of International Affairs. https://www.ui.se/globalassets/ui.se-eng/publications/ui-publications/2022/ui-report-no.-1-2022.pdf

Falling Walls. (2024, February 26). Retrieved from https://falling-walls.com/

FAO. (2021). Los pueblos indígenas y tribales y la gobernanza de los bosques - Una oportunidad para la acción climática en Latina América y el Caribe. In *FAO eBooks*. 10.4060/cb2953es

Federal Foreign Office. (2020). Latin America and the Caribbean: A strategically important region. Regional Focuses. Retrieved from https://www.auswaertiges-amt.de/en/aussenpolitik/regionaleschwerpunkte/lateinamerika/-/227548

Federal Foreign Office. (2024, March 2). Education and Research. Retrieved from https://www.auswaertiges-amt.de/en/aussenpolitik/regionaleschwerpunkte/lateinamerika/-/227554

Federal Foreign Office. (2024. March 23). Science Diplomacy (09.11.2020-Article). Retrieved from https://auswaertiges-amt.de

Federal Ministry of Education and Research (BMBF). (2024. March 2). Science diplomacy. Retrieved from https://www.bmbf.de/bmbf/en/international-affairs/science-diplomacy/science-diplomacy.html

Federal Ministry of Education and Research. (2021) German Research Landscape. Publisher Deutsche Akademische Austauschdienst (DAAD), Section Internationl Research Marketing 86p. https://www.research-in-germany.org/dam/jcr:3e5ad44b-8e6a-4832-b64d-ff037d106018/The-German-research-landscape_2021_barrierefrei.pdf

Federal Ministry of Education and Research. (2021) German Research Landscape. Retrieved from. https://www.research-in-germany.org/en/research-landscape/university-landscape/universities.html

Federal Ministry of Education and Research. (2022, Mach 2). CALAS: Science cooperation with Latin America. Retrieved from CALAS: Science cooperation in and with Latin America | Humanities & Social Sciences (geistes-und-sozialwissenschaften-bmbf.de)

Federal Ministry of Education and Research. (2024, March 2). Funding: Research funding. Retrieved from https://www.bmbf.de/bmbf/en/ministry/research-funding/research-funding_node.html

Federal Ministry of Education and Research. (2024, March 2). International affairs. *Horizon*, 2020, •••. https://www.bmbf.de/bmbf/en/ministry/research-funding/research-funding_node.html

Federal Ministry of Education and Research. (2024, March 2). Research. Hightech and innovation. Cluster-networks-international. Retrieved from https://www.bmbf.de/bmbf/en/research/hightech-and-innovation/clusters-networks-international/clusters-networks-international_node.html

Federal Ministry of Education and Research. (2024, March 2). Worldwide networking. Retrieved February 27, 2024, from https://www.bmbf.de/bmbf/en/international-affairs/worldwide-networking/worldwide-networking.html

Fedoroff, N. V. (2009). Science Diplomacy in the 21st Century. *Cell*, 136(1), 9–11. 10.1016/j.cell.2008.12.03019135879

Feliú Mójer, M. I., Ayala Rosado, E., Ortiz Torres, X., González Otero, R., & López, A. I. (2024). Creating risk know-how amongst marginalized communities in Puerto Rico during Covid-19. *Risk know how*. Retrieved March 2, 2024, from https://riskknowhow.org/risk-know-how-around-the-world/stories/aqui-nos-cuidamos/

Feliú-Mójer, M. I. (2020, March 19). The Boricua science community responds to Covid-19. *Cienciapr*. Retrieved March 21, 2024, from https://www.cienciapr.org/en/blogs/equipo-informa/boricua-science-community-responds-covid-19

Feliú-Mójer, M. I. (2020, May 26). Scientific organizations ask governor to prioritize purchase of molecular tests for COVID-19. *Cienciapr*. Retrieved from https://www.cienciapr.org/en/blogs/pr-span/scientific-organizations-ask-governor-prioritize-purchase-molecular-tests-covid-19

Feliu-Mojer, M. (2022). Advancing inclusion through culturally relevant science communication: A perspective from Puerto Rico. *JCOM, Journal of Science Communication*, 21(07), C04. 10.22323/2.21070304

Fernández, E. G. (2012). Conocimientos Tradicionales, Propiedad Intelectual y Negociaciones Internacionales: ¿Gestación de Nuevos Derechos? *Revista Propiedad Intelectual*, 11(15), 204–226.

Ferraz da Silva, P. (2023). Nuevos rumbos para la diplomacia científica. In Nuevos mapas para la diplomacia. Ciencia, conocimiento, tecnología e Innovación. Santiago de Chile: Academia diplomática de Chile.

FGV. (2023). Higher quality education associated with higher growth rates, study finds. Retrieved March 1, 2024, from https://portal.fgv.br/en/news/higher-quality-education-associated-higher-growth-rates-study-finds

Fidler, D. P. (2001). "The Globalization of Public Health: The First 100 Years of International Health Diplomacy" Articles by Maurer Faculty. 599. https://www.repository.law.indiana.edu/facpub/599

Figueiredo, M. S. N., & Pereira, A. M. (2017). Managing Knowledge – The Importance of Databases in the Scientific Production. *Procedia Manufacturing*, 12, 166–173. 10.1016/j.promfg.2017.08.021

FILAC & Secretaria General Iberoamericana. (2023). *Diálogo participativo con Organizaciones de Pueblos Indígenas: Rumbo a la Cumbre del Futuro y la COP 28.*

FILAC. (2021a). *Evento Paralelo: Conocimientos tradicionales de los Pueblos Indígenas fundamentales en la transformación de los Sistemas Alimentarios | Fondo para el desarrollo de pueblos indígenas de América Latina y el Caribe*. https://www.filac.org/evento-paralelo-conocimientos-tradicionales-de-los-pueblos-indigenas-fundamentales-en-la-transformacion-de-los-sistemas-alimentarios/

FILAC. (2021c). *Presentaron experiencias de cómo los conocimientos indígenas ayudan en la prevención de riesgos ambientales | Fondo para el desarrollo de pueblos indígenas de América Latina y el Caribe*. https://www.filac.org/presentaron-experiencias-de-como-los-conocimientos-indigenas-ayudan-en-la-prevencion-de-riesgos-ambientales/

FILAC. (2021d). *Taller Global: Aplicación de los Conocimientos Indígenas en la Prevención y Gestión de Riesgos Ambientales | Fondo para el desarrollo de pueblos indígenas de América Latina y el Caribe*. https://www.filac.org/taller-global-aplicacion-de-los-conocimientos-indigenas-en-la-prevencion-y-gestion-de-riesgos-ambientales/

FILAC. (2022a). *FILAC realiza diálogo con mujeres indígenas para la preservación de los conocimientos tradicionales | Fondo para el desarrollo de pueblos indígenas de América Latina y el Caribe*. https://www.filac.org/filac-realiza-dialogo-con-mujeres-indigenas-para-preservar-los-conocimientos-tradicionales/

FILAC. (2022b). *Mujeres indígenas juegan un papel crucial en la transmisión de conocimientos tradicionales | Fondo para el Desarrollo de Pueblos Indígenas de América Latina y el Caribe*. https://www.filac.org/mujeres-indigenas-juegan-un-papel-crucial-en-la-transmision-de-conocimientos-tradicionales/

FILAC. (2022c). *Pueblos Indígenas contribuyen con sus conocimientos a la reducción de riesgos ambientales | Fondo para el desarrollo de pueblos indígenas de América Latina y el Caribe*. https://www.filac.org/pueblos-indigenas-buscan-contribuir-con-sus-conocimientos-a-la-reduccion-de-riesgos/

FILAC. (2023). *Infraestructura de Investigación del Conocimiento Indígena, un aporte valioso para proteger los conocimientos y Sistemas Alimentarios de los Pueblos Indígenas | Fondo para el desarrollo de pueblos indígenas de América Latina y el Caribe*. https://www.filac.org/nfraestructura-de-investigacion-del-conocimiento-indigena-un-aporte-valioso-para-proteger-los-conocimientos-y-sistemas-alimentarios-de-los-pueblos-indigenas/

FILAC. (2023a). *Conocimientos tradicionales de los Pueblos Indígenas para la transformación de los sistemas alimentarios | Fondo para el desarrollo de pueblos indígenas de América Latina y el Caribe*. https://www.filac.org/conocimientos-tradicionales-de-los-pueblos-indigenas-para-la-transformacion-de-los-sistemas-alimentarios/

FILAC. (2023b). *COP 28: FILAC destaca la realidad que enfrentan los Pueblos Indígenas frente a urgencia de la Crisis Climática en América Latina | Fondo para el desarrollo de pueblos indígenas de América Latina y el Caribe*. https://www.filac.org/cop-28-filac-destaca-la-realidad-que-enfrentan-los-pueblos-indigenas-frente-a-urgencia-de-la-crisis-climatica-en-america-latina/

FILAC. (n.d.). *Consejo Directivo | Fondo para el desarrollo de pueblos indígenas de América Latina y el Caribee*. https://www.filac.org/consejo-directivo/

FILAC. (n.d.). *Fondo para el desarrollo de pueblos indígenas de América Latina y el Caribee | FILAC*. https://www.filac.org/

FILAC. (n.d.). *Kolom No'ooj | Fondo para el desarrollo de pueblos indígenas de América Latina y el Caribe*. https://www.filac.org/iniciativas/kolom-noooj/

Fiocruz Foundation. Brazil. Official website https://portal.fiocruz.br/en/production-and-innovation

Fischer-Bollin, P. (2019). Die Welt und wir: Globale Trends und Zukunftsthemen für die deutsche Politik. Konrad-Adenauer-Stiftung. https://www.kas.de/documents/259121/4624147/Die+Welt+und+Wir+Globale+Trends+und+Zukunftsthemen+f%C3%BCr+die+deutsche+Politik.pdf/f3e4b449-ef03-da73-2cf1-8c915d24a740?version=1.2&t=1575999718243

Fjæstad, M. (2010, April). Research institutes in Germany: Basic and applied science institutionalized? CESIS Electronic Working Paper Series, (Paper No. 232). Centre of Excellence for Science and Innovation Studies (CESIS), The Royal Institute of Technology. Retrieved from https://static.sys.kth.se/itm/wp/cesis/cesiswp232.pdf

Flink, T., & Rüffin, N. (2019). The current state of the art of science diplomacy. In Handbook on Science and Public Policy (pp. 104-121). Edgar Elgar Publishing. 10.4337/9781784715946.00015

Flink, T. (2022). *Taking the pulse of science diplomacy and developing practices of valuation. Science and Public Policy, 49(2), 191–200*. Oxford University Press. 10.1093/scipol/scab074

Flink, T., & Schreiterer, U. (2010). Science diplomacy at the intersection of S&T policies and foreign affairs: Toward a typology of national approaches. *Science & Public Policy*, 37(9), 665–677. 10.3152/030234210X12778118264530

Förderberatung des Bundes. (2024, March 2). Förderung. Retrieved from https://www.foerderinfo.bund.de/foerderinfo/de/foerderung/foerderung_node.html

Fox, N. J. (2011). Boundary objects, social meanings and the success of new technologies. *Sociology*, 45(1), 70–85. 10.1177/0038038510387196

Foyer, J., & Kervran, D. David (2022) Traditional knowledge in the global climate change regime: Narratives for an alternative science diplomacy. In Mays C, Laborie L, Griset P (eds) Inventing a shared science diplomacy for Europe: Interdisciplinary case studies to think with history. Zenodo. 10.5281/zenodo.6582702

Foyer, J., & Kervran, D. D. (2020). Mettre en récit les savoirs traditionnels Une diplomatie scientifique alternative à la COP21 [bonus—Article en ligne]. Terrain, 73. *Terrain*, (73). Advance online publication. 10.4000/terrain.20607

Franco Ferreira, L., Justo, A., Cotelo, C., & Arias, I. (2019). Multi-actor engagement: An open innovation process of knowledge exchange and co-creation. In *Proceedings of the 12th annual International Conference of Education, Research and Innovation*, Seville, Spain. DOI:10.21125/iceri.2019.0969

Franco Giraldo, A. (2015) Academic review of global health approaches: an analytical framework. *Cad. Saúde Pública* 31 (9) • Set 2015 https://doi.org/10.1590/0102-311X00027115

Frauenhofer-Gesellschaft. (2014, March 13). Frauenchofer in Chile. Retrieved from https://www.fraunhofer.de/en/institutes/international/north-southamerica/fraunhofer-chile.html#:~:text=The%20Fundaci%C3%B3n%20Fraunhofer%20Chile%20Research%20FCR%20was%20established,%E2%80%94%20aimed%20at%20attracting%20international%20centers%20of%20excellence

Fulbright Chile. (2024). Quiénes somos. Disponible en: https://fulbrightchile.cl/quienes-somos/

Fulbright Commission Colombia. (n.d.). *Fulbright Commission Colombia—Our history*. Fulbright Colombia – Becas Para Colombianos. Retrieved in March 12, 2024, from https://fulbright.edu.co/nuestra-historia/

Gainza, C. (2023). Ciencia, tecnología, conocimiento e innovación en Chile: problemas complejos, miradas transdisciplinarias. In Nuevos mapas para la diplomacia. Ciencia, conocimiento, tecnología e Innovación. Santiago de Chile: Academia diplomática de Chile.

Gallarotti, G. M. (2011). Soft Power: What it is, Why it's Important, and the Conditions Under Which it Can Be Effectively Used. Division II Faculty Publications. Paper 57.

Galvão, T., Tavares, K. A., Fernandes, M., Souza, J. C. M., & Noll, M. (2021). Case report: Scientific dissemination as a mediator of the research consolidation process: Proposals for institutionalization [version 1; peer review: awaiting peer review]. *F1000 Research*, 10, 886. 10.12688/f1000research.55655.1

Gálvez-Soriano, O. de J. (2020). Could Education Increase the Economic Growth of Mexico? *Análisis económico- SciELO Analytics*.

García, B. (2013). Contaminación transfronteriza en cuenca binacional Puyango Tumbes. http://genderandwater.org/es/biblioteca/enlaces

Garcia, R., Araújo, V., Mascarini, S., dos Santos, E. G., & Costa, A. R. (2018). An analysis of the relation between geographical and cognitive proximity in university-industry linkages. *Anais do XLIV Encontro Nacional de Economia*, 1–20.

Garrido, R., & Hernández, J. (2008). *Agua: recurso único. Conclusiones y propuestas. Expo Agua Zaragoza 2008.* Caja Azul de la Tribuna del Agua.

Gavilanes (2016). Estudio de la concentración del plomo en el agua del río Tumbes, periodo 2012 - 2015 como causa de la minería aurífera y su relación con la salud de los pobladores del caserío de Rica Playa - Tumbes.

GDC. (2023). Welcome to the Global Diaspora Confederation. GDC.

Georghiou, C. A. (2018). Decolonization and Diplomacy. In The Encyclopedia of Diplomacy (pp. 1–9). John Wiley & Sons, Ltd. 10.1002/9781118885154.dipl0071

German Center for Research and Innovation (DWIH) (2024. March 2). Science Diplomacy. Retrieved from Science Diplomacy (dwih-newyork.org)

German universities of technology (TU9). (2024, March 2). Retrieved from https://www.tu9.de/

German University Alliance (UA) 11+. (2024, March 2). Retrieved from https://www.ua11plus.de/

Germann, J. (2022). Global rivalries, corporate interests and Germany's 'National Industrial Strategy 2030'. *Review of International Political Economy*. Advance online publication. 10.1080/09692290.2022.2130958

Germany Trade & Invest. (2024, March 12). Start-up location Germany. Retrieved from https://www.gtai.de/en/invest/hot-topics/start-up-location-germany

Gerring, J. (1999). What makes a concept good? A criterial framework for understanding concept formation in the social science. *Polity*, 31(3), 357–393. 10.2307/3235246

Gibbs, G. (2012). *El análisis de datos cualitativos en investigación cualitativa.* Ediciones Morata.

Gittens, R. A., Lopez-Verges, S., Collado, T., Pimentel, J., Vazquez, A., Pulido-Salgado, M., & Torres-Atencio, I. (2021). Science Diplomacy as an Umbrella Term for Science Advisory in Public and Foreign Relations in Small Developing Countries: The Case of Panama. *Frontiers in Research Metrics and Analytics*, 6, 655335. Advance online publication. 10.3389/frma.2021.65533533912787

GLACIER. (2024, March 1). Homepage. Retrieved from https://glacieronehealth.org/

Global Water Partnership (2000) Integrated Water Resources Management. Technical Committee Background Paper No. 4. GWP, Stockholm, Sweden. Gual, M. Diplomacia Científica en América Latina y El Caribe: Estrategias, mecanismos y perspectivas para fortalecer la diplomacia de la ciencia, tecnología e innovación- Foro CILAC UNESCO.

Global Water Partnership -GWP. (2014). *Water Security: Putting the Concept into Practice. Technical Committee Background Paper No. 20.* GWP.

Gluckman, P. D., Turekian, V. C., Grimes, R. W., & Kishi, T. (2017). *Science Diplomacy: A Pragmatic Perspective from the Inside.*

Gluckman, P. D. (2016). Science advice to governments: An emerging dimension of science diplomacy. *Science & Diplomacy*, 21–30.

Gluckman, P. D., Turekian, V. C., Grimes, R. W., & Kishi, T. (2017). Science diplomacy: A pragmatic perspective from the inside. *Science & Diplomacy*, 20–33.

Gluckman, P. D., Turekian, V., Grimes, R. W., & Kishi, T. (2017). Science Diplomacy: A Pragmatic Perspective from the Inside. *Science & Diplomacy*, 6(4).

Gobierno de Argentina. (2018). Argentina y China colaborarán mutuamente en Educación, Cultura y Ciencia. Gobierno de Argentina, Noticias (20/11/2018). Available at https://www.argentina.gob.ar/noticias/argentina-y-china-colaboraran-mutuamente-en-educacion-cultura-y-ciencia. Retrieved in December 2023.

Gobierno de Argentina. (2019). 20 años de cooperación entre Argentina y la Unión Europea en Ciencia y Tecnología. Gobierno de Argentina, Noticias (11/06/2019). Available at https://www.argentina.gob.ar/noticias/20-anos-de-cooperacion-entre-argentina-y-la-union-europea-en-ciencia-y-tecnologia. Retrieved in December 2023.

Gobierno de Argentina. (2023a). El Ministerio de Ciencia participó en un seminario de diálogo sobre política digital entre Argentina y Estados Unidos. Gobierno de Argentina, Noticias (30/06/2023). Available at https://www.argentina.gob.ar/noticias/el-ministerio-de-ciencia-participo-en-un- seminario-de-dialogo-sobre-politica-digital-entre. Retrieved in February de 2024.

Gobierno de Argentina. (2023b). Argentina y China acuerdan crear un centro de investigaciones en RRNN y transición energética. Gobierno de Argentina, Noticias (07/09/2023). Available at https://www.argentina.gob.ar/noticias/argentina-y-china-acuerdan-crear-un-centro-de-investigaciones-en-rrnn-y-transicion Retrieved in December 2023.

Goertz, G. (2006). *Social science concepts.* Princeton university press. 10.1515/9781400842995

Goldman, I. (2022). El rol de la complementación entre ciencia y diplomacia en la Cuestión Malvinas. Revista de Investigación en Política Exterior Argentina, 2(4), 80-100.

Goldman, I. (2022). El rol de la complementación entre ciencia y diplomacia en la Cuestión Malvinas. *Revista de Investigación en Política Exterior Argentina,* 2(4), 80-100. Retrieved from https://politicaexteriorargentina.org/wp-content/uploads/2022/12/Goldman-I.-El-rol-de-la-complementacion-entre-ciencia-y-diplomacia-en-la-Cuestion-Malvinas.pdf

Gonesh, A., & Melissen, J. (2005). *Public diplomacy: Improving practice.* Netherlands Institute of International Relations Clingendael.

Göring, J. Intellectual Property Protection - German Patents and International Competition. ATZ Worldw 125, 50–53 (2023). https://doi.org/, J. (2023). Intellectual property protection - German patents and international competition. ATZ Worldwide, 125(1), 50–53. https://doi.org/10.1007/s38311-023-1537-4Göring

Gorry, C. (2020) Global Collaboration in Times of COVID-19: Cuba's Emergency Medical Contingent. MEDICC Review.

Grabois Gadelha, C. A., Seixas da Costa Braga, P., Mattoso Montenegro, K. B., & Bahia Cesário, B. (2020). Access to vaccines in Brazil and the global dynamics of the Health Economic-Industrial *Complex.Cadernos de Saude Publica*, 36(Suppl 2), e00154519. 10.1590/0102-311x0015451932876097

Granstrand, O., & Holgersson, M. (2020). Innovation ecosystems: A conceptual review and a new definition. *Technovation*, 90–91, 102098. 10.1016/j.technovation.2019.102098

Gregory, B. (2016). Mapping boundaries in diplomacy's public dimension. *The Hague Journal of Diplomacy*, 11(1), 1–25. 10.1163/1871191X-12341317

Gual Soler, M. (2014). El Papel de las Redes Científicas Intergubernamentales en las Relaciones Regionales y la Integración de América Latina. *Science & Diplomacy*, 3(4).

Gual Soler, M. (2021). Science diplomacy in Latin America and the Caribbean: Current landscape, challenges, and future perspective. *Frontiers in Research Metrics and Analytics*, 6, 670001. Advance online publication. 10.3389/frma.2021.67000134222772

Gual Soler, M. (2021). Science diplomacy in Latin America and the Caribbean: Current landscape, challenges, and future perspective. *Frontiers*, 1–9.

Gualini, E., & Fricke, C. (2019). Who governs Berlin's metropolitan region? The strategic-relational construction of metropolitan scale in Berlin–Brandenburg's economic development policies. Environment and Planning C. *Politics and Space*, 37(1), 59–80. 10.1177/2399654418776549

Gustafsson, K. M., & Lidskog, R. (2018). Boundary organizations and environmental governance: Performance, institutional design, and conceptual development. *Climate Risk Management*, 19, 1–11. 10.1016/j.crm.2017.11.001

Guston, D. H. (2001). Boundary organizations in environmental policy and science: An introduction. *Science, Technology & Human Values*, 26(4), 399–408. 10.1177/016224390102600401

Gutiérrez, A. G. (2020). Imbabura and Chimborazo at the cradle of Biogeography in the South American Andes. *Biodiversity International Journal*, 4(5), 198–200. 10.15406/bij.2020.04.00185

Guzman-Holst, A., Rodrigo De Antonio, A., Prado-Cohrs, D., Patricia Juliao, P. (2020) Barriers to vaccination in Latin America: A systematic literature review. Vaccine. Volume 38, Issue 3, 16 January, Pages 470-481

Hafner, M., Yerushalmi, E., Fays, C., Dufresne, E., & van Stolk, Ch. (2020). COVID-19 and the cost of vaccine nationalism. Published by Rand Corporation. e RAND Corporation, Santa Monica, Calif., and Cambridge, UKHotez, P.J. (2014) "Vaccine Diplomacy": Historical Perspectives and Future Directions. *PLoS Neglected Tropical Diseases*, 8(6), e2808. 10.1371/journal.pntd.0002808

Harden-Davies, H. (2018). The next wave of science diplomacy: Marine biodiversity beyond national jurisdiction. *ICES Journal of Marine Science*, 75(1), 426–434. 10.1093/icesjms/fsx165

Harman, G. H. (1965). The inference to the best explanation. *The Philosophical Review*, 74(1), 88–95. 10.2307/2183532

Harrison, N. E., & Singer, D. J. (2006). Complexity is more than systems theory. In Complexity in world politics. Concepts and methods of a new paradigm. State University of New York Press.

Hay, C. (2020). *International Relations Theory and Globalization* (Dunne, T., Kurki, M., & Smith, S., Eds.; 5th ed.). International Relations Theories. 10.1093/hepl/9780198814443.003.0015

Heidelberg Center Latin America (HCLA). (2024, March 2). Retrieved from https://www.hcla.uni-heidelberg.de/en/the-center and https://www.hcla.uni-heidelberg.de/en/research

Hernández, B. & Lebret, P. (2019). Dos décadas de cooperación europea en Chile: una asociación para el desarrollo. *Revista Estudios internacionales* 51(192), 35-56. Chile. 10.5354/0719-3769.2019.52812

Hernández-Mondragón, A. C. (2022). From lab to science policy advisor. *Nature Human Behaviour*, 6(4), 477. 10.1038/s41562-022-01314-w35177810

Hersh, M. A. (1999). Sustainable decision making: The role of decision support systems. *IEEE Transactions on Systems, Man, and Cybernetics. Part C, Applications and Reviews*, 29(3), 395–408. 10.1109/5326.777075

Hodgson, G. M. (2019). Taxonomic definitions in social science, with firms, markets and institutions as a case study. *Journal of Institutional Economics*, 15(2), 207–233. 10.1017/S1744137418000334

Horizon Europe. (2023). *List 3rd country participation horizon-Europe*. Avialable at: https://ec.europa.eu/info/funding-tenders/opportunities/docs/2021-2027/common/guidance/list-3rd-country-participation_horizon-euratom_en.pdf

Hotez, P. J. (2021) Preventing the Next Pandemic: Vaccine Diplomacy in a Time of Anti-science. Edited by Johns Hopkins University Press. DOI??

Huish, R., & Kirk, J. M. (2007) Cuban Medical Internationalism and the Development of the Latin American School of Medicine. *Latino American perspectives* Vol. 34, No. 6, Aggressive Capital and Democratic Resistance (Nov, 2007), pp. 77-92 Published By: Sage Publications, Inc.

Hurtado, D. (2010). *La ciencia Argentina un proyecto inconcluso 1930-2000*.

InnScid. (2019). *São Paulo Framework of Innovation Diplomacy*. Retrieved September 5, 2021, from https://innscidsp.com/framework.pdf

Institute for Latin American Studies (LAI). (2014, March 3). Freie Universität Berlin, Institute for Latin American Studies. Retrieved from https://www.fu-berlin.de/en/einrichtungen/fachbereiche/zentralinstitute/lai/index.html

Instituto de Hidrología. Meteorología y Estudios Ambientales - Ideam. (2023). Informe del estado de los glaciares colombianos 2021. Bogotá: Ideam.

Instituto, C. A. P. A. Z. (2024. March 5). About Us. Retrieved from https://www.instituto-capaz.org/en/about-us-2/

Instituto, C. A. P. A. Z. (2024. March 5). Home. Retrieved from https://www.instituto-capaz.org/en/

Inter-American Development Bank. (2021). *IDB Report Highlights Future Growth Opportunities for Latin America and the Caribbean*. Retrieved from IDB.

International Consortium of Critical Theory Programs. (2021, March 1). Homepage. Retrieved from https://directory.critic

International Council for Science. (2002). Report from ICSU Study Group on Science and Traditional Knowledge. SCIENCE AND TRADITIONAL KNOWLEDGE. https://council.science/wp-content/uploads/2017/05/Science-traditional-knowledge.pdf

International Labour Organization. (2021). Executive Summary "Overview of social protection systems in Latin America and the Caribbean: Progress and setbacks in the face of the pandemic". Available at https://www.ilo.org/wcmsp5/groups/public/---americas/---ro-lima/---sro-port_of_spain/documents/publication/wcms_824905.pdf

International Unión for Conservation for Nature - IUCN. (2017). Gobernanza del Agua desde las bases: un aporte a la diplomacia y a la paz. Gland. Disponible en: https://www.iucn.org/node/27781

Ipiranga. (2024). Museu do Ipiranga-USP. Retrieved March 22, 2024, from <https://museudoipiranga.org.br/en/>

Isaza, J. P. (2020). Breve repaso del contexto institucional de la diplomacia científica, tecnológica y de innovación en Colombia 2004-2006. Innovación y ciencia, 27(2). https://innovacionyciencia.com/articulos_cientificos/breve-repaso-del-contexto-institucional-de-la-diplomacia-cientifica-tecnologica-y-de-innovacion-en-colombia-2004-2006

Islam, S., & Madani, K. (2017). *Water Diplomacy in Action: Contingent Approaches to Managing Complex Water Problems*. Anthem Environment and Sustainability. 10.2307/j.ctt1jktqgh

Iza, A. (2021). *Implementing Nature Based Solutions in Transboundary Basins*. Centro de Derecho Ambiental de la UICN.

Izquierdo, S. (2021). "Binational Commission for Integrated Water Resources Management in the Ecuadorian-Peruvian transboundary basins – Creation process, experience and mandate". Disponible en: https://www.gwp.org/contentassets/092e5da4a6b0454ab98526442b143961/moocevent-16-march-ppts.pdf

Jungbluth, C. (2019) Roadmap 2030: 3 Ways to Ensure the German Economy's Competitiveness in the Coming Decade. Bertelsmann Stiftung. GED New Perspectives on Global and European Dynamics. Retrieved from 3 ways to ensure the German Economy 's competitiveness in the coming decade (globaleurope.eu)

Kalergis, A., González, P., & Rivero, T. (2023). La paradiplomacia científica logra el acceso preferente, prioritario y temprano a vacunas contra el COVID-19. In Nuevos mapas para la diplomacia. Ciencia, conocimiento, tecnología e Innovación. Santiago de Chile: Academia diplomática de Chile.

Kaltofen, C., & Acuto, M. (2018a). Science diplomacy: Introduction to a boundary problem. *Global Policy*, 9(S3), 8–15. 10.1111/1758-5899.12621

Kaltofen, C., & Acuto, M. (2018b). Rebalancing the encounter between science diplomacy and international relations theory. *Global Policy*, 9(S3), 15–23. 10.1111/1758-5899.12620

Karlsson, Á. G. (2023). The Lion, the Spider, and the Laid-Off Janitor: Tales from the World Intellectual Property Organization. Narrative Culture, 10(1), 88–107. *Scopus*. Advance online publication. 10.1353/ncu.2023.0003

Kennedy, B. L., & Thornberg, R. (2018). Deduction, induction, and abduction. In *The SAGE Handbook of Qualitative Data Collection* (pp. 49-64). 10.4135/9781526416070.n4

Keohane, R. O., & Nye, J. S.Jr. (1977). Power and interdependence. *Survival*, 15(4), 158–165. 10.1080/00396337308441409

Kevany, S. (2014). Global health diplomacy, 'smart power', and the new world order. *Global Public Health: An International Journal for Research, Policy and Practice*, 9(7), 787–807. 10.1080/17441692.2014.92121924953683

Khojasteh, D., Davani, E., Shamsipour, A., Haghani, M., & Glamore, W. (2022). Climate change and COVID-19: Interdisciplinary perspectives from two global crises. *The Science of the Total Environment*, 844, 157142. Advance online publication. 10.1016/j.scitotenv.2022.15714235798107

Kickbusch, I. (2011, June 10). Global health diplomacy: How foreign policy can influence health. *BMJ (Clinical Research Ed.)*, 342(jun10 1), d3154. 10.1136/bmj.d315421665931

Kiese, M. (2019). Regional cluster policies in Germany: Challenges, impacts and evaluation practices. *The Journal of Technology Transfer*, 44(6), 1698–1719. 10.1007/s10961-017-9589-5

Klaes, M. (2004). Evolutionary economics: In defence of vagueness. *Journal of Economic Methodology*, 11(3), 359–376. 10.1080/1350178042000252992

Klein, J. T. (2017). Typologies of interdisciplinarity. In Frodeman, R., Klein, J. T., & Pacheco, R. C. (Eds.), *The oxford handbook of interdisciplinarity (págs* (pp. 21–39). 10.1093/oxfordhb/9780198733522.013.3

Krasnyak, O., & Ruffini, P. (2020). *Science Diplomacy*. Oxford Universisty Press., 10.1093/obo/9780199743292-0277

Kriegesmann, K. (2023, March 25). Doing Area Studies: Present and Future Perspectives on the History of Latin American Studies at Freie Universität Berlin. History of Knowledge: Research, Resources, and Perspectives. Retrieved from https://historyofknowledge.net/2023/03/25/history-of-latin-american-studies-at-fu-berlin/

Kumar, S. (2022) Agua: un recurso que puede acelerar el crecimiento verde, inclusivo y resiliente.

Kundnani, H. (2014). *The paradox of German power*. Hurst & Company.

Kunkel, S. (2021). Science Diplomacy in the Twentieth Century: Introduction. *Journal of Contemporary History*, 56(3), 473–484. 10.1177/00220094211006762

Kvale, S. (2011). *Las entrevistas en la investigación cualitativa*. Ediciones Morata.

Lahl, R., Ferse, S., & Bleischwitz, R. (2023). Reflecting on a Research Institute's Role as a Tool for Ocean Science Diplomacy—ZMT's Mission to Collaborate. In Ittekkot, V., & Baweja, J. K. (Eds.), *Science, Technology and Innovation Diplomacy in Developing Countries: Perceptions and Practice* (pp. 311–328). Springer Nature., 10.1007/978-981-19-6802-0_20

Lange Nacht der Wissenschaften. (2024, February 24). Retrieved from https://langenachtderwissenschaften.de

Lanzano, C. (2013). What kind of knowledge is 'indigenous knowledge'? Critical insights from a case study in Burkina Faso. *Transcience*, 4(2), 3–18.

Lascano & Vedia. J. R (2021). Relaciones internacionales, política exterior y diplomacia. En: Colotta, M., Degiorgis, P., Lascano y Vedia, J., Rodriguez, Á. (compiladores) (2021). Manual de relaciones internacionales, Buenos Aires de: https://www.teseopress.com/manualderelacionesinternacionales/chapter/relaciones-internacionales-politica-exterior-y-diplomacia/#footnote-41-1

Lechuga, R. I., & Castro, A. C. (2017, March). Dramatic effects of control measures on deaths from yellow fever in Havana, Cuba, in the early 1900s. *Journal of the Royal Society of Medicine*, 110(3), 118–120. 10.1177/0141076817694583278397

Leff, E. (2004). Racionalidad ambiental y diálogo de saberes. *Polis Revista Latinoaméricana*, 7, 2–29.

Leibniz Universität Hannover. (2019, March 13). Verbundprojekt CALAS erforscht Krisen in Lateinamerika. Presseinformation. Retrieved from https://www.uni-hannover.de/de/universitaet/aktuelles/presseinformationen/detail/news/verbundprojekt-calas-erforscht-krisen-in-lateinamerika/

Leijten, J. (2017). Exploring the future of innovation diplomacy. *European Journal of Futures Research*, 5(1), 1–13. 10.1007/s40309-017-0122-8

Leo, P. D., & Greene, A. J. (2008). Is awareness necessary for true inference? *Memory & Cognition*, 36(6), 1079–1086. 10.3758/MC.36.6.107918927026

Life Sciences Directory. (2023). Excellence in life sciences & healthcare: The German capital region. Life Sciences Report – Contacts 2. Retrieved from https://www.healthcapital.de/files/user_upload/LS_Report_Adressen_Nov23.pdf

Likens, G. E. (2010). The role of science in decision making: Does evidence-based science drive environmental policy? *Frontiers in Ecology and the Environment*, 8(6), e1–e9. 10.1890/090132

Lillo González, M., & Santamaría García, A. (2009). La Unión Europea y Colombia. Historia y desafío. *Anuario de Estudios Americanos*, 66(1), 289–319. 10.3989/aeamer.2009.v66.i1.311

Linkov, I., Basu, S., Fisher, C., Jackson, N., Jones, A. C., Kuklja, M. M., & Trump, B. D. (2016). Diplomacy for science: Strategies to promote international collaboration. *Environment Systems & Decisions*, 36(4), 331–334. 10.1007/s10669-016-9614-5

Lisboa, V. (2023). La producción nacional de vacunas refuerza soberanía y planificación. Article in spanish. Available at https://agenciabrasil.ebc.com.br/es/saude/noticia/2023-09/la-produccion-nacional-de-vacunas-refuerza-soberania-y-planificacion

Lopera, W. (2020). Colombia en la antártida: Aproximaciones a una investigación científico/social … Un llamado a la academia/estado/sociedad. *Revista Boletín Redipe*, 9(12), 12. 10.36260/rbr.v9i12.1145

López de San Román, A., & Schunz, S. (2018). Understanding European Union Science Diplomacy: Understanding EU Science Diplomacy. *Journal of Common Market Studies*, 56(2), 247–266. 10.1111/jcms.12582

López, D., & van Klaveren, A. (2023). La diplomacia científica: una nueva dimensión de la Política Exterior. In Nuevos mapas para la diplomacia. Ciencia, conocimiento, tecnología e Innovación. Santiago de Chile: Academia diplomática de Chile.

López-Calva, L. F. (2022). Finding Unicorns: Investing in innovation-led growth in LAC. *United Nations Development Programme, Latin America and the Caribbean*. Retrieved April 4, 2024, from <https://www.undp.org/latin-america/blog/graph-for-thought/finding-unicorns-investing-innovation-led-growth-lac>

Lopez-Goñi, I. (2020). *Preparados para la Próxima Pandemia*. Editorial Planeta.

López-Vergès, S., Collado, T., Pulido-Salgado, M., Pimentel, J., Vásquez Fábrega, A., & Torres Atencio, I. (2021, February 11). The role of science diplomacy in Panama's response to COVID-19. *Science & Diplomacy*. Retrieved from https://www.sciencediplomacy.org/perspective/2021/role-science-diplomacy-in-panamas-response-covid-19#note6

López-Vergès, S., Macías-Navarro, L., Hernández-Mondragón, A. C., Corrales-Aguilar, E., Soler, M. G., & Guerra, M. (2021). Closing the gap between emerging initiatives and integrated strategies to strengthen science diplomacy in Latin America. *Frontiers in Research Metrics and Analytics*, 6, 664880. 10.3389/frma.2021.66488033912788

Lum, T. (2011). *U.S.-China relations: Policy issues*. Nova Science Publishers, Inc.

Lungeanu, A., & Contractor, N. S. (2015). The Effects of Diversity and Network Ties on Innovations: The Emergence of a New Scientific Field. *American Behavioral Scientist*, 59(5), 548–64. 10.1177/0002764214556804

Lynn, E., & Wolf, A. (2023). Water, Culture and International Institutions Blue Papers2023 (Vol. 2 No. 2), pp. 16-2510.58981/bluepapers.2023.2.01.

Maass, A., Guerrero, G., San Martin, J., & Utreras, F. (2023). Un sistema robusto de infraestructuras críticas para la era de los datos: recurso humano para la investigación, capacidades de super cómputo y conexión internacional. In Nuevos mapas para la diplomacia. Ciencia, conocimiento, tecnología e Innovación. Santiago de Chile: Academia diplomática de Chile.

Manassero-Mas, M.-A., & Vázquez-Alonso, Á. (2019). Conceptualización y taxonomía para estructurar los conocimientos acerca de la ciencia. Revista sobre enseñanza y divulgación de la ciencia, 3104-3116.

Manero, M. L. A. (2018). Actores sociales en la Cumbre del Clima de París: El mensaje de pueblos indígenas, grupos religiosos, mujeres y jóvenes. *Revista Mexicana de Ciencias Politicas y Sociales*, 233(233), 221–254. 10.22201/fcpys.2448492xe.2018.233.59458

Marcelle, G. M. (2016). *Redefining innovation in the Global South: Critical imperatives.* Retrieved April 3, 2024, from https://unctad.org/system/files/non-official-document/CSTD_2015_ppt15_Marcelle_en.pdf

Marck, C., & Rigau-Perez, J. (2009). The World's First Immunization Campaign: The Spanish Smallpox Vaccine Expedition, 1803–1813. *Bulletin of History of Medicine*. Vol. 83, No. 1, Special Issue: Reassessing Smallpox Vaccination, 1789–1900 (Spring 2009)

Marginson, S. (2022). What drives global science? The four competing narratives. *Studies in Higher Education*, 47(8), 1566–1584. 10.1080/03075079.2021.1942822

Martínez, L. M. (2019). La paradiplomacia. Una aproximación a las nuevas teorías. *Oasis*, 30(30), 145–162. 10.18601/16577558.n30.09

Massarani, L. (2018). Estado del arte de la divulgación de la ciencia en América Latina. *Journal of Science Communication in Latin America*, 1(01), A01. 10.22323/3.01010201

Massarani, L., Aguirre, C. C., Reynoso, E., & Lindergaard, L. (2015). RedPOP: 25 años de red en comunicación de la ciencia en América Latina. *Journal of Science Communication*, 14(03), Y06. 10.22323/2.14030406

Mauduit, J. C., & Gual Soler, M. (2020, August). Building a science diplomacy curriculum. *Frontiers in Education*, 5, 138. 10.3389/feduc.2020.00138

Max-Planck-Gesellschaft. (2014, March 13). New laboratory opened in Argentina (newsroom 2014, November 25). Retrieved from https://www.mpg.de/8778738/max-planck-laboratory-argentina

Mayring, P. (2010). *Qualitative Inhaltsanalyse*. Grundlagen und Techniken. Weinheim Basel: Beltz Verlag.

Mazzucato, M., & Penna, C. (2016). *The Brazilian Innovation System: A Mission-Oriented Policy Proposal.*

McClelland, J. L., McNaughton, B. L., & O'Reilly, R. C. (1995). Why there are complementary learning systems in the hippocampus and neocortex: Insights from the successes and failures of connectionist models of learning and memory. *Psychological Review*, 102(3), 419–457. 10.1037/0033-295X.102.3.4197624455

Mejía Navarrete, J. (2020). Epistemología de las políticas de ciencia y tecnología en América Latina. Cinta De Moebio. *Revista De Epistemología De Ciencias Sociales*, (67), 14–25.

Melissen, J. (2005). Wielding soft power: The new public diplomacy. Netherlands Institute of International Relations Clingendael. The Hague: *Clingendael Diplomacy Papers No. 2*.

Mercy Corps. (2021). Rumour tracker programme: A community based approach to address information gaps and misinformation on COVID-19. *World Health Organization*. Retrieved March 11, 2024, from https://cdn.who.int/media/docs/default-source/science-translation/case-studies-1/cs13_rumourtracking.pdf?sfvrsn=829a4b42_4

Miller, C. A. (2006). "An Effective Instrument Of Peace": Scientific Cooperation As An Instrument Of U.S. Foreign Policy, 1938–1950. *Osiris*, 21(1), 133–160. 10.1086/507139

Miller, E. K., & Cohen, J. D. (2001). An integrative theory of prefrontal cortex function. *Annual Review of Neuroscience*, 24(1), 167–202. 10.1146/annurev.neuro.24.1.16711283309

Miller, R. (2018). *Transforming the future: Anticipation in the 21st century*. Taylor & Francis.

MinCiencia. (2019a). Data Observatory: AWS y UAI serán miembros fundadores clave de Chile en la administración de datos. Available at: https://www.minciencia.gob.cl/noticias/data-observatory-aws-y-uai-seran-miembros-fundadores-clave-de-chile-en-la-administracion-de-datos/

MinCiencia. (2019b). Profesores se formarán con científicos norteamericanos para liderar la enseñanza de la ciencia. Available at: https://www.minciencia.gob.cl/noticias/profesores-se-formaran-con-cientificos-norteamericanos-para-liderar-la-ensenanza-de-la-ciencia/

MinCiencia. (2020a). Política Nacional de Ciencia, Tecnología, Conocimiento e Innovación. Available at: https://minciencia.gob.cl/politicactci/documentos/Politica-Nacional-CTCi_Chile-2020.pdf

MinCiencia. (2020b). UC e Instituto Milenio de Inmunología e Inmunoterapia firman alianza con laboratorio chino que está desarrollando vacuna contra el COVID-19. Available at: https://www.minciencia.gob.cl/noticias/uc-e-instituto-milenio-de-inmunologia-e-inmunoterapia-firman-alianza-con-laboratorio-chino-que-esta-desarrollando-vacuna-contra-el-covid-19/

MinCiencia. (2021a). Ministro de Ciencia recibe a representantes de NASA para explorar operación de la misión astronómica SOFIA desde Chile. Available at: https://www.minciencia.gob.cl/noticias/ministro-de-ciencia-recibe-representantes-de-nasa-para-explorar-operacion-de-la-mision-astronomica-sofia-desde-chile/

MinCiencia. (2021b). ESO y Chile firman convenio de cooperación científica y tecnológica con el que se convertirá en el telescopio más grande del mundo. Available at: https://www.minciencia.gob.cl/noticias/eso-y-chile-firman-convenio-de-cooperacion-cientifica-y-tecnologica-con-el-que-se-convertira-en-el-telescopio-mas-grande-del-mundo/

MinCiencia. (2023a). Ministerio de Ciencia e IBM invitan a mujeres de todo el país a cursos de formación gratuita en tecnología. Available at: https://www.minciencia.gob.cl/noticias/ministerio-de-ciencia-e-ibm-invitan-mujeres-de-todo-el-pais-cursos-de-formacion-gratuita-en-tecnologia/

MinCiencia. (2023b). Presidente Boric y ministra Aisén Etcheverry anuncian inicio del proceso para que Chile sea estado asociado del Centro Europeo para la Investigación Nuclear, CERN. Available at: https://www.minciencia.gob.cl/noticias/presidente-boric-y-ministra-aisen-etcheverry-anuncian-inicio-del-proceso-para-que-chile-sea-estado-asociado-del-centro-europeo-para-la-investigacion-nuclear-cern/

MinCiencia. (2023c). Declaración pública sobre Sinovac. Available at: https://www.minciencia.gob.cl/noticias/declaracion-publica/

Ministerio de Ciencia, Tecnología e Innovación. (2020). Libros Misión de Sabios. *Academia (Caracas).*

Ministerio de Ciencia. Tecnología e Innovación. (n.d.-a). *Línea de Tiempo Minciencias*. Retrieved in March 16, 2024, from https://minciencias.gov.co/sites/default/files/ckeditor_files/Linea%20de%20Tiempo%20Minciencias.png

Ministerio de Ciencia. Tecnología e Innovación. (n.d.-b). *Nexo Global*. Minciencias. Retrieved March 12, 2024, from https://minciencias.gov.co/nexo-global

Ministerio de Ciencia. Tecnología e Innovación. (n.d.-c). *MinCiencias y VaxThera firmaron un Memorando de Entendimiento para fortalecer las capacidades en biotecnología del país*. Minciencias. Retrieved March 12, 2024, from https://www.minciencias.gov.co/sala_de_prensa/minciencias-y-vaxthera-firmaron-un-memorando-entendimiento-para-fortalecer-las

Ministerio de Ciencia. Tecnología e Innovación. (n.d.-d). *Conozca por qué Minciencias hace parte del programa Artemis de la Nasa*. Minciencias. Retrieved March 12, 2024, from https://minciencias.gov.co/sala_de_prensa/conozca-por-que-minciencias-hace-parte-del-programa-artemis-la-nasa

Ministerio de Educación Nacional. (2018). *Ministerio de Educación Nacional y el Gobierno de China dan a conocer el programa binacional de becas | Ministerio de Educación Nacional*. Available at: https://www.mineducacion.gov.co/portal/salaprensa/Noticias/377886:Ministerio-de-Educacion-Nacional-y-el-Gobierno-de-China-dan-a-conocer-el-programa-binacional-de-becas

Ministerio de Relaciones Exteriores de Chile. (2015). Informe Anual de Comercio Exterior de Chile 2014-2015. Dirección General de Relaciones Económicas Internacionales. Available at: https://www.subrei.gob.cl/docs/default-source/default-document-library/informe-anual-de-comercio-exterior-de-chile-2014-2015.pdf?sfvrsn=325fde69_0

Ministerio de Relaciones, E., Comercio, I., & Argentina, C. (2023). Candidatura de Bariloche para ser sede de la Exposición Especializada 2027. https://www.cancilleria.gob.ar/es/destacados/candidatura-de-bariloche-para-ser-sede-de-la-exposicion-especializada-2027

Moronta-Barrios, F., Vargas-Domínguez, S., Nuesch-Germano, M., Torres, V., Selvaggi, K., Di Prinzio, C., O'Brien, E., Hernandez, V., & Monteiro, M. (2021). Latin American Network for Scientific Culture (RedLCC): A regional science communication initiative. *Frontiers in Research Metrics and Analytics*, 6, 654022. 10.3389/frma.2021.65402233870070

Morressier. (2022). Women's participation rates are 253% higher at virtual conferences. Here's why. Morressier *Research Team*. Retrieved December 28, 2023, from https://www.morressier.com/post/womens-participation-rates-are-253-higher-at-virtual-conferences-heres-why

MRE. (2024). Innovation Diplomacy Program. Retrieved January 14, 2024, from https://www.gov.br/mre/en/subjects/science-technology-and-innovation/innovation-diplomacy-program

Murhula, P. B. B., & Singh, S. B. (2022). The Impact of COVID-19 Vaccine Nationalism on Global Health and Human Rights to Health Standards. *The Oriental Anthropologist*, 22(1), 93–101. 10.1177/0972558X221096263

Murray, S., Sharp, G., Wiseman, D., Criekemans, D., & Melissen, J. (2011). Criekmans, y J. Melissen. 2011. The Present and Future of Diplomacy and Diplomatic Studies. *International Studies Review*, 13(4), 709–728. 10.1111/j.1468-2486.2011.01079.x

Nadir, M., Siqueira Tybusch, J., & Bonesso de Araujo, L. E. (2020). A diplomacia Ambiental no Sul Global: Um olhar sobre África e América Latina. *JUSTIÇA DO DIREITO*, 34(1), 314–363. 10.5335/rjd.v34i1.11135

Nair-Bedouelle, Sh. (2023). Science diplomacy is part of UNESCO's DNA. In Nuevos mapas para la diplomacia. Ciencia, conocimiento, tecnología e Innovación. Santiago de Chile: Academia diplomática de Chile.

Neghme, A. (1962) La vida y obra de Carlos J Finlay. Edited by Centro de Investigaciones de Historia de la Medicina. Universidad de Chile. Available at https://obtienearchivo.bcn.cl/obtienearchivo?id=documentos/10221.1/55353/1/203253.pdf

Neuhäusler, P., Rammer, C., Frietsch, R., Feidenheimer, A., Stenke, G., & Kladroba, A. (2022). Neue Liste FuE-intensiver Güter und Wirtschaftszweige sowie wissensintensiver Wirtschaftszweige 2021. Studien zum deutschen Innovationssystem Nr. 13-2022. Expertenkommission Forschung und Innovation (EFI). ISSN 1613-4338. Retrieved from https://www.e-fi.de/fileadmin/Assets/Studien/2022/StuDIS_13_2022.pdf

Nieto, M. J., & Santamaría, L. (2007). The importance of diverse collaborative networks for the novelty of product innovation. *Technovation*, 27(6–7), 367–377. 10.1016/j.technovation.2006.10.001

Nilus. (2024). La pobreza no debería ser más cara. Retrieved April 4, 2024, from https://www.nilus.co/

Nolte, D. (2023). China's mask and vaccine diplomacy in Latin America: A success story? *International Politics*. Advance online publication. 10.1057/s41311-023-00525-w

Noticias Argentinas. (2024). La Embajada china en Argentina sostiene que la base en Neuquén es solo "científica". *Diario Noticias Argentinas*, edición del 1/04/2024. Disponible en https://noticiasargentinas.com/politica/la-embajada-china-en-argentina-sostiene-que-la-base-de-neuquen-es-solo--cientifica-_a660b463e2d551ce9d8e1b965

Nye, J. S. (2004). *Soft Power: The Means To Success In World Politics*. New York: PublicAffairs. p. ix, xi. ISBN 9781586482251

Nye, J. S. (2003). *La paradoja del poder norteamericano*. Editorial Taurus Madrid.

Nye, J.Jr. (2008, March). Public Diplomacy and Soft Power. *The Annals of the American Academy of Political and Social Science*, 616(1), 94–109. 10.1177/0002716207311699

OECD. (2020). Science, Technology and Innovation in the New Economy. Retrieved March 1, 2024, from https://www.oecd.org/science/inno/1918259.pdf

OECD. (2021). *OECD Science, Technology and Innovation Outlook 2021: Times of Crisis and Opportunity*. OECD Publishing., 10.1787/75f79015-

OECD. (2022). *Building agility for successful transitions.*, 10.1787/50b32331-

OECD/Eurostat. (2018). Oslo Manual 2018: Guidelines for Collecting, Reporting and Using Data on Innovation, 4th Edition. *The Measurement of Scientific, Technological and Innovation Activities.* OECD Publishing: Paris/Eurostat, Luxembourg. 10.1787/9789264304604-en

Oficina de Enlace Argentina-Unión Europea, n.d. Official Website. Available at https://www.argentina.gob.ar/ciencia/seppcti/enlace_arg-ue/programas-marco

Olšáková, D. (2023). A Review of Science Diplomacy: Theoretical Evolution to a Post-Naïve Approach and Its Relevance for the Czech Republic. *Czech Journal of International Relations.*

ONU. Yearbook of the International Law Commission (2001). Proyecto de artículos sobre prevención del daño transfronterizo resultante de actividades peligrosas, elaborado por la Comisión de Derecho Internacional y adoptado en 2001.

ONU-Agua. (2021). The Sustainable Development Goal 6 Global Acceleration Framework. Ginebra- Suiza. https://sustainabledevelopment.un.org/hlpf/2020/sdg6accelerationframework

Orsini, A. J. (2018). La Diplomatie Environnementale. In *Balzacq, Dans Thierry; Charillon, Frédéric; Ramel, Frédéric. Manuel de Diplomatie* (pp. 263–283). Presses de Sciences.

OSI. (2024). Office of Science and Innovation, Embassy of Sweden. Retrieved July 13, 2024, from https://www.swedenabroad.se/en/embassies/japan-tokyo/about-us/the-office-of-science-and-innovation/

Özkaragöz Doğan, E., Uygun, Z., & Akçomak, İ. S. (2021). Can science diplomacy address the global climate change challenge? *Environmental Policy and Governance*, 31(1), 31–45. 10.1002/eet.1911

PAHO. (2008). Res CD48/15 Health and international relations: linkages with national health development. Available at https://www3.paho.org/english/gov/cd/cd48-15-e.pdf

Pantović, B., & Benavides, S. (2023, August 29). La red de diplomacia científica en Latinoamérica y El Caribe. Retrieved March 1, 2024, from Las habilidades para la diplomacia científica en los futuros investigadores e investigadoras: Reflexiones desde Latinoamérica: https://diplomaciacientifica.org/las-habilidades-para-la-diplomacia-cientifica-en-los-futuros-investigadores-e-investigadoras/

Pantović, B., & Michelini, G. (2018). Ciencia y Cultura de la Memoria en la Diplomacia Serbia / The Science and Culture of Memory in Serbian Diplomacy. *Revista CIDOB D'Afers Internacionals*, (120), 241–258. 10.24241/rcai.2018.120.3.241

Parliament of Australia. (2024). Defining "Public Diplomacy". Retrieved January 14, 2024, from https://www.aph.gov.au/Parliamentary_Business/Committees/Senate/Foreign_Affairs_Defence_and_Trade/Completed_inquiries/2004-07/public_diplomacy/report/c02

Partners of the Americas. (2023). *The 100,000 Strong in the Americas Innovation Fund Announces New Higher Education Partnerships and Training Programs between 17 colleges and universities in the United States and Colombia – 100,000 Strong in the Americas*. https://www.100kstrongamericas.org/c4p-winners-2023/

Pascual, U., Balvanera, P., Christie, M., Baptiste, B., González-Jiménez, D., Anderson, C., Athayde, S., Barton, David., Chaplin-Kramer, R., Jacobs, S., Kelemen, E., Kumar, R., Lazos, E., Martin, A., Mwampamba, T., Nakangu, B., O´Farrel, P., Raymond, C., Subramanian, S., Termansen, M., van Noordwijk, M., & Vatn, A. (2022). Resumen para los encargados de la formulación de políticas de la evaluación metodológica relativa a la conceptualización diversa de los valores múltiples de la naturaleza y sus beneficios, incluidos la diversidad biológica y las funciones y los servicios de los ecosistemas. Plataforma Intergubernamental Científico-Normativa sobre Diversidad Biológica y Servicios de los Ecosistemas (IPBES).

Pastrana Buelvas, E., Castro, R., & González, P. A. (2016). Política exterior colombiana: Escenarios y desafíos en el posconflicto. In E. Pastrana Buelvas & H. Gehring (Eds.), *La proyección de China en América Latina y el Caribe* (Primera edición). Editorial Pontificia Universidad Javeriana.

Paz-Enrique, L. E., Núñez-Jover, J. R., & Hernández-Alfonso, E. A. (2022). Pensamiento latinoamericano en ciencia, tecnología e innovación: políticas, determinantes y prácticas. Desde el Sur, 14(1), e0008. 10.21142/DES-1401-2022-0008

PDI. (2024). Programa Diplomacia da Inovação. *Ministério das Relações Exteriores*. Retrieved July 13, 2024, from https://www.gov.br/mre/pt-br/assuntos/ciencia-tecnologia-e-inovacao/programa-de-diplomacia-da-inovacao

Penca, J. (2018). The rhetoric of "science diplomacy": Innovation for the EU's scientific cooperation? (Working Paper EL-CSID Project). https://aei.pitt.edu/102624/

Pérez Cubillos, C. M. (2022). El Conocimiento Tradicional Ecológico Indígena y su papel en el blindaje de territorios étnicos y en el ordenamiento territorial de Leticia, Amazonas – Colombia. *Espacio y Desarrollo*, 39(39), 1–35. 10.18800/espacioydesarrollo.202201.004

Perez Valerino, D. R. (2020) Health, Science and Diplomacy: Necessity in Epidemic Times – A Cuban Perspective. *Science Diplomacy Review*, Vol. 2, No. 3,November 2020.

Pérez Valerino, D. R. (2021, July 2). Vacunas y diplomacia en tiempos de pandemia. *La red de diplomacia científica en Latinoamérica y El Caribe*. Retrieved from https://diplomaciacientifica.org/vacunas-y-diplomacia-en-tiempos-de-pandemia/

Pérez, C. (2010). Technological revolutions and techno-economic paradigms. *Cambridge Journal of Economics*, 34(1), 185–202. 10.1093/cje/bep051

Perez-Perez, A., & Vallejo, J. R. (2023, June 5). The Smallpox Vaccine in Latin America: A New Approach (1801-1804). *Medicina (Kaunas, Lithuania)*, 59(6), 1093. 10.3390/medicina5906109337374296

Pfotenhauer, S., & Jasanoff, S. (2017). Panacea or diagnosis? Imaginaries of innovation and the 'MIT model' in three political cultures. *Social Studies of Science*, 47(6), 783–810. 10.1177/030631271770611028571510

Piñeros Ayala, R., Echeverría-King, L. F., & Andrade-Sastoque, E. (2022). Política exterior y de CTI en Colombia: la diplomacia científica como frontera. Desafíos, 34(Especial), 1-36.

Piñeros Ayala, R., Echeverría-King, L. F., & Andrade-Sastoque, E. (2022). Política exterior y de CTI en Colombia: la diplomacia científica como frontera. Desafíos, *34*(Especial), 1-36. https://doi.org/10.12804/revistas.urosario.edu.co/desafios/a.11881

Piñeros Ayala, R., Echeverría King, L., & Pantovic, B. (2023). La diplomacia científica frente a los retos globales: Acciones para mitigar el cambio climático en la región de América Latina. In *M. Lascurain Fernández, & A. M. Chípul Castillo, La nueva agenda de la globalización* (pp. 257–277). Universidad de Xalapa.

Piñeros Ayala, R., Echeverría-King, L. F., & Pantović, B. (2024). La diplomacia científica frente a los retos globales: Acciones para mitigar el cambio climático en la región de América Latina. In *La Nueva Agenda de la globalización* (pp. 257–277). Universidad de Xalapa.

PNUD - GEF. (2020). *Análisis de Diagnóstico Transfronterizo y Programa de Acción Estratégica de las Cuencas y Acuíferos Transfronterizos de Puyango - Tumbes, Catamayo - Chira y Zarumilla*. Primera Edición.

Pohl, B. (2014) The Rise of Hydro-Diplomacy - Strengthening foreign policy for transboundary Waters. Climate Diplomacy, Adelphi, Federal Foreign Office. Berlin - Alemania. Disponible en https://climate-diplomacy.org/sites/default/files/2020-10/the_rise_of_hydro-diplomacy_adelphi.pdf

Pohl, B. (2014). Hydro - Diplomacy Can Build Peace Over Shared Waters, But Needs More Support. Wilson Center. Disponible en:https://www.newsecuritybeat.org/2014/09/hydro-diplomacy-builds-peace-shared-waters-support/

Pohl, N., & Luco, N. (2023). Encontrarnos desde el conocimiento: comunicación para la diplomacia en ciencia. In Nuevos mapas para la diplomacia. Ciencia, conocimiento, tecnología e Innovación. Santiago de Chile: Academia diplomática de Chile.

Polejack, A. (2023). Coloniality in science diplomacy—Evidence from the Atlantic Ocean. *Science & Public Policy*, 50(4), 759–770. 10.1093/scipol/scad027

Polejack, A., & Coelho, L. F. (2021). Ocean Science Diplomacy can Be a Game Changer to Promote the Access to Marine Technology in Latin America and the Caribbean. *Frontiers in Research Metrics and Analytics*, 6, 637127. 10.3389/frma.2021.63712733912786

Powell, J. J. W., Fernandez, F., Crist, J. T., Dusdal, J., Zhang, L., & Baker, D. P. (2017). Introduction: The Worldwide Triumph of the Research University and Globalizing Science. In Powell, J. J. W., Baker, D. P., & Fernandez, F. (Eds.), *International Perspectives on Education and Society* (Vol. 33, pp. 1–36). Emerald Publishing Limited.

Pradhan, S. K. (2020). *India's quest for energy through oil and natural gas: Trade and investment, geopolitics, and security*. Springer Singapore., 10.1007/978-981-15-5220-5

Presidencia de la República de Colombia. (2023, September 1). Entre-Vistas con Alma de País hoy: Jorge Luis Ceballos, Glaciólogo IDEAM. Retrieved from https://youtu.be/7KcIBSKgoF4?si=Kvk_nulbw2acN_Bw

Prieto, J., Cumba García, L. M., Echeverría-King, L. F., Alarcón López, C., & Benavides, S. (2024). Recommendations for a science diplomacy strategy for the safeguarding of the Amazon. *Inter-American Institute for Global Change Research & DiploCientífica*. https://zenodo.org/doi/10.5281/zenodo.10695010

Prieto, J., Cumba García, L., Echeverría-King, L., Benavides, S., & Alarcón, C. (2024). Recomendaciones para una estrategia de Diplomacia Científica para la salvaguarda de la Amazonía. DOI:)10.5281/zenodo.10695010

Prieto, J., & Scott, C. A. (2022). Scientific diasporas and the advancement of science diplomacy: The InFEWS US-China program in the face of confrontational "America First" diplomacy. *Frontiers in Research Metrics and Analytics*, 7, 944333. 10.3389/frma.2022.94433336277735

Programa de las Naciones Unidas para el desarrollo (2014). Gestión Integrada de Recursos Hídricos de las Cuecas Transfronterizas y Acuíferos de Puyango - Tumbes, Catamayo - Chira y Zarumilla.

Pugin, M. (2023). Indigenous Australian diplomacy and the United Nations declaration on the rights of Indigenous peoples. *Australian Journal of International Affairs*, 77(6), 625–631. 10.1080/10357718.2023.2268022

Pulido-Salgado, M., & Castaneda Mena, F. A. (2021). Bringing policymakers to science through communication: A perspective from Latin America. *Frontiers in Research Metrics and Analytics*, 6, 654191. 10.3389/frma.2021.65419133981947

Qadri, R. (2021). Global South startups can do better than Silicon Valley. *Reporting Global Tech Stories*. Retrieved April 2, 2024, from https://restofworld.org/2021/grab-indonesia-silicon-valley/

Rachid, A., Moussaoui, M., & Benbba, B. (2023). Rethinking Natural Resource Development: A Sustainable and Equitable Approach to North-South Relations. In Moussaoui, M. (Ed.), *Global Science's Cooperation Opportunities, Challenges, and Good Practices. First* (pp. 50–85). IGI Global. 10.4018/978-1-6684-7874-5.ch003

Ramos, G. (2023). UNESCO's Recommendation on the Ethics of Artificial Intelligence as a reflection of global cooperation in diplomacy, science, technology and policy. In Nuevos mapas para la diplomacia. Ciencia, conocimiento, tecnología e Innovación. Santiago de Chile: Academia diplomática de Chile.

Riaz, M. M. A., Unaiza, A., Anmol, M., dos Santos-Costa, A. C., Khan, H., Maryam Salma Babar, M. S., Hasan, M. M., & Essar, M. Y. (2021). Global impact of vaccine nationalism during COVID-19 pandemic. *Tropical Medicine and Health*, 49(1), 101. 10.1186/s41182-021-00394-034963494

RIS. (2020). Southern Perspectives on Science Diplomacy. Nueva Delhi: Research and Information System (RIS) for Developing Countries. Retrieved from https://cutt.ly/bxDkZq0

Risnock, A. (2016) Historical context and the roots of Jenner's discovery. *Hum Vaccin Immunother*. Aug; 12(8): 2025–2028.

Rispoli, G., & Olšáková, D. (2020). Science and Diplomacy around the Earth. *Historical Studies in the Natural Sciences*, 50(4), 456–481. 10.1525/hsns.2020.50.4.456

Robinson, D. F., Raven, M., & Lumsden, S. (2024). Epistemic injustice: Intellectual property, biodiversity and traditional knowledge. In Amani, B., Ncube, C. B., & Rimmer, M. (Eds.), *The Elgar Companion to Intellectual Property and the Sustainable Development Goals* (pp. 458–480). Edward Elgar Publishing Limited. 10.4337/9781803925233.00027

Robinson, S. (2020). Early Twentieth-Century Ocean Science Diplomacy. *Historical Studies in the Natural Sciences*, 50(4), 384–410. 10.1525/hsns.2020.50.4.384

Rodríguez Cruz, L. A. (2020, October 3). PR-SPAN's science rising initiative to bring visibility and awareness in Puerto Rico's election cycle. *Cienciapr*. Retrieved from https://www.cienciapr.org/en/blogs/pr-span/pr-spans-science-rising-initiative-bring-visibility-and-awareness-puerto-ricos

Roig, A. (2020). *Diplomacia científica y tecnológica*. Secretaría de Relaciones Exteriores de México.

Rojas-Ochoa F., et al. (2011) Vacunas. Cuba. 1959-2008. Editorial Ciencias Médicas (in Spanish)

Rombini, M.F., Mauas, R.P., & Urueña, A. (2019) Ranking de los programas de inmunización en América Latina. *Rev Panam Salud Publica.* 2022;;46:1. Article in spanish

Rotolo, D., Hicks, D., & Martin, B. R. (2015). What is an emerging technology? *Research Policy*, 44(10), 1827–1843. 10.1016/j.respol.2015.06.006

Royal Society & AAAS. (2010). New frontiers in science diplomacy. London, England: Royal Society. Retrieved from https://royalsociety.org/topics-policy/publications/2010/new-frontiers-science-diplomacy

Royal Society of Great Britain. (2010). *The scientific century : securing our future prosperity*. Royal Society.

Roy, J. (Ed.). (2013). *Después de Santiago: Integración regional y relaciones Unión Europea –América Latina*. Miami-Florida European Union Center of Excellence.

Ruckert, A., Almeida, C., Ramírez, J., Guerra, G., Salgado de Snyder, V. N., Orozco, E., Alvarenga, A. A., Nassim de Saboya, A., Martins dos Santos, R., Sepúlveda, D., Rivera Vivian, E. C., Santos, M. D. S., Doriani, K., & Labonté, R. (2022). Global Health Diplomacy (GHD) and the integration of health into foreign policy: Towards a conceptual approach. *Global Public Health: An International Journal for Research, Policy and Practice*, 17(6), 1041–1054. 10.1080/17441692.2021.190031833736572

Ruffini, P. (2020). Collaboration and competition: The twofold logic of science diplomacy. *The Hague Journal of Diplomacy*, 15(3), 371–382. 10.1163/1871191X-BJA10028

Ruffini, P. B. (2017). *Science and Diplomacy. A New Dimension of International Relations*. Springer.

Ruffini, P. B. (2020). Conceptualizing science diplomacy in the practitioner-driven literature: A critical review. *Humanities & Social Sciences Communications*, 7(1), 124. 10.1057/s41599-020-00609-5

Ruffini, P.-B. (2018). The Intergovernmental Panel on Climate Change and the Science-Diplomacy Nexus. *Global Policy*, 9(S3), 73–77. 10.1111/1758-5899.12588

Ruffini, P.-B., & Kransyak, O. (2023). Science diplomacy from a nation-state's perspective: A general framing and its application to Global South countries. *Science & Public Policy*, 50(4), 771–781. 10.1093/scipol/scad023

Rüland, A.-L., Rüffin, N., Cramer, K., Ngabonziza, P., Saxena, M., & Skupien, S. (2023). Science diplomacy from the Global South: The case of intergovernmental science organizations. *Science & Public Policy*, 50(4), 782–793. 10.1093/scipol/scad024

Rungius, C., & Flink, T. (2020). Romancing science diplomacy for global solutions: On narratives and interpretative schemas of science diplomacy. *Humanities & Social Sciences Communications*, 1–10.

Rusieshvili, A. (2020). Role of Science Diplomacy in Russian-Georgian Relations. In Research and Information System for Developing Countries (Coord.), Southern Perspectives on Science Diplomacy. Nueva Delhi: RIS.

Sandroni, L., Ramos Quispe, L., House-Peters, L., Alonso-Yanez, G., Carabajal, M. I., Valentine, M., Schweizer, S., Murisa, M., Roy, N., & de Vernal, A. (2023). Transdisciplinarity 101: Short-term training in knowledge co-production to face global environmental change. *Journal of Science Policy & Governance*. 10.38126/JSPG220207

Sanmartin, A. M., & Bortz, G. (2023) Between regional convergence, and locally rooted technological and cultural aspects: lessons from the case of COVID-19 vaccines in Latin America. UN Report available at https://sdgs.un.org/sites/default/files/2023-05/B8%20-%20Cecilia%20Sanmartin%20-%20Lessons%20from%20the%20COVID-19%20pandemic%20in%20Latin%20America.pdf

São Paulo, D. W. I. H. (2024. March 3). Mission. Bavarian Academic Center for Latin America (BAYLAT). Retrieved March 3, 2024, from https://www.dwih-saopaulo.org/en/dwih-sao-paulo/mission/

Scartozzi, C. (2018). A new taxonomy for international relations: Rethinking the international system as a complex adaptative system. Journal of Policy and Complex Systems, 4(1), 109-132.

Schmeier, S. (2018). What is water diplomacy and why should you care? Global Water Forum. Disponible en https://www.globalwaterforum.org/2018/08/31/what-is-water-diplomacy-and-why-should-you-care/

Science Panel for the Amazon. (2021). Connecting and sharing diverse knowledge systems to support sustainable pathways in the Amazon. In Amazon Assessment Report 2021 (First). Science Panel for the Amazon. Retrieved from https://www.theamazonwewant.org/wp-content/uploads/2022/05/Chapter-33-Bound-May-16.pdf

SENACYT. (2019). Lineamientos de la estrategia de diplomacia científica. Panama City: *Ministerio de Relaciones Exteriores de Panamá y la Secretaría Nacional de Ciencia, Tecnología e Innovación.* Retrieved from https://www.senacyt.gob.pa/publicaciones/?p=21846

Servicio Nacional de Aduanas de Chile. (2023). *Reporte De Comercio Exterior 2023.* Enero – Julio. Available at:https://www.aduana.cl/aduana/site/docs/20181217/20181217125337/reporte_comex_julio_final.pdf

Sharma, J., Pérez-Valerino, D. R., Widmaier, C. N., Lima, R., Gupta, N., & Varshney, S. K. (2022). Science Diplomacy and COVID-19: Future Perspectives for South–South Cooperation, *Global Policy*, London School of Economics and Political Science, vol. 13(2), pages 294-299

Sheridan, M. B. (2022) How Cuba became a pioneer in covid-19 vaccines for kids. The Washington Post. Available at https://www.washingtonpost.com/world/2022/06/18/cuba-coronavirus-vaccine-abdala-soberana/

Shortwave Podcast. (2020, April 7). Puerto Ricans are at risk from the coronavirus and a lack of information. *NPR*. Retrieved from https://www.npr.org/2020/04/06/828439886/puerto-ricans-are-at-risk-from-the-coronavirus-and-a-lack-of-information

SIN. (2024). UK Science and Innovation Network. Retrieved July 13, 2024, from <https://www.gov.uk/world/organisations/uk-science-and-innovation-network>

Singer, J. D. (1971). A general systems taxonomy for political science. General Learning.

SIRIUS. (2024). SIRIUS: Accelerating The Future Of Brazilian Science. Retrieved March 22, 2024, from https://lnls.cnpem.br/sirius-en/

Skiles, M., Yang, E., Reshef, O., Muñoz, D. R., Cintron, D., Lind, M. L., & Rush, A. (2022). *'Conference demographics and footprint changed by virtual platforms'. Nature Sustainability*. Nature Research. 10.1038/s41893-021-00823-2

Skipper, M. (2021). Science lies at the heart of solutions to important problems. In Hazelton, A. (Ed.), Why is scientific collaboration key? 4 experts explain. World Economic Forum. Retrieved from https://www.weforum.org/agenda/2021/06/4-views-on-why-scientific-collaboration-is-key-for-the-future/

Smith, F. L.III. (2014). Advancing science diplomacy: Indonesia and the US Naval Medical Research Unit. *Social Studies of Science*, 44(6), 825–847. 10.1177/030631271453586425608440

Snidjers, D. (2023). Reassessing STI Diplomacy in the age of Techno-politics. *Science & Diplomacy*, 7(2), 16–18.

Soler, M. G. (2020). *Diplomacia científica en América Latina y el Caribe*. United Nations Educational, Scientific and Cultural Organization.

Soler, M. G. (2021a). *Diplomacia científica en américa latina y el caribe*. UNESCO. Retrieved from https://forocilac.org/wp-content/uploads/2020/11/PolicyPapers-DiplomaciaCientifica-ES.pdf

Solimano, G., & Valdivia, L. (2015). Global Health in Latin-American academic institutions: towards a distinctive development and identity. *Dossie Saude Global • Saude* soc. 23 (2) • https://doi.org/10.1590/S0104-12902014000200001

Soto, G. H. (2024). The effects of foreign direct investment on environmentally related technologies in Latin America. *Resources Policy*, 90, 104711. 10.1016/j.resourpol.2024.104711

Spector, J. M., & Ma, S. (2019). Inquiry and critical thinking skills for the next generation: From artificial intelligence back to human intelligence. *Smart Learning Environments*, 6(1), 1–11. 10.1186/s40561-019-0088-z

Srinivasan, R. (2019). Opinion: The Global South is redefining Tech Innovation. *Wired*. Retrieved April 2, 2024, from https://www.wired.com/story/opinion-the-global-south-is-redefining-tech-innovation/

Stake, I. E. (1998). Investigación con estudios de Caso. Veracrúz. Universidad Veracruzana. Retrieved from https://www.uv.mx/rmipe/files/2017/02/Investigacion-con-estudios-de-caso.pdf

Star, S. L. (2010). This is not a boundary object: Reflections on the origin of a concept. *Science, Technology & Human Values*, 35(5), 601–617. 10.1177/0162243910377624

Startup Blink. (2024, February 24). Startup Ecosystem of Berlin. Retrieved from https://www.startupblink.com/startup-ecosystem/berlin-de

State Department. (n.d.). *100,000 Strong in the Americas*. United States Department of State. Retrieved March 12, 2024, from https://www.state.gov/100000-strong-in-the-americas/

Straubhaar, J. (2007). *World Television: From Global to Local*. SAGE Publications, Inc. 10.4135/9781452204147

Subramanian, S., & Pisupati, B. (Eds.). (2013). *Traditional Knowledge in Policy and Practice: Approaches to Development and Human Well-Being (First)*. United Nations Press., 10.18356/0e3bca6c-en

Suhay, E., Cloyd, E., Heath, E., & Nash, E. (2019). Recommended practices for science communication with policymakers. Retrieved from https://www.american.edu/spa/scicomm/upload/recommended-practices-booklet_v17-digital.pdf

Sustainability Times. (2023). Global South needs locally driven innovation to turn tide on UN's SDG agenda. *Sustainability Times*. Retrieved April 2, 2024, from https://www.sustainability-times.com/in-depth/global-south-needs-locally-driven-innovation-to-turn-tide-on-uns-sdg-agenda/

Tamayo Ortiz, C. V., & Dilas Jiménez, J. O. (2021). Conocimientos tradicionales y recursos genéticos: Una revisión conceptual, importancia y marco legal en Ecuador y Perú. *Revista de Investigación Científica y Tecnológica Alpha Centauri*, 2(3), 2–14. 10.47422/ac.v2i3.37

Tassara, C. (2019). La cooperación internacional de la Unión Europea en Colombia. Buenas prácticas y lecciones aprendidas. In E. Tremolada Álvarez (Ed.), *Colombia y la Unión Europea: Una asociación cada vez más estrecha* (Primera edición). Universidad Externado de Colombia.

Taubman, A. (2007). Cereal offenders: Access and equity in trade negotiations on knowledge resources. In J. P. Kesan (Ed.), Agricultural Biotechnology and Intellectual Property: Seeds of Change (First, pp. 97–131). CABI.

Tengö, M., Brondizio, E. S., Elmqvist, T., Malmer, P., & Spierenburg, M. (2014). Connecting diverse knowledge systems for enhanced ecosystem governance: The multiple evidence base approach. *Ambio*, 43(5), 579–591. 10.1007/s13280-014-0501-324659474

Termansen, M., Jacobs, S., Mwampamba, T. H., Ahn, S., Castro, A., Dendoncker, N., Ghazi, H., Gundimeda, H., Huambachano, M., Lee, H., Mukherjee, N., Nemogá, G. R., Palomo, I., Pandit, R., Schaafsma, M., Ngouhouo, J., Choi, A., Filyushkina, A., Hernández-Blanco, M., . . . González-Jiménez, D. (2022). Chapter 3: The potential of valuation. En: Balvanera, P., Pascual, U., Christie, M., Baptiste, B., and González-Jiménez D. (eds). *Methodological Assessment Report on the Diverse Values and Valuation of Nature of the Intergovernmental Science-Policy Platform on Biodiversity and Ecosystem Services*. IPBES secretariat, Bonn, Germany. DOI: 10.5281/zenodo.6521298

Thales. (2024). What is Human Factors? Retrieved April 3, 2024, from <https://www.thalesgroup.com/en/human-factors-design#:~:text=Human%20Factors%20is%20about%20designing,things%20better%20and%20more%20effective>

The Federal Ministry of Education and Research (BMBF). (2023. March 2.) R&D policy framework. Retrieved from https://www.research-in-germany.org/en/research-landscape/why-germany/r-d-policy-framework.html

The Royal Society, & AAAS. (2010). *New frontiers in science diplomacy. Navigating the changing balance of power*. The Royal Society.

The Royal Society. (2010). *New frontiers in science diplomacy*. Retrieved from https://royalsociety.org/-/media/policy/publications/2010/4294969468.pdf

The Royal Society. (2010). *New frontiers in science diplomacy*. Retrieved July 14, 2024, from https://www.aaas.org/sites/default/files/New_Frontiers.pdf

The Royal Society. (2010). *New frontiers in science diplomacy*. The Royal Society.

The Royal Society-AAAS. (2010). New frontiers in science diplomacy. Royal Society.org. https://royalsociety.org/~/media/Royal_Society_Content/policy/publications/2010/4294969468.pdf

Tremolada, E., & Quiroga Chacón, G. A. (2022). La Diplomacia Pública de la Unión Europea en Colombia: ¿Contribuye a una mejora en su percepción? *Revista Internacional de Cooperación y Desarrollo*, 9(2), 81–89. 10.21500/23825014.6071

Turchetti, S. (2020). The (Science Diplomacy) Origins of the Cold War. *Historical Studies in the Natural Sciences, 50*(4), 411–432. Union Europea. (n.d.). *La Unión Europea y Colombia | EEAS*. Retrieved March 15, 2024, from https://www.eeas.europa.eu/colombia/la-union-europea-y-colombia_es?s=160

Turchetti, S., & Lalli, R. (2020). Envisioning a "science diplomacy 2.0": On data, global challenges, and multi-layered networks. *Humanities & Social Sciences Communications*, 7(1), 144. Advance online publication. 10.1057/s41599-020-00636-2

U.S. Department of State. (2024). *Technology and the Transformation of U.S. Foreign Policy* [Government website]. United States Department of State. https://www.state.gov/technology-and-the-transformation-of-u-s-foreign-policy/

UNESCO (2021). Avances en la cooperación en materia de aguas transfronterizas. Situación mundial del indicador 6.5.2 de los ODS y necesidades de aceleración.

UNESCO Science Report: Towards 2030 (2020). Latin America's Place in the World. Available online at: https://en.unesco.org/sites/default/files/latin_americas_place_in_the_world_usr15.pdf

UNESCO Science Report: Towards 2030. (2015). United Nations Scientific, Cultural and Education Organization.

UNESCO. (2016). How science can help to create a sustainable world. News. Retrieved from https://en.unesco.org/news/how-science-can-help-create-sustainable-world

United Nation Office for the Coordination of Humanitarian Affairs (OCHA). (2020) Desastres Naturales en America Latina y el Caribe 2000-2019. Report in Spanish Available at https://reliefweb.int/report/world/desastres-naturales-en-am-rica-latina-y-el-caribe-2000-2019

United Nations Children's Fund. (2018). Immunization Regional Analytics. Latin America and the Caribbean. Available at https://www.unicef.org/timorleste/media/3251/file/Immunization-regional-snapshots-LAC-2020.pdf.pdf

United Nations Economic Commission for Latin America and the Caribbean. (2021). *Emerging technologies and sustainability in Latin America and the Caribbean.* Retrieved from https://www.cepal.org

United Nations General Assembly. (2008) Resolution A/63/33 Global Health and Foreign Policy. Available at http://www.worldlii.org/int/other/UNGA/2008/71.pdf

United Nations Industrial Development Organization (UNIDO). (2021, October 8). Germany and UNIDO strengthen cooperation in Latin America and the Caribbean and boost trade capacity-building worldwide. Retrieved from https://www.unido.org/news/germany-and-unido-strengthen-cooperation-latin-america-and-caribbean-and-boost-trade-capacity-building-worldwide

United Nations. (2022). Science, technologies can transform global challenges, but need to be accessible to all, senior officials stress, as Economic and Social Council forum concludes. Economic and Social Council Plenary Multi-Stakeholder Forum, Am & Pm Meetings ECOSOC/7083. Meetings Coverage and Press Releases. Retrieved from https://www.un.org/press/en/2022/ecosoc7083.doc.htm

Universidad de Luxemburgo. (2023). Conference Indigenous Diplomacy in the Contemporary World: Indigenous Knowledge as a Pivotal Asset for the Achievement of an Effective Global Sustainable Development. https://www.uni.lu/wp-content/uploads/sites 2/2023/11/20231114_AFF_DSOC_Dyplomacy_VF_WEB.pdf

Us, H., & Martin, C. (2021, 2 de agosto). *Pueblos indígenas, ciencia y tecnología. BID Blog.* https://blogs.iadb.org/igualdad/es/los-pueblos-indigenas-la-ciencia-y-tecnologia/

Valderrama, B., Islas, P., Tibaduiza, Y., & Gual-Soler, M. (2021) Vaccine Diplomacy and Political Diversification in Mexico's Science and Technology Stakeholders. Available at https://www.sciencediplomacy.org/article/2021/vaccine-diplomacy-and-political-diversification-in-mexicos-science-and-technology

Välimaa, J. (2019). The emergence of universities in the Middle Ages. In *A history of Finnish higher education from the Middle Ages to the 21st century* (pp. 11–38). Springer., https://link.springer.com/chapter/10.1007/978-3-030-20808-0_210.1007/978-3-030-20808-0_2

van den Hove, S. (2007). A rationale for science–policy interfaces. *Futures*, 39(7), 807–826. 10.1016/j.futures.2006.12.004

Van der Berg, A., Latorre, P., Diaz, R., Vargas, N., Leiva, S., & Olmos, F. (2023). Internacionalización en universidades chilenas: un proceso en construcción. In Nuevos mapas para la diplomacia. Ciencia, conocimiento, tecnología e Innovación. Santiago de Chile: Academia diplomática de Chile.

Van Langenhove, L. (2016). Multilateral EU Action through Science Diplomacy. In Ujvari, B. (Ed.), *The EU Global Strategy: going beyond effective multilateralism?* European Policy Centre.

Vera, N. (2018). *Cooperación nuclear sino-argentina en el nuevo rol de China como potencia emergente del siglo XXI. ¿Lógica de cooperación sur – sur o norte – sur? (2003 - 2015*). XIII Congreso Nacional y VI Congreso Internacional sobre Democracia. Rosario, 10 a 13 de septiembre de 2018. Available at https://es.scribd.com/document/394877936/7-Memorias -Politica-Internacional-Ok

Vera, N., & Echeverría King, L. (2022) Hacia una diplomacia científica para América Latina. En Oregioni (comp.) Derribando estructuras, construyendo puentes, tejiendo Futuro. Aportes para pensar la internacionalización desde una perspectiva situada en el Sur. Editorial-Cooperativa El Zócalo, Buenos Aires, Argentina. https://ceipil.fch.unicen.edu.ar/wp-content/uploads/2023/02/Libro-PICT-final.pdf

Vera, N., & Echeverría King, L. (2022) *Hacia una diplomacia científica para América Latina.* En Oregioni, S. (comp.) Derribando estructuras, construyendo puentes, tejiendo Futuro. Aportes para pensar la internacionalización desde una perspectiva situada en el Sur. Editorial-Cooperativa El Zócalo, Buenos Aires, Argentina, pp. 213-236.

Vera, N., & Guglielminotti, C. (2018). Potencialidades de la agenda de cooperación sur-sur en tecnología espacial: Entre las proyecciones y las posibilidades para la Argentina (2003-2015). *Saber y Tiempo*, 77-94. https://ri.conicet.gov.ar/bitstream/handle/11336/112594/CONICET_Digital_Nro.48-53-PB%20%281%29.pdf?sequence=5&isAllowed=y

Vera, N., Salvati, A., de Angelis, I., & Pantović, B. (2023). Diplomacia Científica Transregional y prioridades en la agenda de cooperación científico-tecnológica de la Comunidad de Estados Latinoamericanos y caribeños (2013 - 2023). Ciencia, tecnología y cooperación internacional en clave de desarrollo: reflexiones sobre Argentina y América Latina, de María Paz López. Tandil: Centro de Estudios Interdisciplinarios en Problemáticas Internacionales y Locales (CEIPIL).

Vera, N. (2021). *Ciencia, tecnología y Relaciones Internacionales: visibilizar la simbiosis. Desarrollo y políticas de ciencia, tecnología e innovación en un mundo en transformación. Reflexiones sobre la Argentina contemporánea.* CEIPIL-UNICEN.

Vera, N., & Colombo, S. (2020). Reflexiones en torno a la Cooperación Internacional en Ciencia y Tecnología y la Diplomacia Científica: los casos de Estados Unidos y la República Popular China. In *López, M. P. (comp.). Perspectivas sobre la cooperación internacional en ciencia, tecnología y universidad: políticas, prácticas y dinámicas a principios del siglo XXI* (pp. 13–38). UNICEN.

Vera, N., & Echeverría King, L. (2022). Hacia una diplomacia científica para América Latina. In Oregioni, S. (Ed.), *Derribando estructuras, construyendo puentes, tejiendo futuro: Aportes para pensar la internacionalización desde una perspectiva situada en el Sur* (pp. 213–236). Editorial-Cooperativa El Zócalo.

Vera, N., & López, M. P. (2023). Diplomacia Científica en la Argentina Contemporánea: Un mapeo de iniciativas en los ministerios de Ciencia, Tecnología e Innovación y de Relaciones Exteriores (2019-2022). *Estudios Internacionales (Santiago)*, 55(204), 153–185. 10.5354/0719-3769.2023.67792

Vera, N., Salvati, A., De Ángelis, I., & Pantović, B. (2023). Diplomacia Científica Transregional y prioridades en la agenda de cooperación científico-tecnológica de la Comunidad de Estados Latinoamericanos y Caribeños (2013-2023). In *López, M. P. (comp.). Ciencia, tecnología y cooperación internacional en clave de desarrollo. Reflexiones sobre Argentina y América Latina* (pp. 73–124). CEIPIL – UNICEN.

Vessuri, H. (1994). La ciencia académica en América Latina en el siglo XX. *Rev. Estudios Soc. Ciencia*, 1(2), 41–76. https://ridaa.unq.edu.ar/handle/20.500.11807/304

Vessuri, H. (2013). El nuevo mantra de la diplomacia científica internacional: ¿Co-diseño de conocimiento? ¿Investigación integrativa? *Universitas Humanistica*, 76, 25–50. https://revistas.javeriana.edu.co/index.php/univhumanistica/article/view/5905

Vickers, R. (2004). The New Public Diplomacy: Britain and Canada Compared. *British Journal of Politics and International Relations*, 6(2), 182–194. 10.1111/j.1467-856X.2004.00133.x

Vila-Viñas, D., & Crespo, J. M. (2015). Saberes y conocimientos ancestrales, tradicionales y populares. In Buen Conocer / FLOK Society: Modelos sostenibles y políticas públicas para una economía social del conocimiento común y abierto en Ecuador (Digital, pp. 551–616). Asociación aLabs. https://book.floksociety.org/ec/3/3-2-comunidades-saberes-y-conocimientos-ancestrales-tradicionales-y-populares/

Virchez, J., Zepeda, R., & Marquina-Márquez, A. (2017). Gobernanza y Paradiplomacia en el Ártico canadiense: El caso del Consejo Circumpolar Inuit. *Revista Internacionales*, 3(6 (julio-diciembre)), 14–37.

Wagner, C. S. (2002). The elusive partnership: Science and foreign policy. *Science & Public Policy*, 29(6), 409–417. 10.3152/147154302781780741

Wagner, C., & Simon, D. (2023). China's use of formal science and technology agreements as a tool of diplomacy. *Science & Public Policy*, 50(4), 807–817. 10.1093/scipol/scad022

Water, U. N. (2024). *The United Nations world water development report 2024: Water for prosperity and peace*. UNESCO.

Weiss, C. (2015). How do science and technology affect international affairs? Minerva, 411-430.

Whitesides, G. (2012). *Science and American Foreign Relations since World War II*. Cambridge University Press.

WHO. (1978) Declaration of Alma Ata. Available at https://www.who.int/docs/default-source/documents/almaata-declaration-en.pdf

WHO. (2020a) Report of the WHO-China Joint Mission on Coronavirus Disease 2019 (COVID-19). Available at https://www.who.int/publications/i/item/report-of-the-who-china-joint-mission-on-coronavirus-disease-2019-(covid-19)

WHO. (2020b) Report. Origin of the SARS-Cov-2 Available at https://iris.who.int/bitstream/handle/10665/332197/WHO-2019-nCoV-FAQ-Virus_origin-2020.1-eng.pdf

WHO. (2020c) Timeline COVID-19. Available at https://www.who.int/news/item/27-04-2020-who-timeline---covid-19

Widmaier, C., Echeverría King, L., & Aquino, K. (2020). La diplomacia científica y la internacionalización solidaria: Catalizadores en tiempos de pandemia para las universidades. In G. Mutti (Coord.), *El tiempo que vivimos: COVID-19 y su impacto en nuestras sociedades* (pp. 83-89). UNR Editora. https://cutt.ly/ghO7uRV

Wikipedia contributors. (2024, March 3). Deutsches Wissenschafts- und Innovationshaus New York. Wikipedia. Retrieved from https://en.wikipedia.org/wiki/Deutsches_Wissenschafts-_und_Innovationshaus_New_York

Wilder, M. (2019). *U. S. Mexico Hydrodiplomacy: Foundations*. Change, and Future Challenges.

Wissenschaftsstandort Berlin. (2024, February 23). Retrieved from https://digital.zlb.de. https://www.startupblink.com/startup-ecosystem/berlin-de

Wolf, A. (2005). Gestionando conflictos por el agua y cooperación. Centro de Investigación para la Paz, 2005. Barcelona - España.

World Bank Report (2023) Building resilience against disasters in Latin America and the Caribbean. Available at https://www.worldbank.org/en/results/2023/11/19/building-resilience-against-disasters-in-latin-america-and-the-caribbean

World Intellectual Property Organization (WIPO). (2021). World Intellectual Property Indicators report: Worldwide trademark filing soars in 2020 despite global pandemic (PR/2021/883). Retrieved from https://www.wipo.int/pressroom/en/articles/2021/article_0011.html

World Intellectual Property Organization. (2023). Global Innovation Index 2023. Retrieved from https://www.wipo.int/edocs/pubdocs/en/wipo-pub-2000-2023/de.pdf

Yin, R. K. (2018). *Case study research and applications: Design and methods* (6th ed.). SAGE.

Yiu, E., Kosoy, E., & Gopnik, A. (2023). Imitation versus Innovation: What children can do that large language and language-and-vision models cannot (yet)? *arXiv preprint arXiv:2305.07666*. 10.31234/osf.io/kt9es

Zárate, M. S., Sierra, D., & Goldflam, M. (2022). Orígenes y desarrollo de una política científica nacional en Chile: Conicyt, 1967-1981. *Historia, Ciencias, Saude—Manguinhos*, 29(4), 953–972. 10.1590/s0104-5970202200040000536542032

Zhou, Y. R. (2022) Vaccine nationalism: contested relationships between COVID-19 and globalization. GLOBALIZATION, VOL. 19, NO. 3, 450–465https://<ALIGNMENT.qj></ALIGNMENT>10.1080/14747731.2021.196320

Zydney, J. M., & Warner, Z. (2016). Mobile apps for science learning: Review of research. *Computers & Education*, 94, 1–17. 10.1016/j.compedu.2015.11.001

About the Contributors

Luisa F. Echeverría-King holds a PhD in Education from Universidad de Murcia and is researcher and consultant on issues related to international higher education, science diplomacy and the development of science, technology and innovation capacities in Latin America. In 2020, takes a Course on Science Diplomacy in India; she is also a fellow from the Warsaw School of Science Diplomacy and the Brazilian School of Science and Innovation Diplomacy. Fellow of the AAAS and TWAS for a course on Science Diplomacy with a perspective on Southern countries and DAAD-fellow for a course on Management of Internationalization. Executive Director of Diplocientifica. Funding member of the Colombian National Chapter of the Organization for Women in Science in the Developing World from UNESCO (OWSD). Member of the Liaison Committee for the Regional Focal Point for Latin America and the Caribbean of the International Science Council Member of the Global Young Academy since 2023.

Branislav Pantović was born and raised in Serbia, with an interdisciplinary academic background from Argentina. He holds a PhD in international relations and works as an independent researcher, specialized in public diplomacy and international academic affairs. As a trilingual professional, has worked on three continents in different contexts. He is currently serving as a Consultant in the International Relations Department of the National University of Río Negro (Argentina).

Rafael Enrique Pineros-Ayala is Professor of Foreign Policy and International Relations Theory at the Universidad Externado de Colombia, in Bogotá. Member of the International Studies Association (ISA) in the committee of Foreign Policy and Latin America. Interested in relations between Latin American countries and the United States in foreign policy issues.

Pedro Figueroa is President and Co-founder of DiploCientifica, the Science Diplomacy Network for Latin-America and the Caribbean. Multilingual professional with experience as policy counselor and project manager in international partnerships. He is an expert engaged in long-standing endeavours for the promotion of Science Diplomacy. His career focused to the intersection of scientific cooperation, international affairs, policy advice and global challenges. Pedro was head of the International Relations Unit at CONICYT – The Chilean Agency for Research and Development. He studied Political Science and International Relations.

Alejandra Fabiola Flores-Zamora has a Master's Degree in International Higher Education from the University of Guadalajara and Bachelor's Degree in Educational Processes from the Benemérita Universidad Autónoma de Puebla. Interested in the internationalization of Higher Education and the processes developed by different institutions in this field, such as the internationalization of the curriculum and the teacher. With experience in the areas of education, curriculum design and educational evaluation. Currently working as a pedagogical advisor at the Center for Innovation and Teaching at CUCEA-UdeG, where her main tasks are related to research on international educational trends, improvement of the teaching and learning process, and teacher training on active learning, flipped classroom and the transition to a learning-centered approach, to mention a few.

* * *

Hannia Bridg is a dynamic industrial scientist with a combination of studies in biology, biotechnology, genetics, and business management. Her scientific career and professional qualifications enable her to effectively bridge the gap between science and business leadership, seamlessly integrating both fields. Hannia's professional experience in the German life sciences industry seamlessly integrates science and business, enabling her to translate groundbreaking scientific discoveries into consumer-ready products, combining activities that drive innovation and economic growth while respecting the relationship with nature. As founder of Bridg-Networking Consulting, Dr Bridg has expertise in strategic direction, effective communication, and efficient implementation. She specializes in translating complex scientific concepts for diverse audiences and teams by synchronizing communication with the rapidly evolving business landscape. Dr Bridg supports marketing and sales initiatives, developing and implementing strategies that drive business growth and facilitating brand entry into new markets. Her consultancy is particularly adept at establishing cross-border projects for mid-sized companies and start-ups, guiding them to expand their activities abroad and achieve new successes by serving the needs of a particular niche. Her work significantly strengthens business links between Germany and Latin America, fostering collaboration and innovation through her deep understanding of both regions. In addition, Dr Bridg is a committed member of the Organisation for Women in Science for the Developing World (OWSD), advocating for the participation and advancement of senior women in science and contributing to the broader diplomatic goal of creating equitable and inclusive scientific communities.

María Inés Carabajal holds a PhD in Anthropology. She is an assistant researcher at the National Scientific and Technical Research Council (CONICET) and an assistant professor at the University of Buenos Aires, Argentina. Her research focuses on the human dimensions of climate change and the Anthropocene. She is particularly interested in the production, circulation, and use of climate information and how it is applied to decision-making processes. Dra. Carabajal has been involved in national and international projects about the provision of climate services in South America, co-production of knowledge, transdisciplinary science, and science communication. She has published several articles and chapters about these topics in national and international journals.

Luz Cumba Garcia is an immunologist working at the intersection of biomedical science, public policy, and science diplomacy. Currently, Luz is a second-year AAAS Science & Technology Policy Fellow in the Office of Mexican Affairs at the U.S. Department of State. In this role, she leverages her technical expertise to inform and support bilateral policy development between the U.S. and Mexico. In her initial fellowship year, Luz worked with the Middle East Regional Cooperation (MERC) Program at the U.S. Agency for International Development (USAID), contributing to the advancement of Arab-Israeli scientific collaboration and regional development through research grants. Her active role as an Inter-American Institute for Global Change Research (IAI) Science, Technology, and Policy (STeP) Science Diplomacy Fellow underscores her dedication to policy and diplomacy. Dr. Cumba Garcia's influence extends beyond her fellowship roles; she is a founding member of the Science Diplomacy Network in Latin America and the Caribbean and participated in several science diplomacy workshops, including the AAAS Science Diplomacy and Leadership Workshop, the AAAS/TWAS Science Diplomacy Course, and the InnScid SP Science Diplomacy School. Her academic background comprises a Ph.D. and M.S. in Immunology, alongside a B.S. in Cellular and Molecular Biology. Originating from San Juan, PR, she strives to champion advocacy, science communication, and community engagement.

Amâncio de Oliveira has a Bachelor at the Medical School at the University of Sâo Paulo (USP), PhD in Political Science. Post-Doc Researcher at New York University (NYU), 2010. Full Professor at the Institute of International Relations from USP. Coordinator of the Innovation and Science Diplomacy School (InnScid SP).

Martha Isabel Gómez Lee is a PhD in Political Studies, Universidad Externado de Colombia, Outstanding mention cum laude. Scholarship German Academic Exchange Service (DAAD) for Doctoral and Postdoctoral research stays at the Berlin Free University. Research professor at the Faculty of Finance, Government and International Relations, Universidad Externado de Colombia.

Santiago Gómez R. is an International Relations student from the Pontificia Universidad Javeriana.

Cristian Guglielminotti is a Member of the Center for Interdisciplinary Studies on International and Local Issues (CEIPIL), Professor at the Rectorate. Universidad Nacional del Centro de la Provincia de Buenos Aires. Bachelor's degree in International Relations, Universidad Nacional del Centro de la Provincia de Buenos Aires. Master's Degree in Science and Technology Policy and Management from the University of Buenos Aires. Doctoral candidate at the National University of Luján in Social and Human Sciences.

Danev Ricardo Perez Valerino is a Nuclear Chemist (2003) degree from the Higher Institute of Nuclear Sciences and Technologies-University of Havana. Head of International Relations at the Finlay Vaccine Institute (2007-2023). Master in International Relations (2009) from the "Raúl Roa García" Higher Institute of International Relations. Participant in international scientific events in Latin America, Europe and Asia. Lecturer on Science Diplomacy, Vaccine Diplomacy and Health Diplomacy. Specialist in International Collaboration, South-South Collaboration, Technology Transfers and International Negotiation. Participant in Science Diplomacy Courses organized by RIS (India) and AAAS-TWAS (USA) and Sao Paolo University (Brazil). Author/co-author of scientific publications. Since September 2023, began postgraduate studies at Tokyo International University as a scholarship recipient from the Japan International Collaboration Agency (JICA).

Jose Prieto is a Doctoral Candidate in Education Policy and Leadership with Dual Title in Transdisciplinary Studies on the Environment and Society at Penn State University (Fulbright Fellow). Holds a Master of Arts in Development and Good Governance (University of Duisburg Essen-DAAD Fellow), and a Master of Public Policy (Hertie School of Governance - Heinrich Böll Stiftung Fellow), and a bachelor's degree in political science (Universidad de los Andes). He has served as Senior International Officer and Confucius Institute Director at Universidad Jorge Tadeo Lozano, international affairs advisor in the Colombian President's Office and at the National Training Service – SENA, and advisor for the German Development Agency (GIZ).

Juan David Reina-Rozo is a consultant, teacher, and researcher of the Research Group on Technologies and Innovation for Community Development of the National University of Colombia. He has dedicated his professional life to reflecting and acting on the relationship between technology and society, with some experiences in Colombia, Peru, Brazil, Uganda, and Nepal. He has participated in citizen innovation processes in LabICS, particularly in Brazil, Mexico, and Colombia, as well as in the Distributive Labs at the beginning of the global pandemic of Covid-19 in 2020. He has been part of Rural Labs initiatives in the territory of Sumapaz and Gualivá in Cundinamarca, central Colombia. Finally, He develops collaborative Research-Creation processes with rural communities in the Caribbean and Pacific coast of Colombia. In addition to the academic and working world, he spends his time climbing and hiking in the Colombian páramos.

Felix Enrique Torres is a research professor, Doctor in Agricultural Sciences and Master in Water Resources, National University of Cajamarca. Improvement studies in Hydrology and Basin Management at the National Agrarian University of La Molina. Publications on ecological flow, hydrogeology, water footprint, climate change.

Angie Vanessa Vargas Junco is an internationalist focused on political studies at the Universidad Militar Nueva Granada with research so far focused on the study of populism and science diplomacy. Where commitment to the exploration and understanding of these issues in the contemporary context is evident. With interest in generating contributions from new and relevant perspectives in the academic field. Its interest lies in promoting greater understanding and constructive dialogue around these issues, with the aspiration of generating a positive impact on the understanding and approach of these phenomena at a global level.

Nevia Vera is a post-doctoral researcher at Argentina's National Council of Scientific and Technological Research (CONICET), and member of the Center of Interdisciplinary Studies on Local and International Affairs (CEIPIL) at the University of the Center of the Province of Buenos Aires (UNICEN), Argentina. She holds a Ph.D. in Political Science from National University of San Martin (UNSAM).

Index

A

B

C

D

E

F

T

V

www.ingramcontent.com/pod-product-compliance
Lightning Source LLC
LaVergne TN
LVHW080602300926
849162LV00005B/1030